T0378647

Praise for *The Ultimate Vaccine Timeline*

"Shaz Khan's *The Ultimate Vaccine Timeline* provides a *tour de force* overview of the history of vaccines from long before they were known as vaccines to the present. In her trajectory of events, we see the development of the vaccine industry from a pharmaceutical sidebar to today's major moneymaker, culminating (so far) in the catastrophic, but very lucrative, rollout of mRNA gene products, the latter now threatening to become the dominant type of 'vaccine.' It is the vaccines that have made the pharmaceutical industry as powerful as it is today, so powerful in fact that it can dictate to the WHO and to national governments in its endless hunger to expand vaccine markets.

"I doubt that even 1 in 1000 physicians will know even a small fraction of the material in this book, but they should. Indeed, this should be required reading for any in the medical profession who deliver or counsel the use of vaccines. And it is essential reading for the rest of us as well to see how much control of our lives the pharmaceutical industry has seized. If we are to begin the process of taking back control of our own health, this book would be a good place to start."

—Christopher Shaw, neuroscientist and professor of ophthalmology at the University of British Columbia, author of *Dispatches from the Vaccine Wars,* and coauthor of *Down the COVID-19 Rabbit Hole*

"Shaz Khan's aptly titled book *The Ultimate Vaccine Timeline* is an extraordinary compilation of historical details about a particular type of pharmaceutical product that we are supposed to believe has enabled humanity to thrive, but the true history of which is riddled with fraud, corruption, deceit, harm, and state violence against bodily integrity. This book is an incredible reference illuminating why someday in the future, when humankind finally becomes civilized, people will look back on the era of coerced mass vaccination and recognize it as an age of barbarism."

—Jeremy R. Hammond, independent journalist and author of *The War on Informed Consent*

"My career in pediatrics from 1985 to 2022—when I relinquished my medical license due to taking a stand for children's right to informed consent and bodily integrity—awakened me to the challenges we have with regard to vaccine safety. Vaccines are dangerous products with virtually no proper placebo safety testing, no analysis of overall health outcomes, and no official studies comparing the health of the vaccinated to that of the unvaccinated.

"To fully understand how we got here, we now have a treasure in this book. Shaz Khan's detailed masterpiece, *The Ultimate Vaccine Timeline: A Fact-Packed History of Vaccines and Their Makers*, includes a timeline of vaccine-relevant information ranging from AD 570 to the present day. This is both a vital reference work and a fascinating book to browse, containing everything you could possibly need to know about the history of vaccines in one beautiful place."

—Paul Thomas, MD, longtime pediatrician, coauthor of *The Vaccine-Friendly Plan*, author of *Vax Facts,* founder and host of *Pediatric Perspectives,* and cofounder of KidsFirst4Ever.com

"The Ultimate Vaccine Timeline by Shaz Khan is indeed an accurate and timely book at this stage in vaccine history. We are at a point where new and experimental vaccines—experimental in concept, in delivery mechanisms, in immune and other mechanisms of action—have been unleashed large scale on the world's unwitting population under great duress from powers that be, for a disease that was itself questionable. All vaccines are now under scrutiny by many who before wholeheartedly accepted them as 'safe and effective.' In this current time frame, the 'grandfather of vaccines' Dr. Stanley A. Plotkin has himself just released a paper noting that vaccines are not adequately tested for safety: 'Postauthorization studies are needed to fully characterize the safety profile of a new vaccine, since prelicensure clinical trials have limited sample sizes, follow-up durations, and population heterogeneity.'

"Shaz has successfully, without any words herself but simply by documenting events in history, revealed the true story of vaccines, of collusion between interested parties, of governmental controls, of repeated vaccine failures and disease outbreaks following the institution of new vaccination programs.

"This book demands a place on the shelf of every person interested in the truth about vaccines and in making their own informed decisions: which should mean everyone. Having been in the health and wellness industry for over thirty-five years, I have many books on vaccination and this stands amongst the very best, deserving its own place in vaccine history. I heartily recommend it to all."
—Robyn Cosford MBBS (Hons), retired doctor, professor of nutritional and environmental medicine, and chair of Children's Health Defense Australian Chapter

"This book impressively organizes the history of vaccines—from both an industry and public health perspective—all in one visually stunning timeline. Seeing serious reactions in my patients and my own family, I was inspired to research vaccines to determine which were worth the risk to keep my children safe. After studying many resources, I decided no vaccines would be safe for my children. This book would have been an excellent resource to fast-track my understanding of what we are really facing— decades of industry propaganda and medical gaslighting with an appalling lack of true safety studies. Ms. Khan has done all families, doctors, nurses, and scientists a favor by giving us a clear framework to begin our exploration for truth."
—Cammy Benton, MD, family physician

THE
ULTIMATE
VACC NE
TIMEL NE

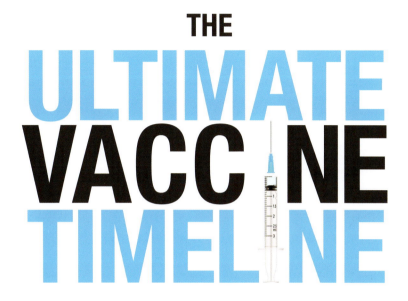

A Fact-Packed History of Vaccines and Their Makers

Research, compilation, and layout design by Shaz Khan

Foreword by Pierre Kory, MD, MPA

Skyhorse Publishing

Children's
Health Defense

Skyhorse Publishing books may be purchased in bulk at special discounts for sales promotion, corporate gifts, fund-raising, or educational purposes. Special editions can also be created to specifications. For details, contact the Special Sales Department, Skyhorse Publishing, 307 West 36th Street, 11th Floor, New York, NY 10018 or info@skyhorsepublishing.com.

Skyhorse® and Skyhorse Publishing® are registered trademarks of Skyhorse Publishing, Inc.®, a Delaware corporation.

Visit our website at www.skyhorsepublishing.com.
Please follow our publisher Tony Lyons on Instagram @tonylyonsisuncertain

10 9 8 7 6 5 4 3 2 1

Library of Congress Cataloging-in-Publication Data is available on file.

Hardcover ISBN: 978-1-64821-067-9
eBook ISBN: 978-1-64821-068-6

Cover design by David Ter-Avanesyan
Cover image by Getty Images

Printed in China

This book is dedicated to my father,
whose death triggered my deep dive into vaccines.

Contents

THE VACCINE TIMELINE

THE VACCINE INDUSTRY

References and source materials are available on
www.TheUltimateVaccineTimeline.com

"Those who cannot remember the past are condemned to repeat it."

—George Santayana (1863-1952)

Spanish-American philosopher, essayist, poet, and novelist, born
Jorge Agustín Nicolás Ruiz de Santayana y Borrás.

"Those who fail to learn from history are doomed to repeat it."

—Sir Winston Leonard Spencer Churchill (1874-1965)

British statesman who served as Prime Minister of the United Kingdom from 1940 to 1945
during the Second World War and again from 1951 to 1955.

Foreword

"Safe and effective" is one of the most successful marketing slogans in history as it implies 100% safety and efficacy without being accountable for demonstrating either. Since that slogan has been repeated so many times by the mass media, most of us never give it a second thought. Had we questioned it, the world likely would not have blindly accepted an experimental vaccine which, as time has shown, was one of the most dangerous and ineffective pharmaceutical products in history.

Because the vaccine industry has been able to hide behind the mantra of "safe and effective," this industry has been shielded from outside scrutiny, and an imaginary history has been created where everyone assumes each previous vaccine was perfect and without issue.

Once you start critically examining the history of vaccination, however, you will make a disturbing discovery. There have been many disastrous vaccines in the past, yet each time a disaster happened, it was swept under the rug and all we heard was "safe and effective," so nothing was learned, and before long, the exact same thing happened again.

After I advocated for treating COVID patients with repurposed drugs which could keep them out of the hospital, I began to be deluged with requests from vaccine-injured patients who had nowhere else to turn to for help. These patients are suffering immensely and are the sickest and most challenging patients that I've dealt with in my career.

Searching for answers, I learned there had been many previous disasters caused by experimental vaccines which should have never been put onto the market, and as I began studying them I discovered many of the unbelievable symptoms I was witnessing had, in reality, been seen many times in the past and were very real side effects of poorly made vaccines no one ever tells us about.

The Ultimate Vaccine Timeline gives a critical window into that lost history of vaccination. As COVID-19 has shown, it is essential that we understand how vaccines work and what can go wrong with them, because if we don't, this damaging history will repeat itself again and again.

—Pierre Kory, MD, MPA,
cofounder and president emeritus of the FLCCC Alliance

Glossary

Abbreviations

AAP	American Academy of Pediatrics
ACIP	Advisory Committee for Immunization Practices
AEFI	Adverse Events Following Immunization
AESI	Adverse Events of Special Interest
AFEB	Armed Forces Epidemiological Board
AMA	American Medical Association
AMRIID	Army Medical Research Institute of Infectious Diseases (at Fort Detrick, US)
ASPR	Administration for Strategic Preparedness and Response
ATCC	American Type Culture Collection
BARDA	Biomedical Advanced Research and Development Authority
BCG	Bacillus Calmette-Guérin—live-attenuated vaccine used to prevent tuberculosis
BLA	Biologics License Application (US-FDA)
BMGF	Bill & Melinda Gates Foundation
BSE	Bovine spongiform encephalopathy
BSL	Biosafety level laboratory—levels 1 to 4 (BSL-4 being the highest security)
CBER	FDA's Center for Biologics Evaluation and Research (US)
CDC	Centers for Disease Control and Prevention (US)
CEPI	Coalition for Epidemic Preparedness
CFR	Code of Federal Regulations (US)
CH	Switzerland—*Confoederatio Helvetica*
CHMP	Committee for Medicinal Products for Human Use (EU-EMA)
CICP	Countermeasures Injury Compensation Program (US-HHS)
CIOMS	Council for International Organizations of Medical Sciences (UN-WHO)
CISA	Clinical Immunization Safety Assessment (US-CDC)
COVID-19	Coronavirus disease 2019
CPMP	European Union Committee for Proprietary Medicinal Products
CRADA	Cooperative Research and Development Agreement
CRISPR-Cas9	Clustered regularly interspaced short palindromic repeats and CRISPR-associated protein 9
CSM	Committee on Safety of Medicines (UK)
DARPA	Defense Advanced Research Projects Agency (US)
DHSS	Department of Health and Social Security (UK)
DTaP / Tdap	Diphtheria, tetanus, acellular pertussis vaccine for pediatric use / **Tdap** is the reduced-antigen formulation (booster)
DTP / DPT / DTwP	Diphtheria, tetanus, whole-cell pertussis vaccine—DPT and DTP are used interchangeably
DOD	Department of Defense (US)
EMA	European Medicines Agency
EPA	Environmental Protection Agency (US)
EUA	Emergency Use Authorization (US)
EUL	Emergency Use Listing Procedure (WHO)

FAO	Food and Agriculture Organization (UN)
FDA	Food and Drug Administration (US)
GACVS	Global Advisory Committee on Vaccine Safety (WHO)
GAO	Government Accountability Office (US)
GAVI	Gavi, the Vaccine Alliance—formerly known as the Global Alliance for Vaccines and Immunization
GISRS	Global Influenza Surveillance and Response System
GMC	General Medical Council (UK)
GRAS	Generally recognized as safe
GVSI	Global Vaccine Safety Initiative (WHO)
HBsAg	Hepatitis B surface antigen
HBV	Hepatitis B virus
hCG	Human chorionic gonadotropin
HHS	Department of Health and Human Services (US)
HIV / AIDS	Human immunodeficiency virus that purportedly causes AIDS (acquired immunodeficiency syndrome)
HPV	Human papillomavirus
HRSA	Health Resources and Services Administration (US)—the agency within HHS that administers VICP
ICAN	Informed Consent Action Network (US)
ICD	International Classification of Diseases
ICRC	International Committee of the Red Cross
IFRC	International Federation of Red Cross and Red Crescent Societies
IHR	International Health Regulations—WHO's legally-binding instrument of international law
IIV	Inactivated influenza vaccine
IOM	Institute of Medicine—known as the National Academy of Medicine since 2015
IPV	Inactivated poliovirus vaccine (developed by Jonas Salk, MD)
JCVI	Joint Committee on Vaccination and Immunisation (UK)
LAIV	Live-attenuated influenza vaccine
LMICs	Low- and middle-income countries
MDL	Multidistrict litigation
MedDRA	Medical Dictionary for Regulatory Activities
MMR	Measles, mumps, rubella vaccine
MRC	Medical Research Council (UK)
MRC-5	Medical Research Council cell strain 5—derived from human fetal lung tissue of a 14-week-old male
MHRA	Medicines and Healthcare products Regulatory Agency—responsible for the safety of medicines licensed in the UK
mRNA / sa-mRNA	Messenger RNA / Self-amplifying messenger RNA
MSD	Merck Sharp & Dohme (Merck & Co., Inc. in USA/Canada and MSD outside)—different from Merck KGaA
MSF	Médecins Sans Frontières / Doctors Without Borders
NATO	North Atlantic Treaty Organization
NIH	National Institutes of Health (US)
NSABB	National Science Advisory Board for Biosecurity
NVIC	National Vaccine Information Center (US)

OPV	Oral live poliovirus vaccine (developed by Dr. Albert Sabin)
P3CO	Potential Pandemic Pathogen Care and Oversight—policy framework for gain-of-function research, governed by HHS
PAHO	Pan American Health Organization
PATH	Program for Appropriate Technology and Health
PCR	Polymerase chain reaction
PHEIC	Public health emergency of international concern
PPP	Potentially pandemic pathogens
PRISM	Post-Licensure Rapid Immunization Safety Monitoring (US-CDC)
SAGE	Strategic Advisory Group of Experts—the principle advisory group to the WHO for vaccines
SARS-CoV-2	Severe Acute Respiratory Syndrome Coronavirus 2
SDGs	Sustainable Development Goals
SIDS	Sudden infant death syndrome
SOC	System Organ Class—terminology distinguished by physiological system or disease origin (MedDRA)
SPEAC	Safety Platform for Emergency vACcines Project (the Brighton Collaboration and CEPI)
TBE	Tick-borne encephalitis
TLR	Toll-like receptor—a class of proteins that are important mediators of inflammatory pathways
TRIPS	Trade-Related Aspects of Intellectual Property Rights—multilateral agreement by the World Trade Organization
TT	Tetanus toxoid vaccine
UAE	United Arab Emirates
UK	United Kingdom
UN	United Nations
UNEP / UNDP	UN Environment Programme / UN Development Programme
UNFPA	United Nations Population Fund, formerly the United Nations Fund for Population Activities
UNICEF	United Nations International Children's Emergency Fund—renamed UN Children's Fund in 1953
US	United States of America
USAID	United States Agency for International Development
USAMRIID	US Army Medical Research Institute of Infectious Diseases
USPHS	US Public Health Service
USSR	Union of Soviet Socialist Republics (Soviet Union)
VAERS	Vaccine Adverse Events Reporting System (US)
VAPP	Vaccine-associated paralytic poliomyelitis
VDPV	Vaccine-derived poliovirus—disease-causing virus that comes from live poliovirus in oral polio vaccines
VICP	Vaccine Injury Compensation Program (US)
VRBPAC	Vaccines and Related Biological Products Advisory Committee (US-FDA)
VSD	Vaccine Safety Datalink—managed by the CDC-Task Force for Global Health
WEF	World Economic Forum
WHO	World Health Organization
WI-38	Wistar Institute cell line #38 derived from the embryonic lung tissue of 3-month-gestation female
WOAH / OIE	World Organization for Animal Health—former OIE Office International des Épizooties
WRAIR	Walter Reed Army Institute of Research
WTO	World Trade Organization

Principles of Vaccination – Topley and Wilson
General principles for any vaccine

1. Vaccination should be harmless to the healthy child.

2. The disturbance caused by the vaccine should not be greater than that of the disease itself.

3. It must be easy to administer.

4. Vaccination must provide herd as well as individual benefit.

5. Vaccine-conferred immunity should be sufficiently solid to obviate need for frequent revaccination.

W.W.C. Topley, and G.S. Wilson, *The Principles of Bacteriology and Immunity* (New York: William Wood & Co.), 1929.

Glossary

Definitions

Active immunity	Refers to immunity acquired when exposed to a disease organism that triggers the immune system to produce antibodies against that disease—it may be acquired through the experience of a natural infection or achieved through the act of vaccination.
Adjuvant	From the Latin adjuvare (to enhance)—a substance that induces and augments an immune response in the body, and is typically added to inactivated and subunit vaccines to improve immunogenicity. The most widely used adjuvants in vaccines are aluminum phosphate and aluminum hydroxide. Merck's proprietary aluminum adjuvant is amorphous aluminum hydroxyphosphate sulfate (AAHS) and there are also oil-in-water or water-in-oil based adjuvants, as well as the most recent CpG-containing oligonucleotide sequence termed 1018 that has been developed by Dynavax Technologies for Heplisav-B®, its hepatitis B vaccine. See page 194 for further details.
Adsorbed	A surface process that leads to transfer of a molecule from a fluid bulk to solid surface.
AEFI Adverse Event Following Immunization	Any untoward medical event that follows vaccination and which does not necessarily have a causal relationship with the vaccine. The adverse event may be any unfavorable or unintended sign, abnormal laboratory finding, symptom or disease, or death. The Brighton Collaboration is commissioned with standardizing these definitions and determining the diagnostic criteria for AEFI, in partnership with the WHO and CIOMS.
Antibody	Also called an immunoglobulin (Ig), an antibody is a protective protein produced by the immune system in response to the presence of a foreign substance, called an antigen.
Antigen	A substance (toxin, virus, bacteria, foreign blood cell, and/or transplant organ cell) that induces an immune response. It is usually the key active ingredient in a vaccine.
Antiserum	Blood serum that contains antibodies against an infective agent (such as a bacteria or virus) or toxic substance (such as snake venom) that can be used to prevent or treat an infection or poisoning. This can only provide passive immunity from disease.
Antitoxins	Providing passive immunity, antitoxins are produced by injecting an animal with a toxin (for example, diphtheria). The animal, most commonly a horse, is given repeated small doses of toxin until a high concentration of the antitoxin builds up in the blood. The blood is then taken to extract the resulting highly concentrated preparation of antitoxins, called an antiserum.
Attenuation	The process by which viruses and bacteria are made less virulent, typically through chemical or heat-based methods, so that they may be used in vaccines.
Bacillus	Rod-shaped bacterium.
Bacterins	Vaccines containing whole, killed bacteria, commonly used in the early 20th century.
Bioinformatics	A hybrid science that links biological data with techniques for information storage, distribution, and analysis to support multiple areas of scientific research, including biomedicine. Bioinformatics uses biology, chemistry, physics, computer science, computer programming, information engineering, mathematics, and statistics to analyze and interpret biological data. The subsequent process of analyzing and interpreting data is referred to as computational biology.
Biologic	A biologically active preparation made from living organisms or their products and used as diagnostic, preventive, or therapeutic agents. Kinds of biologics include antitoxins, serums, products derived from human blood and plasma, and vaccines—which is a form of immunostimulatory therapy.
Biological weapon	A tool of biological warfare that uses biological toxins or infectious agents, bacteria, viruses, insects, and fungi—man-made or natural—with the intent to kill, harm, or incapacitate humans, animals, and/or plants.

Carcinogen	Any agent, process, or action that can cause cancer, notably tumor growth, damage to the genome, and/or disruption of cellular metabolic processes.
Chimeric	Consisting of parts sourced from varied origins—in medicine typically refers to to a person, organ, or tissue that contains cells with different genes than the rest of the person.
Cooperative Research and Development Agreement	A Cooperative Research and Development Agreement (CRADA) is an agreement between a federal laboratory (e.g., NIH, NIAID, DoD etc.,) and a non-federal party (e.g., a pharmaceutical company) to perform collaborative research and development. It is frequently used as a mechanism for formalizing interactions between private industry and government. The goal is to foster mutually beneficial partnerships in order to facilitate cutting-edge research and development for commercialization and technology transfers.
Corporate Integrity Agreement	A document outlining the obligations that a healthcare company in the US makes with a federal government agency like the HHS, as part of a civil settlement. There are a list of obligations that the company must fulfill that typically lasts for five years.
CRISPR-Cas9	A gene-editing technology that research scientists can use to selectively modify the DNA of living organisms, with the help of DNA-slicing enyzme Cas9. This method of genetic engineering can potentially be used in gene therapies, as well as utilized to generate novel biological weapons.
CRM197	A genetically detoxified mutant of diphtheria toxin widely used as a carrier protein for conjugate vaccines, first commercialized in the childhood vaccine HibTITER® approved in the USA in 1990. Characterized by Italian immunologist Rino Rappuoli and his team at Chiron (later Novartis), Cross Reacting Materials-197 is used in Hib, pneumococcal, and meningococcal vaccines.
DDT	Dichloro-diphenyltrichloroethane—a powerful pesticide and neurotoxic agent that was used extensively from the 1940s to control typhus, malaria, and polio. It was banned in the USA and Canada in 1972, with most European countries following suit. It is still used in developing nations for the control of malaria.

DNA / RNA
Deoxyribonucleic acid
Ribonucleic acid

Molecules within a cell that carry genetic information for the development and functioning of an organism, composed of long chains of nucleotides: adenine, guanine, cytosine, and thymine—in RNA, thymine is replaced with uracil.

Efficacy v. effectiveness	According to Gavi, "Efficacy is the degree to which a vaccine prevents disease, and possibly also transmission, under ideal and controlled circumstances—comparing a vaccinated group with a placebo group [which rarely occurs]. Effectiveness meanwhile refers to how well it performs in the real world."
Encephalitis	Inflammation of the brain due to an infection or disease.
Encephalopathy	A disorder of the brain caused by damage to, or dysfunction of, the neurons.
Epidemic	Any sudden increase in disease prevalence in a localized population within a short period of time, within a single country. An outbreak is a disease increase on a regional level.
Epidemiology	The study of disease in a specific population, its determinants and patterns of occurrence and distribution. Though it is an essential cornerstone of public health, being an important tool for shaping policy decisions, epidemiological studies are insufficient to definitively prove causation of a disease.
Escherichia coli	*E. coli* is a gram-negative, facultative anaerobic (does not require oxygen for growth), rod-shaped bacterium that has been extensively used to express antigens for a large variety of vaccines, including bacteria-derived proteins, genetically modified toxins, and virus-derived peptides.
Gain-of-function	Scientific research that genetically alters and intentionally enhances the biological functions of a microorganism—this could mean increasing transmissibility and/or augmenting virulence. This risky gain-of-function research is justified as being essential for scientists to better predict emerging infectious diseases and to develop medical countermeasures—vaccines and therapeutics.
Genome	The genetic information of an organism that consists of nucleotide sequences of DNA (or RNA in RNA viruses), including protein-coding (representing only 2% of the human genome) and non-coding genes in the nuclear genome and the mitochondrial genome. The human genome is estimated to be over 3 billion base pairs (bp) that are distributed across 23 chromosomes.
Gram-negative / positive	Named after its inventor, Danish scientist Hans Christian Gram (1853–1938), this bacteriological laboratory technique was developed in Berlin, Germany, in 1884. This refers to a method of staining used to classify bacterial species into two large groups based on the physical properties of their cell walls: gram-positive bacteria (show blue or pink—thick cell walls) and gram-negative bacteria (show pink or red—thin cell walls). While not all bacterial organisms can be stained and classified this way (such as many species of *Mycobacterium*), the method remains a valuable diagnostic tool. Antibiotic resistance affects mainly gram-negative bacteria.
Human chorionic gonadotropin (hCG)	Human chorionic gonadotropin is a hormone for the maternal recognition of pregnancy produced by trophoblast cells that are surrounding a growing embryo, which then eventually forms the placenta after implantation.
Hemagglutinin	A group of naturally occurring glycoproteins found on influenza viruses that cause red blood cells to clump, or agglutinate. It is estimated that there are 16 forms of hemagglutinin (H), designated H1-H16, associated with influenza type A viruses. Together with various forms of a viral antigenic protein called neuraminidase (N), hemagglutinin is used to distinguish between subtypes of influenza A viruses.
Herd immunity	Also called "community immunity," it is a concept applied to both human and animal vaccination, with the idea that if enough individuals are vaccinated, the collective protection will benefit even those who are unvaccinated. For highly contagious measles, the percentage of vaccine coverage required to acquire herd immunity was estimated at around 55-60% in the 1960s and gradually revised upward to 95% today. Herd immunity cannot be obtained from vaccination against tetanus, diphtheria, COVID-19, polio (cannot prevent its colonization in the gut), influenza (too many subtypes), and even acellular pertussis, owing to the fact that these vaccines cannot prevent infection.

HRSA	The Health Resources and Services Administration is an agency of the US Department of Health and Human Services located in North Bethesda, Maryland. It is the primary federal agency for improving access to healthcare services for people who are uninsured, isolated, or medically vulnerable.
Immunity	Protection from being infected and getting sick from a specific disease.
Immunization	The process and result of being protected from a specific disease or infection.

The words "immunization" and "vaccination" **are not equivalent nor synonymous despite often being used interchangeably.**

Influenza	Also called "the flu," influenza is an acute viral infection of the upper and/or lower respiratory system, typically affecting the nose, throat, and lungs, often accompanied by fever, chills, and fatigue and pain in the muscles. Influenza viruses that are believed to cause illness are primarily categorized into type A and type B (types C and D also exist, but are not used in vaccines), which are divided into subtypes, that are further split into strains. Subtypes of influenza A are identified mainly by two surface antigens (foreign proteins)—hemagglutinin (H) and neuraminidase (N). Examples of influenza A subtypes include H1N1 and H5N1, and are labeled with the place they were identified and the year. For example, A/USSR/92/1977(H1N1) is the influenza strain from the 1977 outbreak in Russia. Influenza B viruses are subdivided into two major lineages, B/Yamagata and B/Victoria. Vaccines mainly include influenza A subtypes, but trivalent and quadrivalent flu vaccines tend to include at least one influenza B lineage.
Inoculation	The act of inoculating—the introduction of an antigenic substance into the body with the aim of producing immunity to a specific disease.
In silico	Research conducted using a computer—such as modeling, simulation, or gene sequencing.
Lipid nanoparticles	LNP are nanoparticles—defined as a particle of matter that is between 1 and 100 nm, or up to 500 nm in diameter—composed of lipids (fatty acids) in which active molecules can be incorporated, that are increasingly used as a novel therapeutic drug and genetic information delivery system. They are a key component of the COVID-19 mRNA products.
Liposomes	Small, artificial vesicles of spherical shape that can be created from cholesterol and phospholipids, and can be used as carriers of pharmaceutically active agents.
Mathematical models	Described as "simplified descriptions of complex phenomena which are used to gain insights into the real world," mathematical models are based on a series of assumptions that can provide a theoretical framework for simulating reality. Mathematical modeling has been given an increasingly important role in infectious disease epidemiology—including health-economic aspects, emergency planning, and risk assessment—by providing information to guide decisions around public health strategies, policies, and programs. Imperial College London, employer of Professor Neil Ferguson (who alerted governments worldwide with his COVID-19 models, justifying nationwide lockdowns), has been a world leader in providing "mathematical modeling of the epidemiology and control of infectious diseases of humans and animals in both industrialized and developing countries for 20 years."
MedDRA	A clinically validated international medical terminology dictionary created by the International Council for Harmonization of Technical Requirements for Pharmaceuticals for Human Use (ICH). It is used by regulatory authorities and the biopharmaceutical industry during the regulatory process, from pre-marketing to post-marketing activities, and for safety information data entry, retrieval, evaluation, and presentation. It is also the adverse event classification dictionary, and is organized with a five-level hierarchy, the highest level of which is the System Organ Class (SOC). See page 284 for further details.

Instruments for Observing Matter
From nanosized to visible with the naked eye

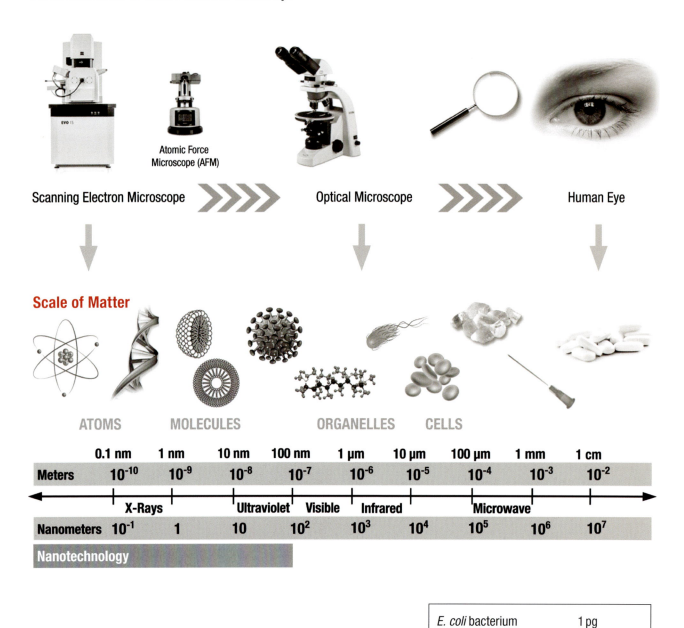

Atomic Force
Microscope (AFM)

Scanning Electron Microscope   Optical Microscope   Human Eye

Scale of Matter

ATOMS   **MOLECULES**   **ORGANELLES**   **CELLS**

	0.1 nm	1 nm	10 nm	100 nm	1 µm	10 µm	100 µm	1 mm	1 cm
Meters	10^{-10}	10^{-9}	10^{-8}	10^{-7}	10^{-6}	10^{-5}	10^{-4}	10^{-3}	10^{-2}
		X-Rays		Ultraviolet	Visible	Infrared		Microwave	
Nanometers	10^{-1}	1	10	10^{2}	10^{3}	10^{4}	10^{5}	10^{6}	10^{7}

Nanotechnology

E. coli bacterium	1 pg
Human hair	100 mcg/µg
Average human cell	1 ng

Microweight Scale

Femtogram	Picogram	Nanograms	Micrograms	Milligrams	Grams
fg	pg	ng	mcg/µg	mg	g
10^{-15} grams	10^{-12} grams	10^{-9} grams	10^{-6} grams	10^{-3} grams	1 gram
1 fg = 0.000000001 mcg	1 pg = 0.000001 mcg	1 ng = 0.001 mcg	1 mcg = 0.000001 g	1 mg = 0.001 g	1 g = 0.001 kg

Metagenomics	Metagenomics is both a field of study—the genomic analysis of microbial DNA from environmental communities—and a research technique that enables population analysis of unculturable or previously unknown microbes via genome sequencing. Viral metagenomics allow scientists to identify new viruses without using traditional methods of isolation, purification, electron photography, and biological characterization.
Monoclonal antibodies (mAbs)	Immunoglobulins that consist of proteins derived "from a cell lineage made by cloning a unique white blood cell." Often engineered from natural sources, mAbs have emerged as a major class of therapeutics used to treat diseases such as cancer, inflammation, or autoimmune conditions, and can be found in hormones, gene therapies, insulin, and plasma-derived medicines.
Mutagenic	Any agent, process, or action that can alter cellular genetic material, causing irreversible and heritable changes in the DNA.
Nanotechnology	The branch of science and engineering devoted to designing, producing, and using structures, devices, and systems by manipulating atoms and molecules at nanoscale, i.e., having one or more dimensions of the order of 100 nanometres or less. The term "nanotechnology" was coined by a Japanese professor at the Tokyo University of Science, Norio Taniguchi—though the concept was first presented by American theoretical physicist Richard Feynman in 1959.
Neuraminidase	A enzyme (protein) that occurs on the surface of influenza-related viruses and is believed to play a crucial role in virus release from infected cells.
New chemical / molecular entity	A new chemical entity (NCE) is, according to the FDA, a novel, small, chemical molecule drug that is undergoing clinical trials or has received a first approval (not a new use) by the FDA. A new molecular entity (NME) is a broader term that encompasses both an NCE or an NBE (New Biological Entity). Vaccine candidates include antigens that are new molecular entities.
Nucleic acid	Large biomolecules, composed of nucleotides, that play essential roles in all cells and viruses, including storing and expressing genomic information.
Nucleotide	Consists of a sugar molecule (either ribose in RNA or deoxyribose in DNA) attached to a phosphate group and a nitrogen-containing base. The bases used in DNA are adenine (A), cytosine (C), guanine (G), and thymine (T). In RNA, thymine is replaced with uracil (U).
One Health	The One Health approach was conceptualized in 2006 as a global public health strategy that encourages interdisciplinary collaboration and communication on health at the human-animal-environmental interface. It is a multisectoral and transdisciplinary approach—working at the local, regional, national, and global levels—with the goal of achieving optimal health outcomes recognizing the interconnection between people, animals, plants, and their shared environment. The current activities are managed and coordinated by the FAO/UNEP/WHO/WOAH.
Palindromic sequence	A palindromic sequence is a nucleic acid sequence in a double-stranded DNA or RNA molecule, whereby reading in a certain direction on one strand is identical to the sequence in the same direction on the complementary strand.
Pandemic	An epidemic of disease that crosses borders and spreads internationally.
Passive immunity	This refers to the transfer of antibodies from one individual to another (for example, a mother to her baby) or from an animal to a human (antiserum). Passive immunity can provide immediate but short-lived protection, lasting several weeks to several months.
Pharmacodynamics	The study of the body's biological reaction to drugs, notably what and how the biological processes in the body respond to, or are impacted by, a drug. Like pharmacokinetics, this study is not required for vaccine clinical trials, hence it is never analyzed.
Pharmacokinetics	The study of the movement of drugs through the body and what happens to the drug in regards to absorption, distribution, metabolism, and excretion—not required for vaccine licensure.

Pharmacovigilance	WHO defines vaccine pharmacovigilance as the "science and activities relating to the detection, assessment, understanding and communication of adverse events following immunization and other vaccine or immunization related issues, and to the prevention of untoward effects of the vaccine or immunization." Pharmacovigilance is essentially a hypothesis-generating activity whereby suspicions of harm spontaneously reported by manufacturers, healthcare providers, and patients in reporting systems give rise to questions of causality between medicines or vaccines and adverse events. Adverse event reports are collected and pooled, and sometimes causality is assessed.
Plasmid	A small, circular DNA molecule found in bacteria and yeast, used extensively in biomolecular research and genetic engineering, notably in vaccine development. It is physically independent of, and can replicate separately from, the bacterium's chromosomal DNA.
Polymerase Chain Reaction procedure	Invented by Kary Mullis, PhD, in 1983, the PCR method is an enzymatic assay that can amplify a specific region of DNA*—making billions of copies—and may be used to identify the presence or absence of a known pathogen or gene, however, it cannot reliably diagnose disease.
PRISM	CDC's Post-Licensure Rapid Immunization Safety Monitoring is a near real-time active surveillance system for monitoring the safety of H1N1 influenza vaccine.
Product Development Partnership (PDP)	PDPs are "public–private partnerships established to develop and provide access to new health products—especially vaccines, therapeutics, and diagnostics—for poverty-related and neglected diseases." PDPs are "nonprofit-making" and generally funded by both public and philanthropic organizations. Their origins can be traced back to the Rockefeller Foundation's 1994 Bellagio conference, where strategies to accelerate HIV vaccine research and development were discussed. Since then, the PDP model has proven its success in advancing scientific research and development for diseases otherwise lacking commercial interest.
Proteins	Molecules made of amino acids, proteins are essential to the myriad of metabolic processes (hormones, enzymes, antibodies) as well as being fundamental to the structure, function, and regulation of the body's tissues and organs.
Public-Private Partnership (PPP)	Conceived in the mid-1990s and early 2000s, PPPs are long-term partnerships and business relationships between international or governmental agencies and commercial pharmaceutical companies. Public sector programs are merged with private-sector participation, creating an incentive environment that can attract private investment. "PPPs combine the deployment of private sector capital and, sometimes, public sector capital to improve public services or the management of public sector assets." The International AIDS Vaccine Initiative founded in 1996 is one of the first examples of a PPP. Their objective is to "ensure the development of safe, effective, accessible, preventive HIV vaccines for use throughout the world."
Recombinant	A term used to describe genetic material that has been lab-created by which genetic material and proteins from different species may be combined.
Recombinant DNA (rDNA)	rDNA are DNA molecules formed by laboratory methods of genetic engineering that bring together genetic material from multiple sources, creating sequences that would not otherwise be found in nature. The technology of recombinant DNA uses restriction enzymes that cleave DNA at specific sites, allowing sections of DNA to be inserted into plasmids or other vectors.
Recombumin®	A recombinant human albumin used in various preclinical, clinical, and marketed vaccines. Human and animal-origin-free, it is produced in the UK by Delta Technology from a proprietary genetically modified yeast strain *Saccharomyces cerevisiae*. Recombumin® is used in MMRVaxPro® and IXCHIQ® vaccines.
Replicons (viral RNA)	Viral RNA replicons are self-amplifying, genetically modified RNA molecules expressing proteins sufficient for their own replication, but which do not produce infectious virions. They are increasingly being investigated as platforms for next-generation vaccine delivery.

*RNA can be amplified using reverse-transcriptase PCR (RT-PCR) where RNA is converted to the more stable DNA. This technique was developed after 1983.

Reverse vaccinology	This process, pioneered by Rino Rappuoli at Chiron in 2000, utilizes bioinformatics to analyze genome sequences to determine potential vaccine antigens for candidate design. "High-throughput screening of the whole genome of pathogens [is used] to identify genes encoding proteins that exhibit excellent immunogenicity." The first pathogen addressed using this method was *N. meningitidis* serogroup B, resulting in the meningococcal B vaccine Bexsero®, approved in 2013.
Safe	Traditionally implies "protected from or not exposed to danger or risk." However, in the medical context, safe means "relative freedom from harmful effect" and that the benefits of a treatment or drug are determined (usually by regulatory agencies) to outweigh the potential risks.
Serum	Blood plasma with clotting factors removed, typically used for transferring immunity (antibodies—immunoglobulins) to another, derived from the blood or tissues of immunized animals (notably horses in the case of diphtheria serum) or humans.
SV40	Polyomavirus simian vacuolating virus number 40 is a known oncogenic—cancer-causing—DNA virus that can induce primary brain and bone cancers, malignant mesothelioma, and lymphomas in laboratory animals. Persuasive evidence now indicates that SV40 is causing infections in humans today and represents an emerging pathogen. It was first discovered by American scientist Bernice Eddy in the Salk polio vaccine, derived from the monkey kidney cell cultures used to produce the vaccine that was injected into around 98 million people. SV40 has been used extensively in cell research, notably in the creation of HEK293T (Human Embryonic Kidney cell line), whose genome contains the SV40 large T antigen, enabling the production of recombinant proteins within plasmid vectors containing the SV40 promoter.
Teratogenic	Any agent, process, or action that is a source of prenatal toxicity characterized by structural or functional defects in the developing embryo or fetus.
Thimerosal **Thiomersal** British spelling **Merthiolate** Trade name **Sodium ethyl mercurythiosalicylate**	An organomercury (mercury-derived) compound that is 49.6% ethylmercury by weight. A 0.01% solution (1 part per 10,000) of thimerosal contains 50 mcg of mercury (Hg) per 1 mL dose or 25 mcg of Hg per 0.5 mL dose. It has been used as a preservative in vaccines since 1929, after pharmaceutical company Eli Lilly discovered its antiseptic and antifungal properties. From 1999–2006, thimerosal in vaccines in industrialized countries has been phased out, replaced largely with 2-phenoxyethanol or phenol. However, thimerosal remains in many vaccines for developing nations, as well as in developed nations in some multidose vials and in "thimerosal-free" vaccines at "trace amount" concentrations of < 0.5 mcg Hg per dose. The WHO refers to "thimerosal-reduced" vaccines as containing only "residual" amounts with a reduction from 50 mcg/mL to 1-2 mcg/mL in the final product. "The effect of vaccines containing thiomersal on concentrations of mercury in infants' blood has not been extensively assessed, and the metabolism of ethylmercury in infants is unknown."
Toxoid	A bacterial toxin that is inactivated by heat or chemical (formalin), but retains its ability to produce antitoxins in the blood and stimulate the ability of the body to produce antibodies.
Trace amounts	

< 0.5 mcg/µg   **micrograms per mL**
< 500 ng   **nanograms**
< 500 ppb   **parts per billion**
< 0.5 ppm   **parts per million**

The Institute of Medicine Committee considers "trace amounts" to be less than 0.5 mcg/µg (500 ng) per dose, left over from the manufacturing process. The Institute for Vaccine Safety at the Johns Hopkins Bloomberg School of Public Health considers "trace amounts" to be less than 0.3 mcg/µg (300 ng). Trace amounts still count for substance that can accumulate and interact with the electrobiochemistry of interconnected systems within the human body, notably via metabolic pathways still not fully understood.

Unavoidably unsafe products	According to a 1985 publication *Vaccine Supply and Innovation* by the National Academy of Sciences: **"There are some products which, in the present state of human knowledge, are quite incapable of being made safe for their intended and ordinary use.** These are especially common in the field of drugs. An outstanding example is the vaccine for the Pasteur treatment of rabies, which not uncommonly leads to very serious and damaging consequences when it is injected. Since the disease itself invariably leads to a dreadful death, both the marketing and the use of the vaccine are fully justified, notwithstanding the unavoidably high degree of risk which they involve. Such a product, properly prepared, and accompanied by proper directions and warning, is not defective, nor is it unreasonably dangerous. The same is true of many other drugs, vaccines, and the like, many of which for this very reason cannot legally be sold except to physicians or under the prescription of a physician."
Vaccination	Vaccination is an attempt to trick the immune system into providing protection from disease without subjecting the host to the actual illness. A live-attenuated, inactivated, or fragmented pathogen is introduced into the body with the objective to induce lasting immunity from infection. It is a process, the intended outcome of which is immunization, although the term "immunization" is most often inaccurately used to allude to vaccination.
Vaccine	A vaccine is a biological product, oral, intranasal, or injected, that aims to provide protection against infectious disease. Vaccines are classified as biologics, as they are active substances made from natural organisms or genetically engineered, consisting of well-characterized proteins. **At present, there are six types of vaccines licensed or in development:** 1. **live-attenuated** *measles, rubella, mumps, chickenpox, influenza, rotavirus, polio (OPV), shingles, BCG, yellow fever, mpox* 2. **inactivated** *polio (IPV), influenza (split-virion), hepatitis A, rabies, Japanese encephalitis* 3. **subunit** (recombinant polysaccharide and conjugate vaccines, virus-like particles, glycoproteins); *pneumococcal, meningococcal, Haemophilus influenzae, HPV, hepatitis B, shingles, acellular pertussis* 4. **toxoid** (inactivated bacterial toxins) *diphtheria, tetanus* 5. **viral vector** (replicating and non-replicating) *Ebola, COVID-19* 6. **nucleic acid vaccines** (mRNA, sa-mRNA and DNA—currently in development) *COVID-19* Smallpox and yellow fever vaccines have also existed as live, non-attenuated vaccines, with smallpox vaccine (ACAM2000®) being the only non-attenuated vaccine still available. No vaccine has ever been evaluated for its carcinogenic or mutagenic potential, or ability to impair fertility in humans.
Vaccine hesitancy	The WHO defines "vaccine hesitancy" as a delay in acceptance or refusal of vaccines despite the availability of vaccination services.
Vaccinia	A double-stranded DNA virus from the poxvirus family, used as a source for modern smallpox vaccines. Vaccinia is also the name given to smallpox vaccine complications in vaccinated people and unvaccinated contacts, involving swollen lymphs and skin lesions that can lead to necrosis.
Vaccinology	The field of science (mostly empirical) that approaches disease prevention and treatment through the research and development of vaccines.
Vaccinovigilance	The activities relating to the monitoring, detection, assessment, understanding, and communication of AEFI and other vaccine-related issues to ensure continued safety of vaccines.
Variolation	The method of inoculation first used to vaccinate individuals against smallpox with material taken from a patient or a recently variolated individual, in the hope that a mild, but protective, infection would result.
Varioloid	In the late 19th century and early 20th century, it meant smallpox of the vaccinated. Today the definition is attributed to a mild form of smallpox that affects people who have already had the disease or have been vaccinated against it.
Zoonosis	Any disease or infection that is naturally transmissible from animals to humans.

What is a vaccine?

"A vaccine, as defined historically, is a suspension of live (usually attenuated) or inactivated microorganisms (e.g., bacteria or viruses) or fractions thereof administered to induce immunity and prevent infectious disease or its sequelae.

"A vaccine, as defined more recently, is a preventative or therapeutic agent that achieves its desired effect by stimulating the immune system of the individual."

US Markets for Vaccines: Characteristics, Case Studies, and Controversies

Ernst R. Berndt, Rena N. Denoncourt, and Anjli C. Warner, 2009

https://www.aei.org/wp-content/uploads/2014/07/-us-markets-for-vaccines_164855346063.pdf

How are vaccines administered?

Vaccines are given as either intramuscular (IM), subcutaneous (SC), or sometimes intradermal (ID) injections—the vast majority of vaccines are given as intramuscular injections in the thigh or upper arm.

Vaccines should never be injected intravenously.

Vaccines may also be administered as oral drops for ingestion—as is the case for Sabin's live polio (OPV), cholera, and rotavirus vaccines, or given as a nasal spray in the case of FluMist®/Fluenz® influenza vaccine. The adenovirus type 4 and type 7 vaccines for military troops are ingested as oral tablets, and the live-attenuated typhoid vaccine Vivotif® is given as oral capsules.

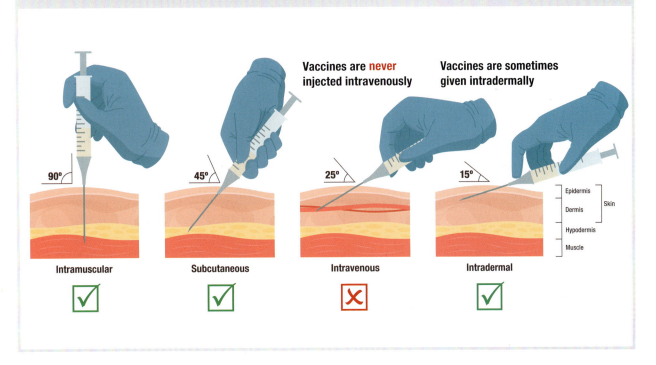

Introduction
A Brief History of Vaccination

While there are many publications that delve into the history of vaccination—covering early inoculation methods through to the novel nanotechnology and bioinformatics era of the 21st century—the goal of this thoroughly researched historical timeline is to provide a chronology of key facts on vaccines. While not all events mentioned in this timeline are specifically related to vaccines, they are included to provide historical context of the medical and societal environment at the time, and as a reminder of some important global events.

Vaccines have been around for over 200 years, and while the technology is significantly more sophisticated, the basic principle remains the same: trick the body into mounting an immune response to a particular pathogen to provide memory and protection from future infection. The efficacy of the vaccine-induced immune response is typically measured by the presence of specific antibodies or T-cells in the blood.

It is interesting to note that the healthcare industry was largely born out of the military, with the first hospitals built for the army during times of war, and health insurance established to cover the medical costs of soldiers and to provide compensation for disability. It is no coincidence that there is much military language embedded within modern medicine. Vaccines have been developed in military laboratories, universities, and companies with ties to governments, including the US Department of Defense. Many vaccines have been developed under biodefense programs (anthrax, plague, botulinum toxin, Ebola) as well as tested on soldiers, often without their informed consent. Biological warfare inherently includes vaccine research and development, as it is the primary countermeasure in case of a biological attack. Dual-use vaccine research can be used for civilian and military applications, for both defensive and offensive purposes.

The initial methods of smallpox variolation consisted of blowing dried smallpox pustule powder up the nostrils, or scaping raw infectious matter from one arm to the next, also called engrafting. These methods are thought to have originated in China and India, spreading throughout Asia in the 1500s, and then to Europe in the early 1700s, thanks to several artistocratic advocates. In the late 1790s, an English country doctor, Edward Jenner, created the first smallpox inoculation using cowpox or horsegrease, which became widely enforced worldwide. The main strategy to circulate and conserve the smallpox vaccine was arm-to-arm vaccination, also known as Jennerian or "humanized" vaccination. Children were often used as vessels to transport the vaccine, and could transmit other human diseases like syphilis through this inoculation method.

When mass vaccination began in the early to mid-1800s, many countries implemented legislation to mandate smallpox vaccine, with non-compliance punishable by fines or imprisonment. Isolation of the sick and the two-week quarantine of exposed individuals became common practice to combat contagion. The invention of the first syringes allowed a direct entry point into the body, and would soon change the administration of vaccines. Another major advance in the vaccination technique took place after 1860, when calves would be used to propagate pox materials, known as "animal vaccine." This safer, more practical approach allowed the production of larger amounts of vaccine and led to the creation of numerous vaccine farms, mostly by medical doctors who saw an

"Popular medical treatments in the 1700s to early 1900s included bloodletting, noxious powders, mercury-based medicine (mercury cyanide*, mercuric iodide, mercury benzoate, calomel or mercuric chloride), arsenical solutions, leeches and blistering with hot irons. As a medicine, mercury was as ubiquitous as aspirin is today."

*used for the treatment of conjunctivitis (pink eye) from the mid-1700s until the 1950s.

—Forrest Maready, *The Moth in the Iron Lung* (2018)

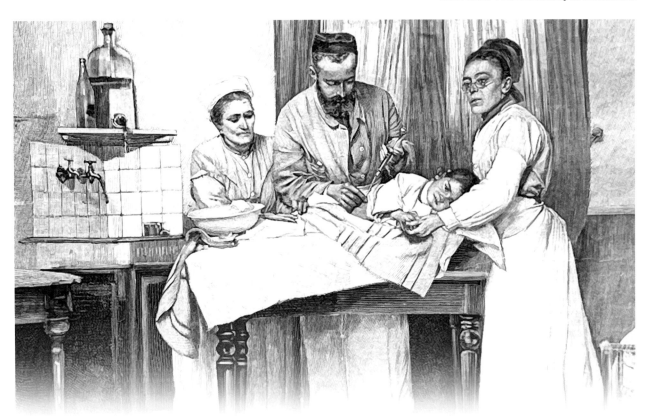

opportunity to respond to the increasing demand and to make a profit. Vaccine farms would also produce and commercialize diphtheria and tetanus antitoxins from the blood of horses and sheep.

The late 1800s saw the explosion of bacteriology and medicine, which coincided with the acceleration of the electrical industrial revolution. Colonial regimes were the first to set up tropical public health programs, primarily to protect their investments, their labor force, upon whom many vaccines were tested and imposed.

By the beginning of the 19th century, vaccine materials were shipped globally. During this time, variolation evolved into live-viral smallpox vaccination and by the end of the century, tetanus and diphtheria antitoxins, plague, cholera, and typhoid antiserums, as well as rabies vaccines, had been developed. Pasteur Institutes were established globally. This was a period characterized by an awakening in public health and recognition of the importance of sanitation. Social reforms were implemented to improve cleanliness through the treatment of waste, access to clean water, and better living and working conditions.

In the early 1900s, laws were enacted to protect the public from contaminated and unsafe products derived from the coal-tar and chemical dye-based

companies that were the predecessors to the modern pharmaceutical industry. Medical education was also transformed during this period, steering the focus away from natural remedies to petroleum-based pharmaceuticals. The First World War was the first international conflict where chemical and biological weapons were used strategically by both sides on the battlefield. The end of this war was swiftly followed by another lethal event with the global outbreak of the Spanish Flu, killing more people than the war.

By the 1920s, many vaccines were marketed and available worldwide for both civilians and the military. This era also saw the expanded use of surgery and radiotherapy. The evolution of sanitary conferences and international collaboration marked the beginning of the first global health organizations, responsible for setting standards and coordinating public health intelligence on mortality and morbidity worldwide—notably for managing shipping ports. This decade also saw the evolution of eugenics into a global movement to improve the human race, leading to the creation of sterilization laws in the following decade in the US and Europe.

In the 1930s, vaccines for pneumonia, influenza, meningitis, yellow fever, typhoid, rabies, tetanus, tuberculosis, diphtheria, and pertussis were available.

1898 Anti-tetanic serum by Parke, Davis & Co. © Division of Medicine and Science, National Museum of American History, Smithsonian Institution

Data on the uptake of these vaccines is unknown, but army troops were often the first to be injected, along with institutionalized adults and children. The electron microscope was invented, and for the first time scientists could observe nanosized organisms, leading to the identification of the influenza H1N1 strain.

The 1940s bore witness to the atrocities of the Second World War with its targeted genocide and abhorrent treatment of humans in the name of science, not only in Germany, but also in Japan. The end of the war inspired the creation of the Nuremberg Code—a set of ethical research principles designed to protect people from inhumane experimentation, highlighting the importance of informed consent. By the late 1940s, the first triple combined vaccine DTP was commercialized as well as the freeze-drying technique, which revolutionized vaccine production and storage. This decade marked the beginning of the Cold War and saw the establishment of the United Nations and the World Health Organization. It was also the the decade of the "Green Revolution" which repurposed the nitrogen-based explosives of WWII into synthetic fertilizer and pesticides for high-yield industrialized agriculture.

By the 1950s, deaths from infectious diseases were in steep decline except in countries where poverty, inadequate access to healthcare, and unsanitary living conditions prevailed. Only five vaccines—smallpox, combined diphtheria, tetanus, whole-cell pertussis, and BCG* tuberculosis—were routinely recommended. Typhoid and yellow fever vaccines were given to troops, as well as to travelers heading to Africa, South America, or Asia. Influenza vaccines were obligatory for soldiers in the US. This decade saw the characterization of DNA and the invention of cell culture methods that would change the landscape of vaccine development. Cell culture methods reduced the need for expensive live animals and paved the way for advances in molecular biology, as well as industrial mass production. Scientists would discover viral contamination in the recently released and highly publicized Salk polio vaccines with the oncogenic monkey virus SV40, and also identified other viral contaminants in mass-produced vaccines. This would influence the shift in research on viruses as the possible cause of cancer in the next decade.

The 1960s saw the development of the first antiviral drugs and the birth of psychiatry. Major advances in vaccine development allowed for the licensing of vaccines against polio (Sabin's live oral version), measles, mumps, and rubella as well as the first inactivated split-virion vaccines against influenza. Following the thalidomide tragedy, the FDA enacted laws that required proof of efficacy from pharmaceutical manufacturers before licensing new drugs in the USA. By the late 1960s, the US military had developed a formidable biological arsenal that included numerous biological pathogens, toxins, and fungal plant pathogens

Though available, BCG was never routinely recommended in the USA.

to induce illness, death, crop failure, and famine. This period also witnessed the mass medication of the American people, mainly women, with benzodiazepine-based neuroleptic drugs like Valium® and Librium®.

The 1970s were defined by the acceleration of biotechnology, creation of recombinant DNA, and President Nixon's declared war on cancer. This initiated an intensive research program into viruses and also advanced biodefense projects using viruses as possible pathogenic agents for weapons. The recombinant DNA era offered wide opportunities for the enhancement of germ warfare technology. By the end of the decade, new vaccines against meningitis and pneumonia were available, and the first "anti-cancer" vaccine, Merck's hepatitis B vaccine, Heptavax® came into existence.

The 1980s introduced the first vaccines made from genetically modified organisms—typically bacteria and yeast—while older vaccines were updated to include synthetic biologically active substances. In the USA, vaccine makers were granted immunity from injury caused by their products in order to ensure a continued supply of vaccines, which had been threatened due to increasing litigation. There was also a paradigm shift in that vaccine manufacturers and academic researchers were now heavily invested in filing patents for their work and receiving royalties. AIDS also arrived on the scene, and despite the declaration of a vaccine being available in less than two years, after billions in funding and forty years of research, there is still no approved vaccine. By the late 1980s, new vaccines against *Haemophilus influenzae* type b were added to the childhood schedule.

By the 1990s, biotech and the recombinant genetic technologies had revolutionized vaccine manufacturing. The ability to file patents on taxpayer-funded inventions encouraged an increasingly aggressive attitude to the protection of intellectual property. This created an ideological move towards vaccines being a liability-free, profitable product—not just a public health tool. Several new and improved vaccines arrived promising protection from meningitis, rotavirus, pertussis, and pneumococcal infections. The childhood vaccination schedule expanded considerably with an ever-growing list of increasing doses given as combination shots. The end of the Cold War also significantly opened previously closed communist countries to vaccine manufacturers looking to expand their global markets.

By the end of the 20th century, gene editing became common practice and most vaccines on the market were genetically engineered and patented.

Several new pentavalent and hexavalent vaccines were added to the vast portfolio of modern and proprietary vaccine products. Together with an accelerating biosecurity agenda, extensive public and private funds were allocated to the development of defensive biological warfare countermeasures, with it advancing research into the creation of completely non-natural pathogens.

The 21st century welcomed the new age of nano-technology with its various applications in medicine and bioengineering. Synthetic biology, artificial intelligence, bioinformatics together with nanotechnology have defined the progress in the early 2000s, promising to advance medical treatments, as well as new biological weapons. With the recently commercialized mRNA platform, the biotech industry—in partnership with vaccine makers—has developed an impressive pipeline of vaccines for both infectious and chronic diseases (cancer) using novel technologies. The first ever synthetic DNA and mRNA vaccines were approved, opening the doorway to a "bio-nano era" of many more synthetic organisms and foreign genetic information to be included in future biologicals. The market for vaccines to prevent and treat cancer, heart disease, and autoimmune disorders is driving investment and innovation in the field of precision vaccinology, using protein engineering and in silico research, designing genetic payloads as active ingredients. These technologies have the advantage of being easily customizable and scaled-up, as well as cheaper to produce than traditional vaccines.

Every effort has been made to make the following vaccine timeline as comprehensive as possible, focusing primarily on the routinely recommended vaccines available for civilians and troops in the USA, Europe, and Switzerland—including some references to Africa, Asia, Australia, and South America.

This timeline is followed by important complementary information on discontinued vaccines, different vaccine types and schedules, along with material on vaccine safety concerns, including a summary of IOM reports. It concludes with the comprehensive illustrated timeline of vaccine manufacturers and pharmaceutical beginnings, acquisitions, and mergers up to mid-2024.

All events directly related to vaccines are highlighted in the color of this background box, and dates showing the introduction of vaccines are marked with a star. ✪

A Chronology of Vaccination

YEAR	EVENT
	VACCINE-RELATED EVENT*

❂ **Vaccines introduced**

570	"Variola" a.k.a. smallpox is named by Bishop Marius of Avenches, Switzerland. The word is derived from the Latin *varius* meaning "stained" or from *varus*, meaning "mark on the skin."
700	First accounts of variolation in China.
910	Persian physician Rhazes distinguishes measles from smallpox and notices the acquired immunity to smallpox in those who survive the disease.
1063	

Smallpox inoculation, or variolation: Chinese text is written by a Buddhist nun describing the practice of grinding scabs of smallpox victims into fine powder and blowing it into the nostrils of non-immune people.

1167	An elaborate sewage system separating running water, rain drainage, and waste is first illustrated as a plumbing diagram by Christchurch monastery in Canterbury, UK.
1346	The Siege of Caffa (a Genoese-controlled seaport) is reported as being the first recorded incidence of biological warfare. Victims of the plague are used as weapons by the Mongol-Tartar army, who catapult plague-infected cadavers into the city. The disease is transmitted to the inhabitants, and fleeing survivors spread the plague from the Crimean city of Caffa (now Feodosia, Ukraine) to the Mediterranean Basin.
1347	Bubonic plague a.k.a The Black Death begins, killing an estimated 75 million people. It originates in Asia, spreading to Crimea, then Europe and Russia.
1357	The term *influenza* is first used in Italy to describe a disease, meaning "influence."

1400s	At the end of the 15th century, the term "small pockes" is first used in England to distinguish the disease from syphilis, which was then known as the great pockes.
1500s	Pertussis (whooping cough) is described for the first time and named in 1670.
1500s	This period is widely acknowledged as the period when variolation was prevalent in China and India.
1507	Smallpox becomes endemic in the West Indies 15 years after Columbus's arrival.
1510	The first recognizable influenza pandemic invades Europe from Africa and proceeds northward to involve all of Europe and then the Baltic States.
1549	*Douzhen Xinfa (Essential knowledge and secrets of pox diseases)* is published in China and believed to be the oldest record of the variolation method.
1596	The flush toilet is invented by Sir John Harington, godson of Queen Elizabeth I, but does not become widespread until 1851 (with modern sewer systems and plumbing).
1557	A highly fatal spring pandemic is the first documented global influenza outbreak which spread westward from Asia to Europe—deaths were recorded as being due to "pleurisy and fatal peripneumony."
1580	Another influenza pandemic appears to sweep over the entire globe, spreading east to west from Asia, noted for its fast progression and high fatality.
1600s	There is still general confusion between syphilis (the Great pox or French pox) and variola (smallpox).
1600	The British East India Company is founded and becomes one of the world's first and most dominant corporations, as well as the most powerful monopoly. It will be instrumental in exporting British imperialism, slavery and pushing opium consumption in China. The company is dissolved in 1874.

*Includes events relating to variolation and inoculation.

1611

Italy issues a health certificate enabling free passage despite quarantines and heavy restrictions on travel due to the plague.

1628 British physician William Harvey discovers the circulation of blood. The first known blood transfusion is attempted and fails.

1657 The first recorded outbreak of measles occurs in Boston, USA.

1659 The first recorded outbreak of the "throat distemper" (also known as quinsy and angina), today called diphtheria, occurs in the USA.

1660s French philosopher René Descartes posits the theory of the body as a machine, separate from the mind. Reductionism and materialism are born.

1661 Royal support for smallpox inoculation is given by the Chinese emperor.

1665 The first successful blood transfusion is recorded in the UK. A dog is kept alive by transfused blood from other dogs.

1665 The Great Plague of London lasts a year, ending with The Great Fire in 1666.

1676 Dutch scientist Antonie van Leeuwenhoek reports the discovery of microorganisms, seen through his single-lensed microscope.

1695 Chinese physician Zhang Lu publishes the first detailed description of variolation.

1701 The US state of Massachusetts passes laws for isolation of smallpox patients and for ship quarantine as needed. A year later, inoculation with material from smallpox scabs would be accepted as an effective means of containing disease when there was a declared epidemic.

1712–1853 England introduces a soap tax, tripling the price of this essentiel item. This makes soap a luxury product, unaffordable for the poor.

1721 Reverend Cotton Mather, the son of Increase Mather, the president of Harvard University, uses variolation during a smallpox epidemic in Boston, USA.

1721 Variolation is brought to England by Lady Mary Wortley Montagu, wife of British ambassador to Turkey, who variolates her two children. Her daughter is allegedly the first person to be variolated in England. Variolation could spread the disease and there was a reported 2–3% fatality rate. Tuberculosis and syphilis are complications arising from this method that is eventually banned in 1840.

1729 An influenza pandemic is first detected in Russia, then spreads westward and southward into Europe. After 1733, there is significant worldwide influenza activity, especially in 1737 to 1738, when America and Europe are invaded in the same month, and in 1742 to 1744, when European deaths associated with influenza-like disease reach extraordinary heights. The period of 1729–1747 is considered one of the more remarkable and epidemiologically chaotic times in the history of pandemic influenza.

1735 An epidemic of "throat distemper" sweeps through Kingston, New Hampshire, as well as Massachusetts and Maine, affecting children, claiming 1,500 lives in New Hampshire alone.

1747 British physician Dr. James Lind leads the first controlled trial on sailors, testing and confirming the protective properties of citrus fruits (vitamin C) against scurvy.

1758 Scottish physician Francis Home inoculates children with morbilous matter or measles.

1761 An influenza pandemic is reported to begin in the Americas and spreads to Europe and around the globe by the next year.

1763 Smallpox is used as a biological weapon, with smallpox-infected blankets distributed to American Indians during wartime. British Army officer and commander-in-chief of the forces in the British Army, Jeffery Amherst, writes to Captain Simeon Ecuyer, a 22-year veteran Swiss mercenary in the British service, "Could it not be contrived to send the smallpox among the disaffected tribes of Indians? We must on this occasion use every stratagem in our power to reduce them."

1767 Chickenpox is distinguished from smallpox by English physician William Heberden.

A Chronology of Vaccination

1774–1812

1774 British Dorset dairy farmer Benjamin Jesty deliberately infects his wife and two sons with cowpox hoping to prevent smallpox. His wife becomes very ill and although she eventually recovers, Jesty is vilified. This is considered the first recorded "vaccination."

1775 Swedish chemist Carl Scheele develops Paris Green, an arsenic-based paint pigment. The highly toxic substance is later widely used to kill rats in Paris sewers.

1775 During the American Revolutionary War, American General George Washington suspects British General William Howe of weaponizing smallpox by deliberately spreading it among American troops. Recently variolated civilians are allegedly sent into Continental Army encampments.

1777 US General George Washington orders the mandatory variolation—inoculation with smallpox—of the entire Continental Army.

1780 An influenza pandemic is recorded as beginning in Southeast Asia, then spreads to Russia and eastward into Europe. It is noted for its high cases and low mortality.

1788 A major outbreak of measles kills thousands in the city of Philadelphia, Pennsylvania, and in New York City.

1788 Another pandemic era is initiated and global influenza activity appears to intensify for almost 20 years until 1806.

1789 English physician Edward Jenner conducts the first experiment on his 18-month-old son, inoculating him with swine-pox.

1793 The first epidemic of yellow fever is recorded in Philadelphia, USA.

1796 Edward Jenner claims protection against smallpox using cowpox (and horse grease, or horsepox). His method involves taking fluid from a cowpox sore and scraping it into someone else's skin. He experiments on 8-year-old James Phipps and his own son, both of whom die from tuberculosis (consumption) in their early 20s.

1796 The *Bibliothèque Britannique,* a monthly magazine, is founded in Geneva and becomes a virtual propaganda sheet for vaccination in Napoleon's expanding French empire. The Swiss, notably the Genevans, come to play an important role in the early transmission to the Continent of knowledge about vaccination.

1798

The US federal government establishes the Marine Hospital Service to provide health services to seamen. To recognize its expanding quarantine duties, Congress changes the service's name to the Public Health and Marine Hospital Service in 1902 and then to the Public Health Service in 1912.

1798 ✪ Jenner publishes an *Inquiry into the Causes and Effects of the Variolae Vaccinae, a Disease Discovered in Some of the Western Counties of England, Particularly Gloucestershire, and Known by the Name of the Cow Pox.* This document outlines Jenner's "success" in protecting James Phipps from smallpox infection with material from a cowpox pustule, in addition to 22 related cases. Initially it receives little attention and is rejected by the Royal Society. This changes when Henry Cline, an associate of Jenner living in London, uses dried vaccine material provided by Jenner to demonstrate that vaccination with cowpox prevents future smallpox infection. Jenner receives £10,000 in 1802, then £20,000 in 1807 as support from the UK government.

1798 The *Act for the Relief of Sick and Disabled Seamen* is signed into US law, authorizing the deduction of 20 cents per month from wages to fund medical care for sick and disabled seamen, and to finance building hospitals to treat seamen.

1799 The first reports of vaccination failure and fatalities in people inoculated against smallpox are documented.

1800–1899

1800s Leukemia is first described as a white blood cell disease, initially named by German physician and biologist Rudolf Virchow in 1847, who is also known as "the father of modern pathology."

1801 It is estimated that 100,000 people in Europe are vaccinated against smallpox.

1801 French medical student and library assistant Jacques Louis Moreau de la Sarthe gathers evidence on vaccination and writes *Traité historique et pratique de la vaccine,* that will be translated into Spanish with hundreds of copies sent with the Royal Smallpox Expedition.

1801 ✪ Benjamin Waterhouse, physician, co-founder, and president of Harvard Medical School begins using "cowpox vaccine," leading Massachusetts to become the first state to encourage smallpox vaccination. Together with US President Thomas Jefferson, they spearhead the US smallpox vaccination campaign.

1801–1802 ✪ Sweden is one of the first countries to adopt Jenner's smallpox vaccine.

1803

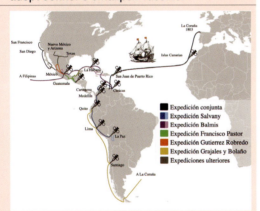

King Carlos IV of Spain commissions the "Royal Philanthropic Vaccine Expedition" to bring smallpox vaccine to the Spanish colonies of the New World. Orphans are shipped, and used as vessels for smallpox vaccine material for arm-to-arm inoculation. In the Americas, the Catholic Church spreads information about the vaccine and coordinates vaccination brigades.

1803 The first act to regulate child labor is passed in the UK.

1803 At the first meeting of the Royal Jennerian Society, Edward Jenner insists that the origin of the term vaccination (Latin for cow, *vacca*) be credited to his friend and fellow physician, Richard Dunning.

1804 ✪ Smallpox vaccine is introduced in Australia, imported from the UK.

1805 Swiss physician Gaspard Vieusseux first describes invasive meningococcal disease. He first differentiates cerebro-spinal fever.

1806 The first regulations requiring smallpox vaccination are passed in the Napoleonic Principalities—today part of Italy.

1807 The UK slave trade is abolished with slave owners compensated for their "losses."

1807 Smallpox vaccination is mandated in Bavaria (present day Germany) and subsequently in Denmark (1810); Sweden (1814); Wurtemberg, Hesse, and other German states (1818); Prussia (1835); Romania (1874); Hungary (1876); and Serbia (1881).

1808 The Act Prohibiting Importation of Slaves takes effect, a federal law forbidding new slaves from being permitted into the USA.

1808 State-supported facilities for vaccination are set up in England with the founding of the National Vaccine Establishment.

1809 The town of Milton, Massachusetts, becomes the first to offer free smallpox vaccination, which is followed the next year with the first US state law mandating vaccination for the general population.

1810 The Karolinska Institutet is founded in Sweden by King Karl XIII as an "academy for the training of skilled army surgeons" to improve hospital care of wounded soldiers. In 1895, Alfred Nobel's testament bequeaths the institute with the right to select the Nobel Prize in Physiology or Medicine.

1810 France requires university students to receive smallpox vaccine.

1810 In the UK, the *Medical Observer* reports on 535 cases of smallpox after vaccination, 97 deaths and 150 post-vaccination injuries.

1812 The US War Department orders variolation to be replaced by smallpox vaccination, and the US Army mandates vaccination for all troops.

A Chronology of Vaccination

1812–1853

1812 The first edition of the medical journal *The New England Journal of Medicine* is published in July.

1813 US President James Madison signs into law An Act to Encourage Vaccination, which creates the first US National Vaccine Agency (now part of the Department of Health and Human Services) led by Baltimore physician James Smith. It would be the first federal law concerning medical products.

1814 Paris Green (copper acetate triarsenate), an arsenic-based toxic paint becomes commercially available, used in clothes, wallpapers, and even toys. By the late 1800s its neurotoxicity is fully appreciated and it is used as a pesticide. Lead arsenate replaces Paris Green in the 1890s and becomes popular in food crops.

1815 The largest volcanic eruption in recorded history occurs in Indonesia. Mount Tambora explodes for months, leading to the 1816 Year Without A Summer, causing massive crop failures and starvation.

1816 Sweden introduces one of the most severe vaccination laws in Europe, making smallpox vaccine obligatory for all children under two years old. This mandate lasts until 1976.

1817 The first in a series of deadly cholera epidemics sweep over India, China, and Japan.

1818 British obstetrician James Blundell uses a syringe to successfully tranfuse human blood in a woman for the treatment of postpartum hemorrhage.

1820 A postindustrial revolution epidemic of hay fever is reported from 1820–1900.

1821 Hay fever is first described by English scientist John Bostock using the term "Summer Catarrh."

1821 Scottish anatomist, neurologist, artist, and surgeon Sir Charles Bell shows that lesions of the seventh cranial nerve produce facial paralysis (now termed Bell's palsy).

1823 The first edition of the medical journal *The Lancet* is published in October.

1826 Diphtheria is named "diphtérite" by French physician Pierre Bretonneau, distinguishing it from scarlet fever.

1827 Boston is the first city to require smallpox vaccination for public school students.

1828 London introduces sand filtration of its water supply.

1832 Fatal cholera outbreaks occur in France, the UK, and the USA, with thousands killed.

1833

The UK Royal Commission, under the first Factory Act, stipulates that children aged 13–18 should work a maximum of 12 hours per day, children aged 9–13 a maximum of 9 hours, and children under the age of nine are no longer permitted to work. No children are allowed to work at night. However, these regulations apply only to children working in the textile factory mills, not for child laborers in coal mines or other industries.

1833 The first US patent for a flush toilet is awarded.

1834 An epidemic of polio is first described by Sir Charles Bell on the Island of Saint Helena.

1835 The Cruelty to Animals Act passes in the UK to protect animals, cattle in particular.

1835 Vaccination and revaccination is made compulsory for all children in Germany.

1836 A method for increasing potency is designed by English physician Edward Ballard, using new strains of cowpox and reintroducing lymph back into cows, enabling sufficient supply for mass smallpox vaccination.

1836 Muscular dystrophy is described by Italian physicians G. Conte and L. Gioja.

1839 The registrar general of Great Britain begins publishing annual reports of births, deaths, and marriages. The first report is mostly compiled by British epidemiologist William Farr, regarded as one of the founders of medical statistics.

1839 Opium is the most traded commodity, controlled by the British East India Company, incorporated by royal charter in 1600.

1840s The first outbreak of meningococcal disease is recorded in Africa.

1840 The British Vaccination Act is established, formalizing the use of optional smallpox vaccine and abolishing variolation. There will be amendments in 1853, 1867, 1871, 1874, 1898, and 1907, with the law repealed by the National Health Service Act in 1946.

1840 In England, local Poor Law institutions are required to appoint qualified medical practitioners and provide free vaccination.

1840 The method of "animal vaccine" is successfully used by Giuseppe Negri in Naples, Italy—he succeeds in continously carrying smallpox vaccine in calves. This technique is confined to Italy for 20 years, until it is discussed at the 1860 Medical Congress in Lyon, France.

1841 Rubella, meaning "little red," is first described in German medical literature, hence the name "German measles."

1847

The American Medical Association (AMA) is founded and incorporated ten years later.

1847 The Factories Act is passed in the UK and limits both adults and children to 10-hour work days.

1848 The American Association for the Advancement of Science is founded in the Academy of Natural Sciences in Philadelphia, Pennsylvania, as the first permanent organization to "promote science and engineering nationally and to represent the interests of American researchers from across all scientific fields." It becomes the world's largest general scientific society.

1848 A severe cholera outbreak in the UK lasts two years and kills over 50,000 people.

1848 The US Navy replaces obligatory variolation with smallpox vaccination.

1848 Hungarian physician Ignaz Semmelweis demonstrates that disinfecting hands and surgical instruments with chlorated lime drastically reduces the number of women dying after childbirth. His recommendations are ridiculed by the medical establishment and he dies in a mental institution in 1865.

1849 A major outbreak of cholera occurs in London, killing around 15,000 people. Early industrialization makes London the most populated city in the world, with its River Thames heavily polluted by untreated sewage. English doctor John Snow proposes that cholera is spread by contaminated water and rejects the commonly held idea that "bad air" or miasma is responsible.

1849 French physician Charles-Paul Diday publishes a report in his journal *Gazette Médicale de Paris* on the development of an anti-syphilitic vaccine.

1851 The French government organizes the first International Sanitary Conference in Paris to discuss cholera and the urgent need for harmonized sanitary maritime laws—twelve European countries are present. Between 1851 and 1912, twelve sanitary conferences take place to address quarantine regulations and similar measures in a series of sanitary conventions—the objective being the protection of the state from external threats, not the well-being of individuals.

1851 Taxes on windows—enforced since the early 1700s—are repealed in the USA. Lack of windows tend to create dark, damp tenements which are a source of disease.

1852 Massachusetts is the first US state to mandate school for children, passing a compulsory public education law.

1852 US Navy doctor Edward Robinson Squibb develops a safe method of steam-distilling ether for use as an anesthetic. Instead of patenting his technique, he publishes his research freely and shares his discovery.

1853 The British Vaccination Act makes smallpox vaccination mandatory for all infants in the first 3 months of life—parents who do not comply are liable to a fine or imprisonment.

A Chronology of Vaccination

1853

Scottish doctor Alexander Wood develops the first all-glass syringe with a plunger and a needle fine enough to pierce the skin. It is first used to inject pain relief medicine. This hypodermic syringe makes mass vaccination possible. That same year, French surgeon Charles Pravaz creates a similar instrument.

1853 The worst yellow fever epidemic strikes New Orleans, with over 12,000 deaths—about 10% of the city's population perish.

1854 During the Soho epidemic, John Snow traces a cholera outbreak to a London water pump, confirming that disease is spread through contaminated drinking water.

1854 Tetraethylpyrophosphate (TEPP) is synthesized as the first organophosphate cholinesterase inhibitor by French chemist Philippe de Clermont and used as pesticide.

1855 Massachusetts passes the first US state law mandating vaccination for schoolchildren.

1855 The Children's Hospital of Philadelphia is founded as the nation's first pediatric hospital.

1856 Deaths from erysipelas—a bacterial infection of the upper dermis—after vaccination is first recorded in England and Wales by the registrar-general (until 1881).

1857 The first edition of the medical journal *The British Medical Journal* is published in January, the combined product of the 1840 established *Provincial Medical and Surgical Journal* that merged with *London Journal of Medicine*—in January 1853 published as the *Associated Medical Journal.*

1857 A two-year smallpox epidemic in England takes over 14,000 lives.

1858 Charles-Paul Diday proposes that everyone, not just travelers and foreigners, should possess a medical certificate of health and disease—essentially a "sanitary passport."

1858 The British Medical Act is passed that introduces a register of all licensed doctors and establishes the General Medical Council. The new Public Health Act replaces the 1848 version, giving power to the Medical Department of the Privy Council to supervise vaccination services.

1858 The Great Stink in London occurs—the River Thames becomes a cesspool of untreated human waste, causing an overpowering stench so putrid that the British House of Commons suspends its sessions. This leads to sanitation reforms and the construction of an underground sewage system.

1858 British anatomist and surgeon Henry Gray first publishes *Gray's Anatomy*, which becomes one of the most influential anatomy textbooks.

1859 The French War Office orders the smallpox vaccination of all recruits on entering service.

1859–1922 In England there are over 1,600 officially reported deaths following smallpox vaccine.

1860 English social reformer and nurse Florence Nightingale promotes "hygiene, sanitation, fresh air and a good diet" during the 1853–1856 Crimean War, returning to London to fund the St. Thomas's Hospital and Nightingale Training School for Nurses.

1862 New York mandates vaccines for schoolchildren, followed by Connecticut (1872), Indiana (1881), Arkansas, Illinois, Virginia, Wisconsin (1882), California (1888), Iowa (1889), and Pennsylvania (1895).

1862 French chemist Louis Pasteur first publishes his research on airborne microbes causing decay and suggests that specific germs are responsible for disease.

1862 Louis Pasteur together with French physiologist Claude Bernard complete the first pasteurization of milk. They discover that heating milk prevents bacterial growth.

1863 Supporters of the Confederate States of America (slave-holding southern states) sell clothing from yellow fever and smallpox victims to Union troops during the US Civil War (1861–1865).

1863

ICRC

The International Committee of the Red Cross (ICRC) is created by five Swiss citizens in Geneva, Switzerland, initiated by banker Jean Henri Dunant. Together with lawyer Gustave Moynier and physician Louis Appia, ICRC's mission is to improve conditions for wounded soldiers during conflicts. For his work, Dunant is awarded the first ever Nobel Peace Prize in 1901.

1863 A two-year smallpox epidemic in England claims over 20,000 lives.

1864 Awareness of the dangers of spreading diseases such as syphilis via arm-to-arm transmission of vaccine virus begins to grow in France. The use of calves for smallpox vaccine material becomes widespread.

1865 Scottish surgeon Joseph Lister elaborates the theory and practice of antiseptic surgery, which includes washing the hands with carbolic acid to prevent infection.

1866 The National Anti-Vaccination League is founded in the UK. They consider vaccination an invasive, unsanitary, and sometimes disfiguring procedure, causing blood-borne diseases, infections, and gangrene.

1866 Scottish Army surgeon Henry Veale names the disease "rubella," a mild and transient illness characterized by a rash with little to no fever.

1866 The last cholera outbreak occurs in the UK.

1866 The Sanitary Act in England gives local authorities new powers to supply clean water and regulate conditions of tenements.

1867 Paris Green is first used as a pesticide.

1867 The British Vaccination Act extends obligatory smallpox vaccination to children 14 years old, with cumulative penalties for non-compliance.

1867 Swedish chemist and businessman Alfred B. Nobel invents and patents dynamite.

1868 In an effort to regulate the sale of opium in the UK, the Pharmacy Act is passed, which requires opium to be labeled as a poison.

1868 French chemist Louis Pasteur suffers a severe brain stroke that paralyzes his left side—he partially recovers.

1869 While working at the laboratory of Felix Hoppe-Seyler, Swiss physiological chemist Johannes Friedrich Miescher discovers the unique macromolecule in the nucleus of every living cell, later called "nucleic acid" or DNA.

1869 The Leicester Anti-Vaccine League is set up.

1869

The first edition of the prestigious scientific journal *Nature* is published in November.

1869 Swiss surgeon Jacques-Louis Reverdin is credited with the first documented surgery of a skin transplant.

1869 The Suez Canal, an artificial sea-level waterway between the Mediterranean and the Red Sea, opens in Egypt after ten years of construction. This transportation milestone initiates the establishment of rigorous public health sanitary protocols for monitoring the passage of migrants and pilgrims to prevent the spread of disease.

1870s The Leicester Method is implemented in England—a method of managing smallpox by prioritizing notification, isolation, quarantine, and disinfection.

1870–1872 One of the worst European smallpox epidemics occurs, lasting two years, coinciding with the Franco-Prussian War.

1870 A condition of receiving a medical degree in the UK includes "proficiency in vaccination."

1871 The New York Department of Health becomes the first municipal agency in the US to produce its own smallpox vaccine.

1873 Financial panic takes place in Europe and the US, marked by the collapse of several banks and 89 American railroad companies.

1874 The Brussels Conference Act of 1890—measures against slavery and importation of firearms and ammunition into Africa—is adapted to prohibit the use of poison weapons and chemical warfare.

THE
LANCET

"...syphilitic infection by vaccination is not only to be expected on theoretical grounds, but it is actually of common occurrence.

I am quite aware that there is now a sort of common consent among medical writers to gloss over the evils that may be attendant upon vaccination, for the sake of its great and manifest benefits. In this course I cannot concur; because I think that the evils should be fairly recognised, and that they should be neutralised by proper and scientific precautions."

—"Vaccination and Syphilis"
Robert B. Carter, *The Lancet*, 13th June 1868
Letter to the Editor of *The Lancet* written by a British surgeon,
obtained from The British Library

The New York Times

"This morning we publish a careful and elaborate letter, which puts the case against vaccination very strongly, and fortifies it by the citation of many remarkable facts and authorities. The writer holds vaccination to be a monstrous infatuation, and an unmixed curse to mankind. He claims that it does not in any degree protect against small-pox, and that it does not even mitigate the severity of the disease. On the other hand, he offers evidence to show how other and very terrible diseases often spring from vaccination. Such are insanity, cancer, scrofula and tubercular diseases.

The facts that he cites on this subject are too appalling, and the opinions he quotes from eminent medical men are too direct and pointed to leave room for doubt of their meaning. One of these physicians declares the 'vaccine has made murder a legal act;' another that 'the vaccine virus is a poison;' a third that 'vaccine does not only not protect, but it also produces other diseases;' a fourth says, 'If I had the desire to describe one-third of the victims ruined by vaccination, the blood would stand still in your veins;' a fifth asserts 'it is a curse to humanity.'"

— **"Is Vaccination Dangerous?"**
***The New York Times*, 26th September 1869**
ProQuest Historical Newspapers obtained from The British Library

A Chronology of Vaccination

1874–1883

1874	Switzerland introduces mandatory public education for children.
1874	The Imperial Vaccination Law makes smallpox vaccine free in the German Empire. Vaccination in infancy and revaccination at age twelve become compulsory. Vaccines are manufactured by state-run production sites and supervised by local authorities.
1874	The National Anti-Compulsory Vaccination League in England is established, leading to acts of civil disobedience and imprisonment.
1874	Smallpox vaccination is made compulsory in Romania and in the state of Victoria, Australia.
1875	A feat of civil engineering, the main drainage system of London, is successfully completed.
1876	The New York Board of Health establishes a vaccine farm with calves to reduce transmission of human illnesses such as leprosy and syphilis, which have been reported as adverse events of smallpox vaccination due to product contamination.
1876	Johns Hopkins is founded as America's first research university.
1876	The British Cruelty to Animals Act is amended, setting limits on the practice of, and instituting a licensing system for, animal experimentation. It is the world's first legislation to regulate the use and treatment of live animals in scientific research.
1876	The government in Japan decides that all people should be vaccinated against smallpox.
1876	German physician Robert Koch links *Bacillus anthracis* to the infectious disease of anthrax.
1877	Louis Pasteur proposes the germ theory, declaring that diseases are caused by microorganisms. By the end of the 1800s, this replaces the miasma theory proposed by Greek physician Hippocrates in the 4th century BC, who attributed illness to foul air and bad vapors emanating from rotting organic matter.
1878	A yellow fever epidemic breaks out in Memphis, USA, spreading to New Orleans.

1878	In an English Parliamentary returns session, it is stated that "25,000 children are slaughtered annually by disease inoculated into the system by vaccination, and a far greater number are injured and maimed for life by the unholy rite."
1879	Louis Pasteur produces the first laboratory-developed vaccine: the bacterial vaccine for chicken cholera Pasteurella multocida.
1879	The Anti-Vaccination Society of America is founded in response to states enacting smallpox vaccine mandates.
1879	*The Vaccination Inquirer* is established by English social reformer, William Tebb.
1879	A first version of the law on epidemics in Switzerland is met with resistance from opponents of compulsory vaccination and is accompanied by the rejection of state intervention in the name of individual freedom. This law is refused by referendum in 1882. Two years later a less ambitious text without mandatory vaccination and with a state of emergency reduced to four diseases (cholera, tuberculosis, typhoid, smallpox) is presented to the population and eventually accepted in 1886.
1880	American army physician (considered the first US bacteriologist) Brigadier-General George Sternberg isolates the bacteria *Streptococcus pneumoniae*.
1880	The first edition of the general scientific journal *Science,* founded by Thomas A. Edison, is published in July by the American Association for the Advancement of Science.
1880	The London Society for the Abolition of Compulsory Vaccination is founded.
1881	 UK adopts the practice of using calves as the source of smallpox vaccine material.

1881 Smallpox vaccine is compulsory for schoolchildren in 10 out of 22 Swiss cantons.

1881 Pasteur publicly demonstrates the efficacy of his anthrax vaccine to the French public in a famous experiment using 50 sheep at Pouilly-le-Fort. In 1995, American historian Gerald Geison reveals, after 18 years of studying hundreds of Pasteur's notebooks, that Pasteur had deceived the public by using a vaccine he had not prepared. Instead of his own published method of air attenuation, Pasteur used a chemical method of attenuation associated with French veterinarian Jean-Joseph Henri Toussaint. Other discrepencies are found, which may explain why Pasteur refused public access to his notebooks, even after his death in 1895.

1881

The International Medical and Sanitary Exhibition takes place in London, showcasing products by Wyeth, Burroughs Wellcome & Co., and many others.

1882

The vaccine farm "Laboratories of the National Vaccine Establishment" is founded in Washington, DC, and managed by Ralph Walsh, MD, and bacteriologist Dr. William F. Elgin—who would later be hired by H.K. Mulford. Most of vaccine farms are "initiated by medical doctors who saw an opportunity to respond to the ever increasing demand of smallpox vaccine from individuals and from health authorities and to make a profit."

1882 The Anti-Vaccination League of America holds its first meeting in New York.

1882 German physician Robert Koch announces his isolation and discovery of the agent that causes mycobacterium tuberculosis.

1883 Swiss pathologist Edwin Klebs identifies the bacterium that causes diphtheria.

1883 German public prosecutor and mayor, August Lürman, reports an epidemic of icterus (jaundice) following smallpox vaccine at a shipyard in Bremen, Germany, and is the first identification of an epidemic of viral hepatitis, type B, a.k.a. hepatitis B.

1883 German chancellor Otto von Bismarck passes legislation to provide workers with insurance, providing both an income and covering medical care in the event of illness or injury. Funded by contributions from both employee and employer, Bismarck's model of health insurance becomes a blueprint for insurance-based welfare services across Europe and beyond.

1883 The American Anti-Vivisection Society is established in Philadelphia by Caroline Earle White, with the goal to eliminate the number of cruel procedures done by medical and cosmetic groups.

1883

Charles Haccius of Lancy near Geneva establishes the Institut Vaccinal Suisse, a vaccine production company. Soon after launching, he secures contracts with nine Swiss cantons, including Geneva, that guarantee acceptance of his smallpox vaccine, which would later become known as "Lancy-Vaxina"—it would remain in the product portfolio for almost 100 years.

1883 The first edition of *The Journal of the American Medical Association,* now known as *JAMA,* is published in July.

A Chronology of Vaccination

1883–1890

1883 Cocaine is synthesized by Merck in Germany and used as the first synthetic local anesthetic, replacing natural substances like opium, mandrake, and cannabis. A year later, Austrian psychologist Sigmund Freud publishes his paper praising cocaine as a miracle drug, and his colleague, Viennese ophthalmologist Dr. Karl Koller, introduces cocaine as an anesthetic for eye surgery.

1884 The International Health Exhibition is hosted in London and displays Victorian developments in sanitation and public health, as well as medicinal drugs and vaccines manufactured by companies like Burroughs Wellcome & Co.

1884 German bacteriologist Friedrich Loeffler first cultivates *Corynebacterium diphtheriae*, the agent claimed to cause diphtheria—its first name was Klebs-Loeffler bacterium.

1884 The New York Cancer Hospital is founded, later named the Sloan-Kettering Institute for Cancer Research in 1945.

1884 Robert Koch and Friedrich Loeffler form the basis for identifying a disease-causing agent called the Koch Postulates:
1. The microbe is present in every case of the disease.
2. The microbe can be taken from the host and grown independently.
3. The disease can be produced by introducing a pure culture of the microbe into a healthy host.
4. The microbe can be isolated and identified from the host infected in Step 3.

1885 Louis Pasteur and French bacteriologist Émile Roux test a rabies vaccine made from rabbit spinal cord tissue on nine-year-old Joseph Meister, who was bitten by a dog suspected of being rabid. It is later revealed that the vaccine had never been previously tested on animals infected with rabies, a detail Pasteur never made public.

1885 Spanish physician Jaime Ferrán develops a live, virulent cholera vaccine using *V. cholerae*. It is the first vaccine for humans against a bacterial disease.

1885 A peaceful mass protest of 80,000–100,000 people takes place against smallpox vaccine mandates in Leicester, England.

1885 Yellow fever vaccine is tested on over 6,000 people in Rio de Janeiro, Brazil.

1885 In a highly publicized story, four boys from Newark, New Jersey, are bitten by a rabid dog. A fundraising effort, partially helped by industrialist Andrew Carnegie, amasses US$ 1,000 to send the boys to France for rabies treatment with Pasteur's new rabies vaccine. The boys are reported as returning home healthy in January 1886.

1886 In Switzerland, a less ambitious Epidemics Act without obligation of vaccination and with a state of emergency reduced to four diseases (cholera, plague, typhoid, smallpox) is presented and accepted. A federal health referent is appointed and signs an international treaty to combat cholera, which makes it possible to establish the Federal Office of Public Hygiene.

1886 New Jersey 7-year-old boy Harold Newell is the first in the US to receive rabies vaccine at American Pasteur Institute's space in the Carnegie laboratory in New York City.

1886 The American Pasteur Institute vaccinate about 12 people against rabies, but closes in 1887 due to lack of support. Other Pasteur Institutes soon open in the US; in New York City (the most famous) and Chicago during 1890, in Baltimore during 1897, and in Atlanta, Pittsburgh, and St. Louis in 1900. About twenty-five more institutes are opened after 1900 in the US and worldwide.

1886 Russian microbiologist and student of Pasteur, Nikolai Gamaleya, opens the first vaccination station for rabies in Russia.

1887 Based on the success of the rabies vaccine, funding is initiated to finance an institute for "the treatment of rabies according to the method developed by Pasteur." Over two million French francs are raised, enabling the purchase of land and construction of what becomes the first official "Pasteur Institute."

1887 One of the first bacteriological laboratories, the Hygienic Laboratory, is established within the Marine Hospital Service in the USA. Its functions expand beyond the system of Marine Hospitals, moving into quarantine and research programs.

1887 Respected Scottish physician—most learned medical scholar and historian of his time—Charles Creighton publishes *History of Cowpox and Vaccinal Syphilis.*

1888 Émile Roux and Alexandre Yersin are the first to demonstrate that the bacteria causing diphtheria releases a deadly toxin. Émile Roux develops passive serum therapies.

1888 Creighton writes the section for vaccination in the ninth edition of *Encyclopedia Britannica*. After thorough investigation on the subject, he ends up questioning the validity and success of smallpox vaccine.

1888 Nikola Tesla invents the polyphase AC motor, bringing in the era of electricity for not just light, but also power. From 1889, the world is electrified on a worldwide scale.

1888 Italy mandates smallpox vaccination and revaccination for children entering school.

1888 American pioneer in public health research, Charles Value Chapin, establishes one of the earliest municipal public health laboratories in Providence, Rhode Island. He launches a New Public Health Campaign to eliminate diphtheria through the isolation of both victims and healthy carriers of the disease, combined with compulsory diphtheria antitoxin treatment and the disinfection of victims' homes. However, he observes "that during the period of strictest control the disease was at its height."

1889 By this date, over 20 institutes to administer rabies vaccines have been set up worldwide.

1889–1890 A deadly influenza pandemic known as the "Asiatic flu" or "Russian flu" kills about one million people.

1889 Japanese physician and bacteriologist Kitasato Shibasaburō isolates tetanus toxin from a human and shows that the toxin can be neutralized by antibodies.

1889
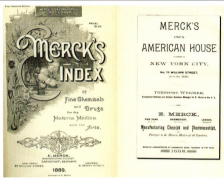
Merck's Chemical Index is first published.

1889 The UK Royal Commission on Vaccination is established. The 13-member commission issues six reports with recommendations between 1892 and 1896, including allowing conscientious exemptions, the abolition of cumulative penalties and the use of a safer smallpox vaccine.

1889 English physician Edgar M. Crookshank publishes a thorough historical investigation of vaccination in *History and Pathology of Vaccination: A Critical Inquiry.*

1889 Creighton publishes his critical views of the smallpox vaccine in *Jenner and Vaccination* and he is branded as an "anti-vaccinator."

1889 As a result of the Infectious Disease Act, the first obligatory communicable disease notification system for physicians and hospitals is set up in England and Wales.

1889 The Inter-Parliamentary Union (IPU) is founded as the world's first multilateral political organization, encouraging cooperation and dialogue between all nations. The IPU helps lay the foundations for the creation of the League of Nations in 1919 and the United Nations in 1945.

1890 The US Sherman Antitrust Act is established by federal government to outlaw monopolies and cartels. Any combination "in the form of trust or otherwise that was in restraint of trade or commerce among the several states, or with foreign nations" is declared illegal.

1890 Cold Spring Harbor Biological Naval Laboratory opens in the USA.

1890 Emil von Behring and Kitasato Shibasaburō vaccinate guinea pigs with heat-treated diphtheria toxin. They name it an "antitoxin" and their treatment, serum therapy.

A Chronology of Vaccination

1890

Glycerine is added to vaccines as an antibacterial agent. English physician S. Monckton Copeman shows that adding glycerin to lymph acts as a germicide, reducing transmission of harmful microbes via the lymph.

1890 Tetanus toxin is purified by German physiologist Emil von Behring and work begins on a tetanus antitoxin serum.

1890 The American Pasteur Institute (mainly to produce rabies vaccines) is opened in New York City and inaugurates a new building three years later.

1891 Lumbar puncture is introduced as a therapeutic procedure for relief in cases of hydrocephalus (water on the brain).

1892 ✪ Widescale use of horse-derived diphtheria antitoxin begins in Europe and a year later is mass produced by city and state health departments in the USA.

1892 The seventh International Health Conference in Venice makes progress towards international health regulations, implementing clear and rigorous sanitary protocols for monitoring the Suez Canal in Egypt.

1892 German bacteriologist Richard Pfeiffer, once a student of Robert Koch, first isolates a bacterium from influenza patients. He claims to have found the causative agent of influenza and names the bacterium *Pfeiffer influenza bacillus*. In the 1930s, it is established that influenza is in fact caused by a virus, not a bacterium. Pfeiffer's *Bacillus influenzae* would eventually be named *Haemophilus* (blood-loving) *influenzae* because of its long-standing, though incorrect, association with influenza.

1892 Russian-French bacteriologist Waldemar Haffkine develops a cholera vaccine at the Pasteur Institute, Paris.

1892 The Wistar Institute in Philadelphia, USA, is founded and named after Caspar Wistar, MD, becoming the nation's first independent biomedical research institution. From the 1950s, under the leadership of Polish-American immunologist Dr. Hilary Koprowski, the Wistar Institute becomes a leader in vaccine research thanks largely to the development of their WI-38 fetal cell line.

1893 Johns Hopkins University opens its School of Medicine by founding physicians William Henry Welch, William Osler, William Stewart Halsted, and Howard Atwood Kelly. The school becomes the model for American medical education, supported by the AMA.

1893 The International Statistical Institute adopts the first international classification of diseases. The system is based on the Bertillon Classification of Causes of Death, developed by French statistician and demographer Jacques Bertillon. In 1898, the American Public Health Association recommends that Canada, Mexico, and the USA use that system and that it be revised every decade. In the following years, Bertillon's classification becomes known as the International List of Causes of Death.

1893 Japan establishes the Institute of Infectious Diseases, that will manage national vaccine production.

1893 The first open water filter system using the slow sand method is implemented in Massachusetts, USA.

1894 The Wellcome Physiological Research Laboratories (officially and consistently named as such from 1898), part of pharmaceutical company Burroughs Wellcome & Co., is established in London. The laboratories focus on biological experimentation and the production of early forms of vaccines, notably diphtheria antitoxin.

1894 The first trials with tetanus antitoxin begin using serum supplied by Professor Tizzoni of the Bologna Institute, imported to the UK by Allen & Hanburys.

1894 A plague pandemic that began in China in 1855 spreads to Hong Kong, India, and then globally. The causative bacterium of plague (or pest) is described and cultured by Swiss-French physician and bacteriologist Alexandre Yersin (with the help of Shibasaburō) in Hong Kong. Treatment with antiserum is initiated two years later, but this therapy is supplanted by sulphonamides in the 1930s and by streptomycin starting in 1947.

1894 The Commonwealth of Massachusetts Board of Health establishes MassBiologics to make diphtheria antitoxin. Two decades later, they introduce diphtheria vaccine into general use in Massachusetts. The organization would go on to develop and manufacture biologics for smallpox, typhoid, scarlet fever, tetanus, and pertussis, among many others. By 2022, MassBiologics becomes the business unit of University of Massachusetts Medical School and the only nonprofit, FDA-licensed manufacturer of vaccines in the US.

1894 The first recorded major polio outbreak in the US occurs from June to September in Otter Creek Valley, Vermont. Eighteen deaths and 132 cases of permanent paralysis are reported, mainly in children.

1894 ✪ Émile Roux at the Pasteur Institute in Paris, France, begins large-scale production of diphtheria antitoxin in immunized horses.

1894 American pathologist Dr. Anna W. Williams isolates a strain of the diphtheria bacteria *Corynebacterium diphtheriae* from a case of tonsillar diphtheria. Together with American bacteriologist and New York City Board of Health laboratory director, William H. Parks, work is begun on inoculating horses to produce diphtheria antitoxin in New York City, developing the first antitoxin outside Europe.

1895 German mechanical engineer and physicist Wilhelm Conrad Röntgen produces and detects electromagnetic radiation in a wavelength range known as X-rays or Röntgen rays, an achievement that earns him the inaugural Nobel Prize in Physics in 1901. The first X-ray ever taken was an image of Röntgen's wife's hand.

1895

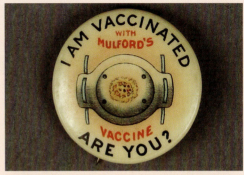

✪ The first commercial diphtheria antitoxin is manufactured in the USA, first by the New York City Health Department, produced in guinea pigs and horses. Soon after, the companies H.K. Mulford Company of Philadelphia (acquired by Sharp & Dohme in 1929) and Parke, Davis and Co., begin producing diphtheria antitoxin.

1895 Alfred B. Nobel, Swedish inventor of dynamite, signs his last will that transfers 94% of his wealth to establish five Nobel Prizes; in physics, chemistry, physiology or medicine, literature, and peace—to be selected by the Karolinska Institutet. The first awards are presented in 1901.

1895

Every dose furnished in Aseptic Glass Syringe with sterile needle, ready for instant use.

Antipneumococcal serum first appears in the H.K. Mulford product catalogue—where it remains until the 1940s.

1895

✪ The Hygienic Laboratory of the US Marine Hospital Service produces its first diphtheria antitoxin from the blood of goats or horses.

A Chronology of Vaccination

1896–1899

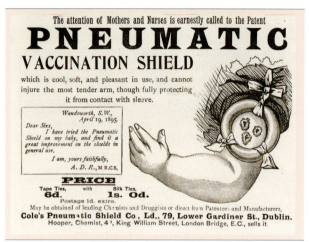

The attention of Mothers and Nurses is earnestly called to the Patent

PNEUMATIC

VACCINATION SHIELD

which is cool, soft, and pleasant in use, and cannot injure the most tender arm, though fully protecting it from contact with sleeve.

Wandsworth, S.W.,
April 19, 1895.

Dear Sirs,
I have tried the Pneumatic Shield on my baby, and find it a great improvement on the shields in general use.

I am, yours faithfully,
A. D. R., M.R.C.S.

PRICE

Tape Ties, with Silk Ties,
6d. 1s. 0d.
Postage 1d. extra.

May be obtained of leading Chemists and Druggists or direct from Patentees and Manufacturers,

Cole's Pneumatic Shield Co., Ld., 79, Lower Gardiner St., Dublin.
Hooper, Chemist, 4?, King William Street, London Bridge, E.C., sells it.

1896 American physician and one of Johns Hopkins School of Medicine founders William Henry Welch launches a journal for the publication of original American research, *Journal of Experimental Medicine.*

1896 A healthy two-year-old infant—son of German pathologist Robert Langerhans— dies following injection with Behring's diphtheria serum—most likely due to anaphylaxis, a serious allergic reaction.

1896 German researchers Richard Pfeiffer and Wilhelm Kolle demonstrate that inoculation with killed typhoid bacteria results in protection against typhoid fever.

1896 German bacteriologist Wilhelm Kolle develops a heat-inactivated cholera vaccine that comes to serve as a model for cholera vaccines in the next century.

1896 ✪ The first typhoid vaccine (phenol-treated) is developed by Almroth Edward Wright and mass produced for military use in Britain. Wright is knighted and supervises the production of bacterial vaccines for Parke, Davis & Co., in the early 1900s.

1896 In the US, variola minor, a much milder and less deadly form of smallpox with a case fatality less than 1%—also known as alastrim—is first identified. This milder illness largely replaces the severe form of smallpox, variola major, in the US and Europe. Alastrim was first described in South Africa in the late 19th century.

1896 European teams develop the first inactivated cholera vaccine and plague vaccine.

1897 German scientist Paul Ehrlich, who wins the Nobel Prize in 1908 for his work on immunity, develops a standardized unit of measure for diphtheria antitoxin.

1897 A Japanese student of Dr. Kitasato Shibasaburō, microbiologist Dr. Kiyoshi Shiga, cultures the bacteria believed to cause dysentery during a "red diarrhea" epidemic that kills 22,000 people. The bacteria is named *Shigella dysenteriae.*

1897 ✪ Bubonic plague vaccine is developed by Russian-French bacteriologist Waldemar Haffkine, who tests it on himself before inoculating millions in India from 1902–03. In 1902, a batch contaminated with tetanus kills nineteen people.

1897 Vaccine farms in the USA:
The Lancaster County Vaccine Farms
The Franklin County Vaccine Farm
The Jenner Vaccine Farm
The Pennsylvania Vaccine Company
The National Vaccine Establishment
The Chicago Vaccine Stables
The Dr. F. C. Martin Vaccine Farm
The New England Vaccine Company
The Missouri Vaccine Farm
The Columbia Vaccine
The Fond du Lac Vaccine Company
The Doctor Henry McNeel Company
The Dr. H. Welker Company

1898 Arm-to-arm transmission of smallpox "vaccine" is banned in the UK.

1898 ✪ H.K. Mulford markets a smallpox vaccine in the USA produced with calf lymph.

1898 Heroin is used as an non-addictive substitute for morphine and marketed as a cough syrup for children.

1898 The British Vaccination Act provides a conscience clause to allow exemptions to mandatory smallpox vaccination. This clause gives rise to the term "conscientious objector" which later refers to those opposed to military service.

1898 Scientists Dmitri Iosifovich Ivanovski in Russia and Martinus Beijerinck in the Netherlands identify the first virus, a plant virus they name the tobacco mosaic virus. They demonstrate that the substance associated with diseased tobacco plants passed through the then-smallest pore-sized filter used in the Pasteur–Chamberland apparatus, without losing infectivity. The material at the bottom of the filter-containing flask (smaller than bacteria) was shown to remain infective when transferred to an uninfected plant of the type from which it was first retrieved.

1898 German microbiologists Friedrich Loeffler and Paul Frosch—utilizing a similar approach as Iosifovich and Beijerinck—conclude that the agent causing foot-and-mouth disease of cattle also passed through porcelain filters and induced symptoms of disease when inoculated back into healthy cows. These observations provide the basis for defining viruses as subcellular entities that could cause specific diseases. At this time, infectious virus particles are too small to see and could not grow on the culture media available.

1898 Following the victory of the Americans during the Spanish-American war, the US gains control of Puerto Rico, Cuba, and the Philippines from Spain. Natives are forcibly vaccinated in an effort to implement American sanitary standards. It is reported that natives refusing vaccination in Puerto Rico faced US$10 fines for not getting vaccinated and US$5 daily fines for every day they remained unvaccinated, and even imprisonment. Mandates are enforced in these nations to vaccinate all children before six months old as well as a general vaccination order for adults.

1898 In Switzerland, the Institut Bactério-Thérapique et Vaccinal Suisse in Berne is founded as a merger between Institut bactério-thérapique, Haefliger & Co., and the renowned Institut Vaccinal Suisse Haccius, Lancy-Geneva. They are the country's leading vaccine supplier and known worldwide for their quality calf lymph.

1898 ✪ The first commercial antitetanus serum is produced in the US by Parke, Davis & Co.

1899 The London School of Tropical Medicine is founded by Sir Patrick Manson using the donation of £6,666 by Parsi philanthropist Bomanjee Dinshaw Petit and is initially located at the Albert Dock Seamen's Hospital in the London Docklands. In 1924, the London School of Hygiene & Tropical Medicine is established by Royal Charter.

1899 German pharmaceutical company, Bayer, patents Aspirin, a derivative of salicylic acid—a synthesized version of willow bark. Aspirin will become widely used to reduce fever and pain, as well as treat influenza.

1899 The French officially abandon their efforts to build the Panama Canal and transfer the rights of the project to the US, in part because of the high volume of yellow fever and malaria deaths among workers. The US military officially take over the project in 1904 and open the canal in 1914, having succeeded in eradicating mosquitos contributing to the diseases.

1899 George H. Simmons takes over leadership of the cash-strapped AMA. He transforms the association into a lucrative business by granting the AMA's "seal of approval" to drug companies that take out advertising spaces in *JAMA* and other affiliate publications. The "seal of approval" is abandoned in 1955.

1899 X-ray is used as treatment for cancer.

1899 Almroth E. Wright's heat- and phenol-treated typhoid vaccine is used by the British Army during the Anglo-Boer War in southern Africa.

1899 Dr. Kiyoshi Shiga is the first to attempt to develop a vaccine against dysentery (shigellosis) by creating a heat-killed whole-cell anti-dysenteric vaccine that he tests on himself. The local reaction to the vaccine is so severe that it requires "incision and drainage." Dr. Shiga then develops a serum-based passive inoculation and later an oral vaccine that is given to thousands of Japanese citizens, though it ends up being abandoned.

INSTITUT

BACTÉRIO-THÉRAPIQUE ET VACCINAL SUISSE, BERNE

FONDÉ LE 1er NOVEMBRE 1898 PAR LA FUSION DE

L'INSTITUT BACTÉRIO-THÉRAPIQUE HÆFLIGER & Cº, BERNE

ET DE

L'INSTITUT VACCINAL SUISSE CH. HACCIUS, LANCY-GENÈVE

SOUS LE CONTROLE DE L'ÉTAT

Direction scientifique : Prof.-Docteur E. TAVEL

SERVICE DES SÉRUMS

Sérum antidiphtérique renfermant 250-400 unités antitoxiques dans 1 cm³, très actif; il se recommande aussi pour l'injection prophylactique en cas d'épidémies à la dose de 500 U. A. — **Sérum antitétanique** se recommande soit prophylactiquement dans les cas de traumatismes graves, soit curativement quand la maladie a éclaté. — **Sérum antistreptococcique** contre la fièvre puerpérale et autres maladies streptococciques, telles que l'érysipèle, les phlegmons, etc. — **Sérum antityphique.** — **Sérum antipesteux.** — **Sérum de sang aseptique** pour application rectale dans les cas d'anémie, de faiblesse et chez les convalescents. — **Vaccins charbonneux.** — **Seringue aseptique** du Dr Beck. — **Tuberculine** (préparation diluée au 1/10 et prête pour l'usage, pour le diagnostique de la tuberculose du bétail ; suivant la grandeur des animaux, aux doses de : 3 cm³ pour ceux de petite taille, 4 cm³ pour ceux de moyenne taille, 5 cm³ pour ceux de grande taille. — **Toxines de l'érysipèle** contre les tumeurs malignes : on les emploie souvent avec succès dans les maladies infectieuses, aigües et chroniques. — **Solutions médicamenteuses** aux doses usuelles pour injections sous-cutanées en tubes de verre et absolument stériles : **Morphine** 1, 2, 5 % : **Ergotine** 25 % ; **Eucaine** 2-10 % ; **Apomorphine** 1 % ; **Solution de Schleich** en tubes de 2, 5 et 10 cm³.

Nos sérums se distinguent par une très longue conservation. — *Adresse télégraphique :* SÉRUM-BERNE.

SERVICE DE LA VACCINE

Vaccin animal (Cow-pox), marque LANCY-VAXINA, cultivé soigneusement de génisse à génisse. — 5 médailles d'or. — Nous garantissons un vaccin absolument pur. — Livré en tubes pour 2 à 3 vaccinations ; Etuis de 5 tubes pour 2 à 3 vaccinations ; Plaquettes pour 2 à 3 vaccinations ; Plaques pour 5 à 6 vaccinations ; Flacons pour 25 à 30 ; 50 à 60 et 100 vaccinations.
H16Y

Adresser les commandes à Berne. — *Adresse télégraphique :* VACCIN-BERNE.

Institut bactério-thérapeutique et vaccinal Suisse, Berne – Antitoxin, serum, and vaccine advert in
Revue Médicale de la Suisse Romande – 20ᵗʰ January 1900

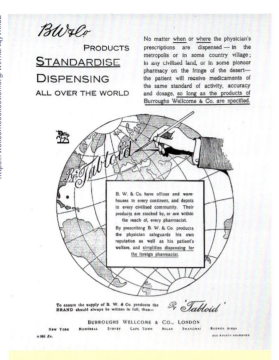

1898–1922
WPRL develops and prepares many vaccines, including this leprosy vaccine.

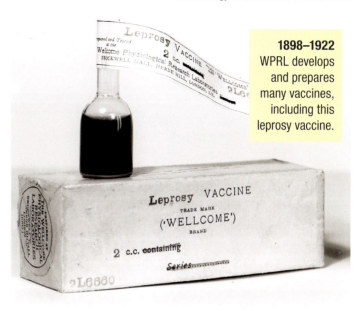

WPRL: Wellcome Physiological Research Laboratories

Early 1900s
At the turn of the century, British pharmaceutical company Burroughs Wellcome delivers medicinal products worldwide. BW has offices in New York, Montreal, Sydney, Cape Town, Milan, Shanghai, and Buenos Aires, and is one of the leading vaccine manufacturers in the world.

Burroughs Wellcome & Co., London
1909 – Situations in which a Tabloid medicine chest would be useful.

Burroughs Wellcome & Co., London
1885 – Pharmacy sign advertising for "new pharmaceutical preparations."

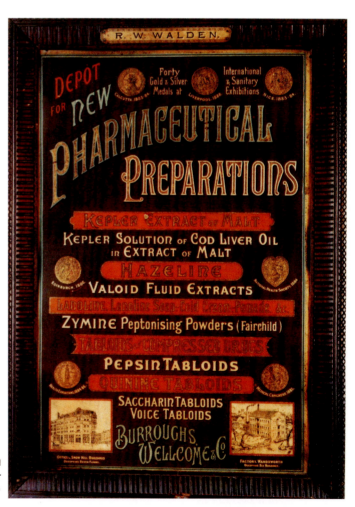

A Chronology of Vaccination

1899–1903

1899 The International Peace Conference takes place at The Hague to create tools for "peacefully settling international crises, preventing wars, and codifying rules of warfare." The Conference also adopts the *Convention for the Pacific Settlement of International Disputes* and establishes the Permanent Court of Arbitration, which begins work three years later.

1900–1909

1900

⭐ Antitetanus serum is prepared by Pasteur Institute (FR), Burroughs Wellcome & Co., (UK) and Parke, Davis & Co. (US).

1900 The first International Classification of Death (ICD) lists smallpox vaccine as a cause of death. The French government holds the first International Conference for the Revision of the Bertillon, or International Classification of Causes of Death. A detailed classification of causes of death are adopted, consisting of 179 groups and an abridged classification of 35 groups.

1900 US Senator Jacob H. Gallinger introduces a proposal for the regulation of human experimentation in the District of Columbia, under Senate Bill 3424. The bill would require the written consent of subjects and forbid scientific experimentation on newborns, people under age 20, pregnant women, the insane, the feeble-minded, and people with epilepsy. This bill would not pass and it is not until decades later that legislation is passed to require informed voluntary consent.

1900 Burroughs Wellcome advertises five different varieties of protective serum against diphtheria, streptococcus, tetanus, typhoid, and snake venom, as well as a range of diagnostic tests for diphtheria, gonococcus, tuberculosis, and typhoid.

1900 Cases of post-vaccinal tetanus are observed after smallpox vaccination in Philadelphia, USA.

1901 Thirteen children die from contaminated diphtheria antitoxin in St. Louis, USA. Investigations show that a horse used in diphtheria antitoxin production for the St. Louis municipal health authority died of tetanus. Rather than being discarded, some of the antitoxin produced from the diseased horse was sent to physicians. This incident, along with a New Jersey tetanus outbreak linked to contaminated smallpox vaccine that kills nine children, leads to the federal regulation of biologic products called The Biologics Control Act of 1902.

1901 Emil von Behring is awarded the first ever Nobel Prize in Physiology or Medicine for his work on serum therapy, notably diphtheria, which has "opened a new road in the domain of medical science and thereby placed in the hands of the physician a victorious weapon against illness and deaths." With the money awarded, Behring establishes the company Behringwerke in 1904 to produce and sell antitoxins, located near the University of Marburg in Germany.

1901 Austrian physician Karl Landsteiner, discovers the first three human blood groups: A,B,O. As a result of this discovery, he is awarded the Nobel Prize in Physiology or Medicine in 1930.

1901 Switzerland experiences an epidemic of cerebro-spinal meningitis with 264 deaths reported from 1901–1905 and 172 deaths between 1906 and 1907.

1901 The Tenement House Law is established in New York State, which outlaws the construction of the dumbbell-shaped-style tenement housing and sets minimum size requirements for tenement housing. It also mandates the installation of lighting, better ventilation, and indoor bathrooms.

1901 *The New York Times* reports on a smallpox vaccination raid where 250 men enter Little Italy in New York City, forcibly vaccinating people and removing children with smallpox.

1901

THE ROCKEFELLER INSTITUTE
FOR MEDICAL RESEARCH
NEW YORK

The Rockefeller Institute for Medical Research is founded by Standard Oil petroleum magnate and the world's first billionaire, John D. Rockefeller. It would be America's first home-grown biomedical institute, like France's Pasteur Institute and Germany's Koch Institute. The name is changed in 1958 to the Rockefeller Institute and in 1965 to Rockefeller University. From 1912–2020, it would boast an impressive 29 Nobel laureates.

1901 US Army physician Walter Reed confirms Cuban doctor Carlos Finlay's theory that yellow fever is transmitted by a particular species of mosquitos, and not by contact.

1902 The first permanent sanitary agency, the International Sanitary Bureau, later renamed the Pan American Sanitary Bureau, is established in Washington, DC. From 1958, it is known as the Pan American Health Organization (PAHO), a specialized agency of the UN and regional office for the WHO in the Americas.

1902 While investigating the toxins produced by jellyfish and and sea anemones, French physiologists Charles Richet and Paul Portier discover the principle of anaphylaxis. Richet coins the term to mean the opposite of protection—instead of inducing tolerance, dogs injected several times with toxins develop a severe reaction.

1902 The Imperial Vaccination League is established in London to evaluate the British Vaccination Act of 1898, and to promote the value of smallpox vaccination and revaccination to those "sections of society now most prejudiced against it."

1902 France mandates smallpox vaccination in the first year of life, and revaccination at the ages of 11 and 21. Conscientious exemptions are not permitted.

1902 British medical doctor Sir Ronald Ross is awarded the Nobel Prize in Physiology or Medicine for his work on the transmission of malaria, which lays the foundation for research and development against the disease. He is the first British national and first person born outside Europe (India) to receive the prestigious prize.

1902 In Mulkowal, India, 19 people die of tetanus following injection with Haffkine's inactivated whole-cell plague vaccine. Initially Waldemar Haffkine is held personally responsible and suspended from his position. Thanks to Sir Ronald Ross, he is exonerated in 1907.

1902 The Statens Serum Institut is founded in Denmark to secure production of diphtheria antitoxin. During the 1930s, fueled by epidemics of tuberculosis and typhoid, the SSI establishes an Epidemiological Department which is initially financed by the Rockefeller Foundation.

1902 The US Congress passes "an act to regulate the sale of viruses, serums, toxins and analogous products" later referred to as the Biologics Control Act (even though "biologics" appears nowhere in the law). This is the first modern federal legislation to control the quality of drugs. As a result, 30% of the companies making antitoxins and vaccines go out of business. The Act creates the Hygienic Laboratory of the US Public Health Service to oversee manufacture of biological drugs. The Hygienic Laboratory is transferred to the FDA and renamed Bureau of Biologics in 1972 and transferred back as the Center for Biologics Evaluation and Research (CBER) under the National Institutes of Health (NIH) over a decade later.

1903 ✪ The first US approval is granted to Gilliland Labs for live smallpox vaccine. Cutter Labs also gets approval for its smallpox vaccine.

1903 ✪ Diphtheria antitoxin prepared by the Massachusetts Public Health Biologic Laboratories is licensed in the USA.

1903 Six companies supply sera and vaccines in the UK: Parke, Davis & Co.; the Royal Veterinary College; Allen & Hanburys; Meister, Lucius & Bruning; and the Jenner Institute.

A Chronology of Vaccination

1903–1909

1903 The US General Education Board is created after John D. Rockefeller makes a donation of US$1 million (US$35M equivalent in 2023). The Rockefellers donate more than US$150 million over the years, with the GEB playing a pivotal role in shaping American education.

1904

The Rockefeller Institute for Medical Research opens its first laboratories in New York.

1904 Meningococcal meningitis sweeps through New York City, killing more than 3,000 people as part of a global pandemic.

1904 In Brazil, the first smallpox vaccine mandate (for school, work, travel, and weddings) sparks popular uprising, escalating into a five-day riot with twelve deaths—it is named the Vaccine Revolts. The mandate is repealed.

1904 Under Simmons, the AMA establishes a Council on Medical Education with a mission to upgrade medical education by grading medical colleges and developing guidelines.

1904 The first attempt to vaccinate American troops against typhoid disease with a dead typhoid bacilli oral vaccine ends in disaster—publication of the results is prohibited by the US Army for reasons of "public policy."

1905 Russian zoologist Ilya Ilyich Mechnikov—also spelled Élie Metchnikoff—first describes the concept of cellular immunity.

1905 Renowned Russian bacteriologist Eugène Gabritschevsky—who worked with both Louis Pasteur and Robert Koch—introduces a scarlet fever vaccine that is prepared from broth cultures of hemolytic streptococci isolated from scarlet fever. Mild attacks of scarlet fever with rashes are described following administration of the vaccine, which would not be widely used.

1905 German physician Robert Koch is awarded the Nobel Prize in Physiology or Medicine for his work on tuberculosis.

1905 The last yellow fever epidemic in North America occurs in New Orleans, Louisiana.

1905 Austrian pediatrician Clemens von Pirquet and Hungarian-born American pediatrician Béla Schick notice that patients vaccinated against smallpox using horse serum react quickly and severely to a second dose. They correctly deduce that the symptoms of what Pirquet called "serum sickness" were being caused by the immune system producing antibodies to fight antigens or foreign substances contained in the serum. Understanding this new form of disease helps pave the way for defining and understanding immunologic diseases. In 1906, Pirquet coins a new term for this antibody-antigen interaction: allergy.

1905 In the case of *Jacobson v. Massachusetts* the US Supreme Court upholds the constitutionality of mandatory smallpox vaccination programs to preserve the public health "for the common good." The US$5 fine—which is what Jacobson was contesting—is imposed upon non-compliance, not forced vaccination. At the time, 11 states have compulsory vaccination laws, though 75% of them have no legal penalties for non-compliance.

1905 The AMA creates the Council on Pharmacy and Chemistry to evaluate medical treatments, including vaccines—the results of which are published in *JAMA*, and then later in book form as *New and Nonofficial Remedies*.

1906 The Pure Food & Drug Act is signed into law by US President Theodore Roosevelt authorizing the government to monitor the purity of transported foods and the safety of medicines—now a responsibility of the FDA.

1906 Harvard- and Hopkins-trained tropical medicine professor Richard Pearson Strong, director of the Biological Laboratory of the American-occupied Philippine Bureau of Science, injects 24 Filipino prisoners with a plague-contaminated cholera vaccine resulting in 13 vaccine-related deaths.

1906 Rocky Mountain spotted fever is first identified as a tick-borne disease by American pathologist Dr. Howard T. Ricketts, from the University of Chicago.

1906 Ernst Lederle, who was formerly New York City Health Commissioner, creates Lederle Laboratories in New York City to make diphtheria antitoxin. Lederle Laboratories later becomes part of Wyeth Laboratories.

1906 Belgian scientists at the Pasteur Institute of Brussels, Jules Bordet and Octave Gengou, first culture the bacterium *Bordetella pertussis* (whooping cough), initially observed in 1900. It would later be named *Bordet-Gengou bacillus*.

1907

Predecessor to the WHO, the permanent Paris-based health agency named the Office International d'Hygiène Publique (OIHP) is founded following an agreement signed in Rome by twelve nations (Belgium, Brazil, Egypt, Spain, USA, France, UK, Italy, Netherlands, Portugal, Russia, Switzerland) on a budget of CHF 150,000.

1907 Von Pirquet develops a skin test requiring a small amount of tuberculin—a combination of proteins used in the diagnosis of tuberculosis—under the skin and measures the body's reaction. Two years later he invents the term "latent TB infection." French physician Charles Mantoux updates the skin test method in 1908 by using a needle and syringe to inject the tuberculin.

1907 ✪ Tetanus antitoxin prepared by Parke, Davis & Co., is introduced in the USA.

1907 The Eugenics Education Society is founded by young widowed social reformer Sybil Gotto, inspired by the work of Francis Galton. In 1924, it is renamed the Eugenics Society—supporting scientific research, public education, and legislation advocating state-forced sterilization. A staunch believer in racial hierarchy, Winston Churchill is an honorary vice-president for the organization.

1907 The American Association for Cancer Research is launched.

1907 ✪ Antimeningococcal serum by the Rockefeller Institute is introduced in the USA, as a therapy initially based on antibodies derived from the blood of horses.

1907 The Rockefeller Institute for Medical Research purchases a farm in New Jersey for the production of laboratory animals. Two years later, following reports of animal cruelty, the farm is burned down by arsonists.

1908 During a polio epidemic in Vienna, Austrian physicians Karl Landsteiner and Erwin Popper obtain spinal cord material from a nine-year-old victim and inject it into two Old World monkeys, who then display polio-like symptoms. This experiment is claimed to be the first isolation of the infectious agent (a virus) of poliomyelitis.

1908 Ilya Ilyich Mechnikov and Paul Ehrlich are awarded the Nobel Prize in Physiology or Medicine for their work on immunity.

1908 The first chemotherapeutic agent, salvarsan (arsphenamine), a toxic organoarsenic, is discovered by Paul Ehrlich and Japanese microbiologist Sahachiro Hata. Salvarsan is used to treat syphilis and trypanosomiasis "sleeping sickness" from 1910.

1908 Chlorination of an American urban water supply is introduced in Jersey City, New Jersey, and rapidly adopted from 1913.

1908 ✪ Typhoid vaccine is licensed in the USA, produced by Burroughs Wellcome & Co., and Swiss Serum and Vaccine Institute Berne—who also commercializes a plague vaccine.

1908 French bacteriologist Charles Jules Henri Nicolle at the Pasteur Institute in Tunis, Tunisia, discovers that lice are the carrier of typhus rickettsia.

1909 The Bureau of Science in the American-controlled colony of the Philippines, prepares and offers vaccines for cholera, anthrax, gonococcus, and *Staphylococcus aureus* and *Staphylococcus albus*, as well as sera for diphtheria, typhoid, plague, and dysentery.

A Chronology of Vaccination

1909–1912

1909 Helen Dean King becomes The Wistar Institute's first female scientist and first female research professor in the country. She participates in breeding the Wistar rat, a strain of genetically homogeneous albino rats for biological and medical research—the first ever standardized lab animal.

1909 Major Frederick Russell of the US Army Medical School adapts European vaccine production methods to produce an inactivated whole-cell typhoid vaccine that is given to American soldiers.

1909 The Rockefeller Sanitary Commission is established and offices open in Washington, DC, in 1910. Their first priority is the eradication of hookworm infection, or Uncinariasis, also called "the disease of laziness"—the understandings of which are deeply influenced by racist ideologies.

1909 The Senate of the Commonwealth of Massachusetts introduces Bill No. 8; "An Act to Prohibit Compulsory Vaccine. Sec. 1. It shall be unlawful for any board of education, board of health, or any public board acting in this state, under political regulations or otherwise, to compel by resolution, order or proceedings of any kind, the vaccination of any child or person of any age, by making vaccination a condition precedent to the attending of any public or private school, either as pupil or teacher." The bill is buried and never submitted for vote.

1909 The first intrauterine device (IUD) made of silkworm gut is used in clinical medicine, invented by German medical practitioner Richard Richter.

1909 ✪ Pneumococcal vaccines are marketed in the USA, manufactured by H.K. Mulford Co., Eli Lilly & Co., and Parke, Davis & Co.

1910–1919

1910 A killed whole-cell pneumococcal vaccine is created by British bacteriologist Sir Almroth Wright. The vaccine is tested on more than 50,000 South African gold miners from 1911–1912. Data from the trials is found to be inconclusive and the vaccine is abandoned.

1910 French physician and Inspector of Hygiene and Epidemiology in the French Army, Jean Hyacinthe Henri Vincent, develops typhoid and paratyphoid vaccines. They are approved by the Académie Nationale de Médecine the next year, and made compulsory for the French Army in 1915. Pasteur Institute in Paris develops heat inactivated vaccines that will also be used.

1910 The AMA publishes *Essentials of an Acceptable Medical College.* This would mark the beginning of the phasing out of homeopathic colleges and the elimination of all competition to the allopathic medical model of healthcare. American medical sociologist Paul Starr would write in his 1984 Pulitzer Prize-winning book *The Social Transformation of American Medicine:* "The AMA Council became a national accrediting agency for medical schools, as an increasing number of states adopted its judgments of unacceptable institutions. Even though no legislative body ever set up the AMA Council on Medical Education, their decisions came to have the force of law."

1910 American pathologist Peyton Rous, working at the Rockefeller Institute, shows that chicken sarcoma—a malignant tumor—can be reproduced by a filterable virus, which lays the foundation of modern virology and oncology.

1910 The sixty-bed Rockefeller Institute Hospital is opened, devoted entirely to clinical research in conjunction with the Rockefeller Institute for Medical Research.

1910 The Carnegie-Rockefeller-sponsored Flexner Report is released by the American Medical Association (AMA). This report is key to the reformation of a medical curriculum based on surgery, radiation, and pharmacology—drug-based medicine derived from petroleum—largely excluding natural remedies and homeopathic medicine.

1910 Mandatory vaccinations against smallpox and typhoid (from 1911) are enforced on military personnel in the USA.

1910

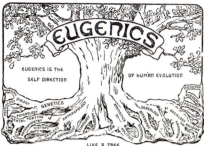

EUGENICS

EUGENICS IS THE SELF DIRECTION OF HUMAN EVOLUTION

LIKE A TREE EUGENICS DRAWS ITS MATERIALS FROM MANY SOURCES AND ORGANIZES THEM INTO AN HARMONIOUS ENTITY.

Located in Cold Spring Harbor, New York, the Eugenics Record Office (ERO) is established as a eugenics and human heredity research department for "Experimental Evolution" by Charles B. Davenport and Harry H. Laughlin (who advocates forced sterilization) of the Carnegie Institution—subsequently administered by its Department of Genetics. Financed primarily by the widow of railroad baron E.H. Harriman, the Rockefeller Foundation, and the Carnegie Institution, in 1938–1939, Laughlin is removed and ERO is renamed the Genetics Record Office (GRO) when American engineer Vannevar Bush takes over direction of the Carnegie Institution of Washington. The GRO will be closed in 1944.

1911 Belgian bacteriologists Jules Bordet and Octave Gengou prepare a whooping cough vaccine from killed whole-cell *B. pertussis* preparations while working at the Pasteur Institute, but it proves to be ineffective.

1911 The Swiss Serum and Vaccine Institute in Berne produces diphtheria antitoxin, antidysenteric serum, antimeningococcic serum, antipneumococcic serum, antiplague serum, antistreptococcic serum, tuberculins, cholera vaccine, plague vaccine, typhoid vaccine, and antitetanic serum.

1911 National Health Insurance is introduced in Britain, providing workers with cash benefits and medical care in the event of sickness and disability. Hospital care is not included, except for treating tuberculosis.

1911 American businessman John Harvey Kellogg establishes the Race Betterment Foundation, sets up a "pedigree registry," and organizes national conferences on the subject, aligning with ideologies rooted in eugenics.

1911 The US Supreme Court rules that Rockefeller's company must be dismantled for violation of federal antitrust laws. It is broken up into 34 separate entities which include companies that become ExxonMobil, Chevron Corporation, and others—some of which still have the highest level of revenue in the world. In the end it turns out that the individual segments of Standard Oil are worth more than the entire company as one entity—the sum of the parts are worth more than the whole—as shares of these doubled and tripled in value in their early years. Consequently, Rockefeller becomes the country's first billionaire with a fortune worth nearly 2% of the national economy.

1911 According to the US Public Health Report, there are 35 vaccine manufacturers worldwide. Over 17 different disease products are available: antidiphtheric serum; antistreptococcic serum; antimeningococcic serum; diphtheria antitoxin; antivenom serum; antigonococcic serum; antitetanic serum; antidysenteric serum; antistaphylococcic serum; anticolon bacillus serum; antipneumococcic serum; typhoid vaccine; rabies vaccine; tuberculin; cholera vaccine; plague vaccine; and other bacterins.

1911 The *American Journal of Public Health*, the official journal of the American Public Health Association, is published for the first time.

1911 St. Elizabeth's Hospital, the first government-funded institution for the insane, opens in the USA. At its peak in 1955, there are 8,000 patients.

1912 The US Federal Public Health and Marine Hospital Service name is changed to the US Public Health Service.

1912 The Eugenics Education Society organizes the first International Congress of Eugenics that takes place in London, UK.

1912 The first pertussis vaccine clinical trials begin in Europe and North Africa.

1912 The first unsuccessful attempts to develop a yellow fever vaccine occur after the opening of the Panama Canal.

US pharmaceutical manufacturer H.K. Mulford Company is one of the leaders of bacterial vaccine production early in the 20th century.

H.K. Mulford Company – Bacterial Vaccines advertisement in
Bulletin of the Medical and Chirurgical Faculty of Maryland – **February 1912**

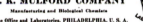

The bestselling H.K. Mulford products during this era are tetanus antitoxin and influenza vaccine.

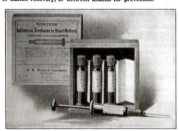

H.K. Mulford Company – Tetanus Antitoxin and Influenza Serobacterin advertisements in
American Medicine – **February 1916**

Parke, Davis & Company, Index of Therapeutics 1913–14
Chief offices and laboratories in Detroit, Michigan, USA
With branches worldwide in Chicago, New York, London, Montreal, Sydney, St. Petersburg, Bombay.

THIS PRICE LIST CANCELS ALL PREVIOUS QUOTATIONS.

1913-14.

Index of Therapeutics
and
Descriptive Catalogue

of the products of
the laboratories of

PARKE, DAVIS & COMPANY,

Manufacturing Chemists,

LONDON, ENGLAND.

Telegraphic and Cable Address: "Cascara, London."
Telephones: 8636 Gerrard (four lines).

Offices and Warehouse:

50-54 BEAK STREET, REGENT STREET, W.
Laboratory: HOUNSLOW.

Chief Offices and Laboratories: DETROIT, MICH., U.S.A.

BRANCH HOUSES:
Canadian Laboratory: WALKERVILLE, ONT.;

NEW YORK,	KANSAS CITY,
BALTIMORE,	NEW ORLEANS,
BOSTON,	CHICAGO,
ST. LOUIS,	MINNEAPOLIS,

SEATTLE, U.S.A.
MONTREAL, CANADA; SYDNEY, N.S.W.;
ST. PETERSBURG, RUSSIA; BOMBAY, INDIA,

It is worth noting that all influenza vaccines at this time are solely bacterial. The only live viral vaccines on the market are against smallpox and yellow fever.

Document in the public domain obtained from the British Library in London, UK.

Vaccines on the market in the mid-1910s

Acne bacillus vaccines, catarrhalis vaccine, hay fever vaccine, gonococcus vaccine, influenza bacillus vaccine, mixed vaccine for colds, streptococcus vaccine, neoformans (anti-tumor) vaccine, smallpox vaccine, typhoid vaccine...

Prepared under the supervision of Sir Almroth E. Wright.

252 PART II.

BACTERIAL VACCINES

Prepared under the supervision of Sir Almroth E. Wright, M.D., F.R.S., etc.

The amount of vaccine required is obtained by drawing up into a graduated inoculating syringe either one full c.c. or such fraction of 1 c.c. as may correspond to the desired dose.

N.B.—Parke, Davis & Co. alone have the concession to sell these vaccines which are prepared under the direct personal supervision of Sir A. E. Wright, at the Vaccine Laboratory of St. Mary's Hospital.

The proceeds which accrue to the Laboratory from the sale of the vaccines are devoted to the upkeep of the In- and Out-patients' Departments of this Branch of the Hospital.

Third Syllables for Code Words.		Second Syllables for Code Words.							First Syllables for Code Words appear after the respective packages.
	OT	OV	OW	OX	UB	UF	UJ	UL	
Dozen	1-12	1-6	1-4	1-3	1-2	3-4	1	2 Dozen.	
	(For other numerals, etc., see page ix.)								

Acne Vaccines,

Prepared from microbes obtained from cases of acne. Acne is associated with the presence of the acne bacillus of Unna and Sabouraud in the sebaceous glands. In uncomplicated and comparatively mild cases, the infection is strictly limited to these glands, and it is brought to notice only by come-dones, a greasy condition of the skin and occasional minute superficial pustules. In the more serious cases, definite retention-cysts may be formed, or inflammatory changes may supervene. Such inflamed acne nodules can be distinguished from ordinary furuncles by the fact that they take origin in the subcutaneous tissues under an uninflamed skin, and redden the skin only when they approach the surface, and also by the fact that they furnish instead of ordinary pus, a pus with an admixture of sebum or simply a caseous mass of sebum. Not unfrequently the staphylococcus infection plays from the outset a very prominent and even a dominant part in acne, giving us the subfuruncular and pustular type.

The different therapeutic requirements of these cases are met by the provision of an Acne Bacillus Vaccine, a Mixed Vaccine (a mixture of acne and staphylococcus albus vaccines), and a Staphylococcus Vaccine.

While the first two of these are issued primarily for use in connexion with cases of acne, they may be used also for the treatment of greasiness of the skin and itchiness and dandruff in the scalp. (See also under Staphylococcus Vaccine.)

Acne Bacillus Vaccine,

Applicable where the acne bacillus is predominant, and the staphylococcus, if present, is quite a subordinate factor, as in *non-pustular cases* in which *comedones* form the principal feature. Issued in dilutions as follows :

VXRG 10 millions of acne bacilli per c.c.
VXPM 20 millions of acne bacilli per c.c.

A suitable initial dose is 5 millions, and this may afterwards be increased to 10 millions.

VXST Bulbs of about 1 c.c. (either dilution).. EBB .... per bulb 2/9
Bottles of 25 c.c. (either dilution). .... EFM .... per bot. 27/9

Mixed Vaccine for Acne,

Applicable where the staphylococcus and the acne bacillus are both present as pathogenetic agents, as in cases in which the lesions assume a *sub-furuncular* form. Issued in a dilution containing 1,000 millions of staphylococcus albus and 10 millions of acne bacilli per c.c.

A suitable initial dose for an average case is 0·25 c.c., and this may afterwards be increased to 0·5 c.c.

Bulbs of about 1 c.c. .. EBB .... per bulb 2/9
Bottles of 25 c.c. .... EFM .... per bot. 27/9

Note.—In cases of suppurating acne the use of Staphylococcus Vaccine is indicated.

Catarrhalis Vaccine, see Mixed Vaccine for Colds.

Gonococcus Vaccine,

Prepared from cultures of gonococcus obtained from cases of gonococcal urethritis and arthritis, and issued for use in connexion with these infections. In *acute* gonococcal urethritis, inasmuch as the use of large doses might here be followed by the development of gonococcal arthritis, the dose which is to be employed is the *minimum effective dose* repeated at short intervals.

For information as to use of Telegraphic Code, see page viii.

PART II. 253

BACTERIAL VACCINES BACTERIAL VACCINES.

Prepared under the supervision of Sir Almroth E. Wright, M.D., F.R.S., etc.

Third Syllables for Code Words.		Second Syllables for Code Words.							First Syllables for Code Words appear after the respective packages.
	OT	OV	OW	OX	UB	UF	UJ	UL	
Dozen	1-12	1-6	1-4	1-3	1-2	3-4	1	2 Dozen.	
	(For other numerals, etc., see page ix.)								

Gonococcus Vaccine—*concluded.*

In the case of gonococcal arthritis it is also well to feel one's way very carefully beginning with very small doses. The results which are obtained in these cases by long-continued treatment are very striking.

VXGC Gonococcus Vaccine is issued in two dilutions as follows :—
VXAL 5 millions of gonococci per c.c.
50 millions of gonococci per c.c.

The minimum effective dose is 1 to 2 millions ; and a medium dose from 5 to 25 millions.

Bulbs of about 1 c.c. (either dilution) ... EBB .... per bulb 5/6
Bottles of 25 c.c. (either dilution) ...... per bot. 55/6

Hay Fever Vaccine, see 'Pollaccine,' page 254.

Influenza Bacillus Vaccine,

Prepared from cultures of the influenza bacillus isolated from cases of acute influenza or chronic bronchitis, and applicable to *those cases of 'influenza' in which the presence of the bacillus* has been ascertained ;

VXIN issued in two dilutions as follows :
VXIS 10 millions of bacilli per c.c.
25 millions of bacilli per c.c.

The minimum effective dose is 5 millions ; and a medium dose from 10 to 15 millions.

VXMC Bulbs of about 1 c.c. (either dilution) .. EBB .... per bulb 5/6
Bottles of 25 c.c. (either dilution) .... EFM .... per bot. 55/6

Mixed Vaccine for Colds,

This vaccine is made from cultures obtained from acute coryzas and 'colds.' An endeavour has been made to include in it a representative of every strain of microbe which is frequently found in association with these, and the vaccine prepared therefrom has appeared to do useful service in this connexion. The vaccine contains the following varieties of micro-organism in the following doses :—

30 millions of micrococcus catarrhalis per c.c.
30 millions of pneumococci per c.c.
30 millions of bacillus septus per c.c.
30 millions of bacillus influenzæ per c.c.
12 millions of streptococcus of the mouth per c.c.
300 millions of staphylococci per c.c.

Inasmuch as the dose of each of these microbes corresponds to a medium dose, it is suggested that the dose should in the case of an acute 'cold' be 0·5 c.c. and in a subacute case 1 c.c. This dose (1 c.c.) may also be employed for the second dose if the cold has been benefited from the first injection of 0·5 c.c.

VXCR Bulbs of about 1 c.c. .. EBB .... per bulb 2/9
Bottles of 25 c.c. .... EFM .... per bot. 27/9

Neoformans Vaccine,

Prepared from cultures of *micrococcus neoformans* (Doyen), which is present in most, if not all, malignant tumours, and, though not to be regarded as the causal agent, is probably responsible for inflammatory changes and the pain associated therewith, as well as for much of the cachexia. Issued in a dilution containing 30 millions of cocci in each c.c. Neoformans Vaccine is not suggested as a substitute for surgical procedures or as a remedial agent, but as *an auxiliary in the treatment of cancer*, which is found to relieve pain by suppressing local inflammation, whilst it appears to diminish the cachexia and thereby to prolong life.

A suitable initial dose in an average case is 15 millions, injected hypodermically in any convenient site on the body—not into the tumour. This dose may be cautiously increased to 30 millions.

Bulbs of about 1 c.c. ... EBB .... per bulb 2/9
Bottles of 25 c.c. .... EFM .... per bot. 27/9

...graphic Code, see page viii.

Parke, Davis & Company, Index of Therapeutics and Descriptive Catalogue 1913–1914

BACTERIAL VACCINES

254   PART I

BACTERIAL VACCINES

Prepared under the supervision of Sir Almroth E. Wright, M.D., F.R.S., etc.

Third Syllables for Code Words.	Second Syllables for Code Words.								First Syllables for Code Words appear after the respective packages.
	OT Dozen 1-12	OV 1-6	OW 1-4	OX 1-3	UB 1-2	UF 3-4	UJ 1	UL 2 Dozen.	
				(For other numerals, etc., see page ix.)					

VXPS

Parodontal Streptococcus Vaccine.

This vaccine is prepared from cultures of streptococci obtained from cases of pyorrhœa alveolaris, dental abscess and other infections of the teeth and gums.

The vaccine is issued primarily for use in connexion with cases of pyorrhœa alveolaris, in the majority of which the streptococci of the mouth appear to be the dominant organisms, but it must be remembered that in these cases there may be other microbes involved, and that it is important to provide for the evacuation of any accumulation of pus. Something may be done towards this end, and towards the washing out of the pockets, by inducing an increased outflow of lymph from the gums. This will be obtained by the use of a 1 per cent. sodium citrate and 4 per cent. sodium chloride solution, which solution may be conveniently obtained by dissolving three Hypertonic Tablets (P., D. & Co., No. 476) in two ounces of water.

The Parodontal Streptococcus Vaccine may be usefully applied in the treatment of small blisters and ulcers of the mucous membrane of the tongue, cheeks and lips; in cases of toothache due to trouble round the roots of the teeth or to inflammation of the pulp cavity; also in connexion with the treatment of nausea, and of the dyspepsias which are associated with definite pyorrhœa, or with the slighter forms of infection which are indicated by sponginess of the gums and a red line round the margin of the teeth. It would seem that the nausea and dyspepsia which are so often a feature of phthisis may be associated with this microbic infection of the gums.

Lastly, the vaccine may be usefully be employed in connexion with those cases of rheumatism which are associated with the presence of pyorrhœa. In typical cases of 'pyorrhœic rheumatism' it will be found that the least overdose of the vaccine will aggravate the symptoms, while a small dose will be followed by a temporary amelioration.

Issued in a dilution containing 30 millions of streptococci per c.c.
As a medium dose, from 6 to 10 millions may be given.

Bulbs of about 1 c.c. . . . . . EBB . . . . per bulb
Bottles of 25 c.c. . . . . EFM . . . . per bot.

Pneumococcus Vaccine.

This vaccine is prepared from cultures of pneumococcus isolated from cases of croupous pneumonia, and is issued for use in that affection. The vaccine finds a large application also in connexion with acute 'bronchial colds' and chronic bronchitis. This latter in particular is generally associated with the presence of pneumococci.

It should be used only in those cases where pneumococci are present.

Issued in two dilutions as follows:—

VXPC
VXPL

           20 millions of pneumococci per c.c.
           50 millions of pneumococci per c.c.

The minimum effective dose is 10 millions, a medium dose, from 20 to 30 millions. In the case of pneumonia, an appropriate initial dose would be 20 millions. Where that dose brings down the temperature it may with advantage be repeated in two or three days, the curve of the opsonic index, or in default of this, the curve of the temperature, being here taken as a guide. If the initial dose of 20 millions leaves the patient's condition unaltered, a further dose of 30 millions may with advantage be administered next day.

Bulbs of about 1 c.c. (either dilution) . . EBB . . . . per bulb
Bottles of 25 c.c. (either dilution) . . . . . EFM . . . . per bot.

'Pollaccine'—Pollen Vaccine for Hay Fever.

This vaccine is prepared from the pollen of grass (*phleum pratense*).

It may be employed both for the prophylaxis and for the treatment of hay fever. The prophylactic treatment gives incomparably the better results, it being possible at times other than the pollen season to advance to larger and much more effective doses without incurring any risk of aggravating the attacks of hay fever.

PART II   255   BACTERIAL VACCINES.

BACTERIAL VACCINES

Prepared under the supervision of Sir Almroth E. Wright, M.D., F.R.S., etc.

Third Syllables for Code Words.	Second Syllables for Code Words.								First Syllables for Code Words appear after the respective packages.
	OT Dozen 1-12	OV 1-6	OW 1-4	OX 1-3	UB 1-2	UF 3-4	UJ 1	UL 2 Dozen.	
				(For other numerals, etc., see page ix.)					

'Pollaccine'—Pollen Vaccine for Hay Fever—*concluded.*

For use in connexion with 'Pollaccine,' a set of graduated dilutions of the pollen toxin are supplied (see 'Hay Fever Reaction Outfit,' below). By instilling these (beginning with the weakest dilution and going on till slight itching and reddening of the conjunctiva is produced) into the patient's eye, the proper initial dose is arrived at. In like manner, the patient's progress can be re-investigated at each subsequent sitting, and the proper dose for carrying on the immunisation can be determined.

'Pollaccine' is issued in four dilutions as follows:—

VXPT
VXPU
VXPW
VXPX

       4 units of pollen toxin per c.c.
     20 units of pollen toxin per c.c.
   100 units of pollen toxin per c.c.
   500 units of pollen toxin per c.c.

Bulbs of about 1 c.c. (any dilution) . . . . EBB . . . . per bulb   2/9
Bottles of 25 c.c. of the 20-unit or 100-unit strength . EFM . . per bot.   27/9

Hay Fever Reaction Outfit.

VXRO

Contains seven capillary tubes of pollen toxin in graduated dilutions, with apparatus for instilling into the patient's eye as above described.

In case, with instructions for applying the test . . ADB . . . . . complete   11/3

Capillary Tubes of Pollen Extract for Diagnostic Use.

VXPE

Any of the several strengths of pollen toxin contained in the Hay Fever Reaction Outfit can be supplied as follows:—

Any one strength, in packages of 12 capillary tubes . ELW . per package   11/3
Any one strength, in packages of 2 capillary tubes . ODG . per package   2/2

Staphylococcus Vaccine.

This vaccine is prepared from cultures of staphylococcus obtained from cases of furunculosis, etc., and may be applied in any of the various forms of staphylococcus infection. It finds useful application (both phylactic and prophylactic) in staphylococcic infections of (a) the subcutaneous tissue—*furunculosis, carbuncle,* and *styes in the eye;* (b) hair-follicles of the beard—*sycosis;* (c) deep-lying structures—*whitlow* and *osteomyelitis;* (d) sinuses and fistulas; (e) moist, lacerated or denuded skin surfaces—*pruritus ani, reddening and pustular affection of anterior nares, eczema, small-pox vaccination, ophthalmia tarsi, dandruff* and *irritation of scalp;* and (f) some cases of *discharge from nose and ear.*

Staphylococcus Vaccine may also be used (either alone, or in combination with Streptococcus Vaccine) as a prophylactic measure against the septic infections which may follow upon surgical operations and all wounds—in particular, serious wounds, such as those associated with compound fractures, but also very slight wounds such as those made in anti-small-pox vaccination, which very readily become infected.

Issued in three dilutions as follows:—

VXQN
VXQP
VXQR

    100 millions of staphylococci per c.c.
    500 millions of staphylococci per c.c.
  1000 millions of staphylococci per c.c.

The minimum effective dose is 50 to 100 millions; and a medium dose 250 millions at the outset, increasing to 750 or 1,000 millions.

Bulbs of about 1 c.c. (any dilution) . . EBB . . . . per bulb   2/9
Bottles of 25 c.c. (any dilution). . . . . EFM . . . . per bot.   27/9

Streptococcus Vaccine.

This vaccine is prepared from cultures of streptococcus isolated from cases of erysipelas and cellulitis.

It is applicable in the various forms of localised streptococcus infections, in particular, in poisoned wounds associated with streptococcal lymphangitis, in onychia, in erysipelas, and in streptococcal infections of sinuses and operation or small-pox vaccination wounds.

It may also advisably be used (alone or in combination with Staphylococcus Vaccine) as a prophylactic measure against the septic infections which may follow upon surgical operations and all kinds of wounds.

For information . . . Code, see page viii.

Parke, Davis & Company, Index of Therapeutics and Descriptive Catalogue 1913–1914

BACTERIAL VACCINES

Prepared under the supervision of Sir Almroth E. Wright, M.D., F.R.S., etc.

Third Syllables for Code Words.	Second Syllables for Code Words.							First Syllables for Code Words appear after the respective packages.
	OT Dozen 1-12	OV 1-6	OW 1-4	OX 1-3	UB 1-2	UF 3-4	UJ 1	UL 2 Dozen.
			(For other numerals, etc., see page ix.)					

Streptococcus Vaccine—*concluded.*

Issued in two dilutions as follows :

5 millions of streptococci per c.c.
20 millions of streptococci per c.c.

The minimum effective dose is 2 millions ; and a medium dose from 5 to 15 millions.

VXSA
VXNP

Bulbs of about 1 c.c. (either dilution). . EBB . . . . per bulb   2/9
Bottles of 25 c.c. (either dilution). . . . EFM . . . . per bot.   27/9

Tuberculin (Bacillary Emulsion),

Issued principally for employment in localised tubercle infections which are not complicated by auto-inoculations. In such cases it is desirable to regulate the dosage by means of the opsonic index, but where this cannot be carried out, tuberculin treatment may nevertheless be usefully employed. In cases of phthisis, it will be well to begin with the minimum effective doses mentioned below for use in extensive infections, and never to advance beyond the doses prescribed as the minimum doses for the slighter infections. In tubercular joints and tubercular glands, it is permissible to begin with the minimum effective doses for the slighter infections, and to advance very gradually to the medium effective doses. (For further particulars, see pamphlet entitled " Tuberculin Therapy.")

Issued in five dilutions as follows :

VXTC
VXTL
VXPZ
VXKO
VXTI

$1/100000$ mgm. comminuted tubercle culture per c.c.
$1/25000$ mgm. comminuted tubercle culture per c.c.
$1/5000$ mgm. comminuted tubercle culture per c.c.
$1/2000$ mgm. comminuted tubercle culture per c.c.
$1/200$ mgm. comminuted tubercle culture per c.c.

Minimum effective doses are from $1/500,000$ to $1/50,000$ mgm. in extensive infections, and $1/15,000$ to $1/10,000$ in slight infections ; medium doses are from $1/6,000$ to $1/3,000$ mgm.

Sets of 3 bulbs of about 1 c.c. (any one dilution) . . EBK . . . . per set   3/4
Bottles of 25 c.c. (any dilution) . . . . EFM . . . . per bot.   22/6

VXTY

Typhoid Vaccine,

Prepared from cultures of the typhoid bacillus. Issued for preventive inoculation. (The most recent statistics from the British Army in India show that the incidence of typhoid fever has been reduced in the inoculated troops by seven-eighths, and the mortality from typhoid by more than nine-tenths by anti-typhoid inoculation.)

It is recommended that, in the case of the ordinary healthy adult, 1,000 millions should be given for the first dose, and 2,000 millions should be given for the second, which should follow the first after an interval of ten days. Young children may receive 2 to 5 minims (0·1 to 0·3 c.c.), according to age, of the weaker vaccine for the first injection, and double that quantity ten days later.

With a view to minimising constitutional disturbance, and the tendency to faintness which may supervene where the patient exerts himself immediately after the inoculation, it is recommended that the inoculation shall be carried out in the late afternoon or early evening, and that the patient should be instructed to go to bed, in particular after the first inoculation, as soon as he feels any malaise. It is advisable, but not necessary, that the patient should remain in bed during the next forenoon.

This vaccine should be used within three months of the date of preparation.

Sets of 2 bulbs, as required for the first and second inoculations. ADB . per set   11/8

*   *   *   *

FUYMI

Sterile Carbolised Salt Solution,

This solution is prepared in the Laboratory of St. Mary's Hospital for use in diluting bacterial vaccines.

Bottles of 25 c.c. . . . ADB . . . . per bot.   1/8

For information as to use of Telegraphic Code, see page viii.

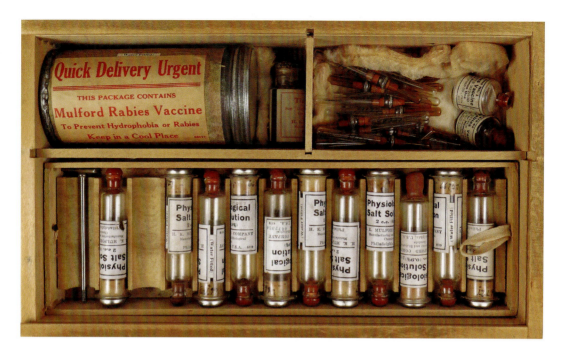

Mulford Rabies Vaccine Kit – early 1900s

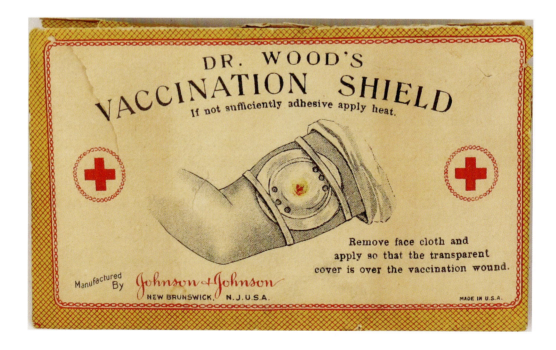

Johnson & Johnson Dr. Wood's Vaccination Shield Kit – early 1900s

Parke, Davis & Co., US Smallpox Vaccine Advert in *The Saturday Evening Post* **– 1928**

A Chronology of Vaccination

1912–1916

1912 Establishments licensed in the USA to manufacture smallpox vaccine include Parke, Davis & Co., H.K. Mulford, Dr. H.M. Alexander & Co., Fluid Vaccine Co., Slee Laboratories, Cutter Laboratories, the Health Department of New York City, and the National Vaccine and Antitoxin Institute Washington, DC.

1913 The UK Medical Research Council (MRC) is founded with funds generated by the National Health Insurance and is immediately charged with establishing the National Institute for Medical Research (NIMR).

1913 The Wellcome Bureau of Scientific Research is established to unify all the scientific research projects, bringing together the work of Wellcome's Chemical and Physiological Research Laboratories.

1913 Hungarian-born American pediatrician Béla Schick develops the Schick test—used to determine whether or not a person is susceptible to diphtheria.

1913 French physiologist Charles Richet wins the Nobel Prize for Physiology or Medicine for his work on anaphylaxis (a term he coined in 1902 to mean the opposite of protection), an acute, severe, sometimes fatal allergic reaction to a second injection of an antigen.

1913

The Rockefeller Foundation, founded by the world's first billionaire and petroleum magnate John D. Rockefeller and his son, is chartered in New York with an influential International Health Division (IHD). It will continue the work of the Rockefeller Sanitary Commission. The foundation backs the General Education Board who gives millions to medical schools to exclude naturopathy and chiropractic teachings, effectively creating a medical monopoly.

1913 Burroughs Wellcome has offices in London, New York, Montreal, Cape Town, Milan, Shanghai, Buenos Aires and Bombay. By this date, the company's catalogue includes concentrated diphtheria antitoxin, tetanus antitoxin (for veterinary use only), anti-colon Bacillus serum, sera against dystentery, gonococcus, meningococcus, staphylococcus and typhoid, as well as six different sera against streptococcus made from horses.

1913 American physician Karl von Ruck announces the successful vaccination against tuberculosis of hundreds of children. As a result, the US Navy Surgeon General considers vaccinating all soldiers and marines with the vaccine, however in the end, the vaccine is abandoned.

1913 The American Society for the Control of Cancer, later the American Cancer Society, is created to educate the public about the importance of early detection and surgery.

1914 The Kitasato Institute is established, Japan's first private medical research facility named after Japanese doctor Kitasato Shibasaburō.

1914 World War I breaks out in Europe, ending in 1918 with millions dead.

1914 ✪ Experimental pneumococcal vaccine is tested on military troops during World War I, without much success.

1914 ✪ Typhoid vaccine by Dr. H.M. Alexander & Co., is first licensed in the USA for civilians.

1914 ✪ Rabies vaccine is first licensed in the US and Europe, though it has been available since 1886.

1914 On the basis of a revision of two articles of the Swiss Constitution in 1911 that was accepted by parliament in 1913, the Swiss Federal Council extends the Epidemics Act to six new diseases (diphtheria, tuberculosis, puerperal fever, cerebro-spinal meningitis, rabies, foot and mouth disease). In 1921, the Chambers will accept a revision that gives federal authorities even more power in exceptional circumstances.

1914 ✪ In France, typhus vaccines are mandated for French soldiers and offered to civilians.

1914

3 vials 2 cc. size
PERTUSSIS ANTIGEN
(DETOXIFIED)
Lederle
BACTERIAL ANTIGEN MADE FROM
Pertussis Bacillus
A formalized filtrate recommended for the early treatment of Whooping Cough, and as a prophylactic for directly exposed contacts; also indicated for general prophylaxis.
Physician's Sample
LEDERLE LABORATORIES, INC.,   NEW YORK

✪ Whole-cell pertussis vaccine containing formaldehyde and Merthiolate (thimerosal) is licensed in the USA, prepared by Lederle Laboratories. Pertussis vaccines are also manufactured by Massachusetts Public Health Laboratories, Upjohn Company, Greeley Laboratories, and E.R. Squibb & Sons.

1914–1918 During WWI, Pasteur Institute prepares and supplies vaccines and serums to the American Army and American Red Cross, with 800,000 doses provided for free.

1914–1918 During World War I, it is reported that the Germans attempt to ship horses and cattle inoculated with *Bacillus anthracis* (anthrax) and *Pscudomonas pseudomallei* (glanders), to the US and other allied countries. These agents are also used to infect Romanian sheep designated for export to Russia. Other allegations include Germany spreading cholera in Italy and plague in St. Petersburg in Russia. Germany denies all allegations.

1914–1918 During WWI, the Medical School at the Royal Naval College Greenwich in England, both research and produce vaccines and sera against typhoid, cholera, influenza, and gas gangrene.

1915 In Ypres, Belgium, 400 tons of chlorine gas is released by the Germans against the British, under the direction of chemist Fritz Haber. Highly toxic and lethal, 6,000 men are killed—it is the world's first recorded large-scale use of chemical warfare. Soon after, phosgene (carbonyl dichloride) is used, ten times as toxic. Phosgene, chlorine, and mustard gases are used by both sides during WWI.

1916 ✪ The first triple typhoid vaccine against typhoid, paratyphoid A, and paratyphoid vaccine (TAB) is produced by the British Medical School at the Royal Naval College.

1916 American physician William Henry Welch and Johns Hopkins University receive US$267,000 from the Rockefeller Foundation to establish the Johns Hopkins School of Hygiene and Public Health. It is the first of its kind in the US and becomes enormously influential in the field.

1916 French researchers Charles Nicolle and Ernest Conseil show that measles patients have specific protective antibodies in their blood. The researchers then demonstrate that serum from measles patients could be used to protect against the disease.

1916 In Columbia, South Carolina, a contaminated batch of typhoid vaccine causes 68 severe reactions, 26 abscesses, and 4 deaths.

1916 ✪ Serum antitoxin for gas gangrene, *Clostridia perfringenes* is licensed in Europe.

1916 Porton Down in Wiltshire, England, is established as the War Department Experimental Station for testing chemical and biological weapons. It then develops into a science and technology campus, home to the Ministry of Defense's most secretive military research facilities. Tear gas and VX nerve agent are developed and Porton Down will house some of the deadliest pathogens.

1916 ✪ In the USA, typhoid vaccines by Cutter Labs and Parke, Davis & Co., are licensed.

1916 A polio epidemic occurs with almost 9,000 cases of paralytic polio and 2,448 deaths in New York City alone. Across the USA polio kills 5,000-6,000 people, mainly children, leaving thousands more paralyzed. Italian immigrants are blamed for the outbreak, as well as dogs and cats—around 72,000 cats are culled to "control polio." Respected microbiologist and biochemist Dr. H.V. Wyatt writes in a 2011 paper that merely three miles from the epicenter of the outbreak, scientists at the Rockefeller Institute "had been passaging spinal cord tissue containing poliovirus, from one Rhesus monkey spinal cord to another. These experiments continued with the passaged virus which at times was reinforced with newly acquired virus from patients."

A Chronology of Vaccination

1916–1922

1916	Smallpox vaccine is produced in Australia by the newly established state-owned Commonwealth Serum Laboratories (CSL).
1916	✪ The US Army Medical School supplies the army with its first paratyphoid A vaccine. The standard routine includes administration of 6 doses of vaccine, 3 doses of antityphoid, and 3 of antiparatyphoid A. The following year a triple typhoid vaccine is released, including antityphoid, antiparatyphoid A, and antiparatyphoid B—known as TAB vaccine
1917	US President Woodrow Wilson establishes the Edgewood Arsenal in Maryland to develop both offensive and defensive chemical weapons; chlorine, chloropicrin, phosgene, and mustard agent. It becomes the home of the nation's chemical and biological weapons program, and the first location for the Chemical Warfare School.
1917	Mustard gas (bis(2-chloroethyl) sulfide) is used by the Germans against British troops.
1917	During WWI, the Rockefeller Institute sends antimeningococcal serum—produced in horses—to England, France, Italy and other countries.
1917	✪ The first chemically inactivated cholera vaccine is approved in the USA.
1917	New Zealander Sir Harold Gillies, considered the father of modern plastic surgery, performs the first reconstructive facial surgeries on severely injured soldiers.
1917	US financier John Pierpont Morgan purchases 25 of America's leading newspapers through his banking firm J.P. Morgan & Co. He inserts his own editors, effectively controlling the media and propagandizing US public opinion in favor of corporate and banking interests.
1917	Bayer's patent for aspirin expires, sparking a large production by competitors.
1917	✪ In Australia, tetanus antitoxin is introduced for the armed forces.
1917–1928	A mysterious epidemic of encephalitis lethargica spreads around the globe.

1918	An experimental pertussis vaccine is used on children at Boys' and Girls' Industrial Home and Farm, Lytton, California, USA.
1918	Horse antiserum is used to treat meningococcal meningitis for the next decade.
1918	From January 21 to June 4, Rockefeller Institute's experimental horse-cultured bacterial meningococcal vaccine is injected into over 4,000 soldiers at Fort Riley, USA.
1918	✪ A bacterial influenza vaccine is given to Swiss army troops made by Swiss Serum and Vaccine Institute.
1918	✪ Triple typhoid lipovaccine (bacteria in oil solution which alone acts as an adjuvant, boosting immune response) is administered to US Army troops during WWI.
1918	Towards the end of the year, the US Army Medical School makes two million doses of pneumococcal types I, II, and III vaccine.
1918	Spanish Influenza has endangered the prosecution of the WAR in Europe. There are 1500 cases in the Navy Yard 30 deaths have already resulted SPITTING SPREADS SPANISH INFLUENZA DONT SPIT

Worldwide reports of a deadly influenza virus is called "Spanish Flu" despite the fact that the first cases occur at US military base Fort Riley, Kansas. Spain is one of the few neutral countries not involved in WWI. Most of the countries involved in the war censored their press to influence public opinion. Free from censorship concerns, the earliest press reports of people dying from disease in large numbers came from Spain. The warring countries did not want to additionally frighten the troops, so they were content to scapegoat Spain. An estimated 95% (or higher) of the deaths were caused by secondary bacterial pneumonia not influenza. Mask wearing and lockdowns are enforced as a response to control the epidemic. Aspirin is widely used against symptoms of fever, and prescribed in high, potentially toxic doses.

1918 ✪ Canada's first domestic supply of pertussis vaccine is prepared and distributed by the Ontario Board of Health Laboratory.

1918 A variety of killed whole-cell bacterial vaccines are tested; these vaccines include *Bacillus influenzae* (now known as *Haemophilus influenzae*) and strains of pneumococcus, streptococcus, staphylococcus and other bacteria.

1918–1919 Early tests of pneumococcal polysaccharide vaccines take place at Camp Upton in New York and at Camp Wheeler in Georgia, USA.

1919 Belgian bacteriologist Jules Bordet is awarded the Nobel Prize for Physiology or Medicine for his discoveries on immunity.

1919 Dozens of Dallas children are sickened and five die from a contaminated batch of diphtheria toxin-antitoxin mixture (TAT). The TAT is manufactured by H.K. Mulford, Philadelphia, who pay damages to the afflicted families.

1919

A mass protest against compulsory smallpox vaccination takes places in Toronto, Canada.

1919 Designed to keep the peace following WWI, delegates at the Paris Peace Conference create the League of Nations under the Treaty of Versailles.

1919 Russia makes smallpox vaccination compulsory for the entire population.

1919 ✪ An influenza vaccine is prepared at the British Medical School at the Royal Naval College, consisting of *B. influenzae* 60 million, streptococci 80 miillion, and pneumococci 200 million per cubic centimeter.

1920–1929

1920s In the early 1920s, French biologist and veterinarian Gaston Ramon devises a new method for chemically inactivating diphtheria and tetanus toxins, using formaldehyde to create an anatoxin (also called toxoid) for use in vaccines.

1920

The League of Nations Health Organization (LNHO), predecessor to the WHO, is founded in Geneva, largely funded by the Rockefeller Foundation and the British government.

1921 Franklin Delano Roosevelt, former New York state senator, assistant secretary to the Navy, and future US president, falls ill with what is assumed to be polio, but it just might be Guillain-Barré syndrome.

1921 Indian lawyer, politician, and social activist Mahatma Gandhi writes *A Guide to Health* and states that "vaccination is a violation of the dictates of religion and morality."

1921 American Public Health Service special expert M.A. Barber discovers that Paris Green is an effective larvicide. It will become widely used by the Health Boards as a new weapon against mosquitos to control malaria.

1921 Burroughs Wellcome and Allen & Hanburys are among the first in the UK to reproduce insulin for commercial use.

1921 At an international conference, it is agreed to cooperate in standardization efforts of active substances, notably sera and vaccines. This responsibility is given to Danish physician Thorvald Madsen of the Danish State Serological Institute, marking the creation of the LNHO Standardization Commission.

1921 Insulin is discovered as a way of managing diabetes by Eli Lilly in the US. Commercial distribution begins in 1923.

1921 French scientists Albert Calmette and Camille Guérin begin their first tests—on children at Charité hospital in Paris—of their attenuated tuberculosis bacilli in humans. Their oral preparation is called Bacillus Calmette-Guérin, or BCG vaccine, which is a weakened form of a tuberculosis bacterium that causes the disease in cows.

1922 The LNHO publishes the first issue of a series entitled *Epidemiological Intelligence* and *Epidemiological Reports* containing data summaries, mainly for typhus.

A Chronology of Vaccination

1922 Following the success of the Second International Congress on Eugenics, the American Eugenics Society is established. It is renamed The Society for the Study of Social Biology in 1973.

1922 A landmark decision by the US Supreme Court, *Zucht v. King*, upholds the constitutionality of excluding unvaccinated children from school, even if there is no outbreak in progress.

1922 *Encyclopedia Britannica* replaces Dr. Charles Creighton's critical entry on vaccination with a favorable article on "Vaccine Therapy."

1922–1927 The UK Ministry of Health records 106 cases of encephalitis, with 52 deaths following smallpox vaccination.

1923 In close collaboration with the University of Toronto, Eli Lilly makes biosynthetic human insulin commercially available.

1923 Allen & Hanburys produces 95% of insulin (derived from cow pancreas) in the UK.

1923 French scientist J. Descombey, working with Gaston Ramon, is credited with developing tetanus toxoid, which becomes widely used during WWII.

1923 ✪ The first diphtheria toxoid vaccine is developed by Gaston Ramon at the Pasteur Institute. British scientists Alexander T. Glenny and Barbara E. Hopkins at the Wellcome Research Laboratories in London also begin experimenting with vaccines created from inactivated toxins (toxoids).

1923 American eugenicist nurse Margaret Sanger opens the Birth Control Clinical Research Bureau in Manhattan to provide birth control to women. She also incorporates the American Birth Control League, "an ambitious new organization that examines the global impact of population growth, disarmament, and famine." The two organizations eventually merge to become Planned Parenthood.

1923 The term "herd immunity" is used for the first time by English bacteriologists William W.C. Topley and Sir Graham S. Wilson.

1923 Aluminum is described by Gaston Ramon and used to flocculate toxins by adsorption.

1924 According to a US Public Health Report, post-vaccinal encephalitis is first brought to the attention of US medical professionals. By 1932, 700 cases with a 40% case fatality rate are reported in the USA and Europe.

1924 The broad economic consequences of rinderpest, also called cattle plague, lead to the creation of the Office International des Epizooties (OIE), renamed the World Organization for Animal Health in 2003.

1924 The Rolleston Committee on Vaccination is established in the UK and chaired by prominent English physician Sir Humphrey Rolleston (later president of the Eugenics Society 1933–1935). The committee is tasked with investigating "observations on the epidemiology and the clinical and pathological character of post-vaccinal nervous disease," though no neurologist is on the committee. By 1931, 90 cases of acute nervous disease within 4 weeks of smallpox vaccination are investigated.

1924 The Malaria Commission of the LNHO is established, in close partnership with the Rockefeller Foundation.

1924 ✪ Diphtheria antitoxin is introduced in Australia and the first mass vaccination program takes places in Victoria.

1924 ✪ American husband and wife medical research team George and Gladys Dick discover scarlet fever is caused by *streptococcus pyogenes* bacteria. They subsequently develop a "Dick" skin test to determine a patient's susceptibility to the disease. They also create a scarlet fever antitoxin and vaccine for treatment and prevention, which is used until the mid-1940s when it is replaced by penicillin. They are nominated for the Nobel Prize in 1925, but there will be no winners that year.

1925 By the end of the year, LNHO receives regular epidemiological intelligence from 76 ports in 27 countries. By 1927, this data is sent from 140 ports.

1925 The American Type Culture Collection (ATCC) is founded by a committee of seven scientists to serve as a national biological resource center for holding and distributing microbiological specimens—proteins, viruses, yeasts, fungi, and bacteria. Their founding collection consists of 175 strains of microorganisms, with the first catalogue being published in 1927. The Rockefeller Foundation gifts US$ 10,000 in 1930 to the ATCC, at which time it begins charging for cultures to cover expenses. ATCC will enjoy a tax-exempt status from March 1947.

1925 Danish physician Thorvald Madsen is the first to test whole-cell pertussis vaccine on a wide scale. Trial data suggests it is a success, but a 1933 report states that two children may have died from the vaccine.

1925 The *Protocol for the Prohibition of the Use in War of Asphyxiating, Poisonous or other Gases, and of Bacteriological Methods of Warfare*, also called the "Geneva Protocol," prohibits chemical and biological weapons in war, however not all countries comply. Belgium, Canada, France, the UK, Italy, the Netherlands, Poland, Japan, and the Soviet Union begin to develop biological weapons soon after they ratify the Protocol. The US signs the treaty, but takes 50 years to ratify it. During that time, the US uses biological and chemical weapons in both the Korean War and Vietnam War (dioxin-based Agent Orange, napalm).

1926 The Metropolitan Life Diphtheria Campaign is launched by US insurance company, who donate US$ 15,000 to promote vaccination.

1926 Developed by Glenny and his team at Wellcome Research Laboratories, aluminum salts are used as "precipitates" or adjuvants to increase effectiveness of toxoid vaccines. Aluminum phosphate $AlPO_4$ and aluminum hydroxide $Al(OH)_3$ are used. Parke, Davis & Co., and Sharp & Dohme soon begin to use alum in their toxoid vaccines.

1926 ✪ Alexander T. Glenny and Barbara E. Hopkins at the Wellcome Research Laboratories in London develop a toxoid-based tetanus vaccine.

1926 ✪ The University of Toronto's Connaught Laboratories begins to supply Canadian public health departments with a standard strain pertussis vaccine, under the direction of Dr. Donald T. Fraser. Connaught also exports this pertussis vaccine to the US, UK, Asia, Central America, and New Zealand.

1926 ✪ Alum-precipitated diphtheria toxoid vaccine is licensed in the USA.

1926 ✪ In the USA, a pertussis vaccine by Merrell-National Laboratories is licensed.

1926 ✪ American Professor of bacteriology and immunology W.P. Larson creates a scarlet fever vaccine based on a ricinoleated toxoid. Due to severe reactions it is not widely used and will be replaced with antibiotics.

1926 ✪ In Australia, tetanus toxoid and pertussis vaccines are made available.

1927 ✪ The Rockefeller Foundation develops a yellow fever vaccine using the African *Asibi* strain.

1927 US Supreme Court mother-daughter case *Buck v. Bell* rules that it is in the state's best interest to forcibly sterilize people deemed unfit and feeble-minded. The Virginia Sterilization Act of 1924 is upheld as not being a violation of the Constitution, legitimizing compulsory sterilization until it is repealed in 1974.

1927 The Chemical Spray Residue Act is passed in the USA making it illegal to pack, ship, or sell fruits or vegetables with harmful pesticide residues.

1927 ✪ Bacilius Calmette-Guerin BCG vaccine is formulated at the Pasteur Institute. This tuberculosis vaccine is given to newborns, and becomes the most widely administered vaccine in the WHO Expanded Program on Immunization (EPI) from 1974. It is the first vaccine based on live-attenuated bacteria.

1927 In Germany, the Kaiser Wilhelm Institute for Anthropology, Human Heredity, and Eugenics is founded in Berlin, partially financed by the Rockefeller Foundation.

1927 ✪ In Australia, diphtheria vaccine (toxoid) is made available to the public.

A Chronology of Vaccination

1927 Perhaps the most celebrated Council of Physics, the Fifth Solvay International Conference on Electrons and Photons, brings the newly formulated theory of quantum physics to center stage. Leading figures Albert Einstein and Niels Bohr famously debate quantum mechanics, and 17 of the 29 attendees will go on to become Nobel Prize winners.

1928 The Office International des Epizooties (OIE) holds its first conference in Geneva, consisting of eight experts who establish a basis for an international sanitary police.

1928 During a diphtheria vaccine campaign in Queensland, Australia, staphylococcal contamination of diphtheria antitoxin-toxin (TAT) serum—manufactured by the federal government's Commonwealth Serum Laboratories (CSL)—leads to the septic deaths of 12 vaccinated children. Events like these lead to the introduction of preservatives in vaccines, notably thimerosal, to prevent microbial infection.

1928 English microbiologist Alexander Fleming discovers penicillin by accident while studying *Staphylococcus aureus* at St. Mary's Hospital in London. The purified version becomes available in 1941.

1928 The Soviet biological weapons program (precursor to the 1974 "civilian" research and pharmaceutical front company Biopreparat), launches its first project to weaponize typhus.

1928 French researcher Charles Nicolle is awarded the Nobel Prize in Physiology or Medicine for his work on typhus.

1928 At the Conference of the Health Committee of the League of Nations, the BCG vaccine is recommended against tuberculosis, declared "safe and effective" and its widescale use strongly promoted.

1928 The first iron lung is used to preserve respiratory function in polio-afflicted patients.

1929 The Great Depression plunges the world into recession, driving millions of people globally into unemployment and poverty.

1929–1931 The first bacterial polysaccharide antigen-based, chemically conjugated vaccine is synthesized by American-Canadian physicians Oswald T. Avery and Walther F. Goebel at the Rockefeller Institute. They improve polysaccharide immunogenicity by the tool of conjugation to proteins. This technique will be applied in future *Haemophilus influenzae* type b (Hib), pneumococcal and meningococcal vaccines, using diphtheria and tetanus toxoid as carrier proteins.

1929 Eli Lilly & Company patents the mercury-derived product Merthiolate, or thimerosal, formulated at the University of Maryland with support of a Lilly research fellowship. It is used from 1930 in medicine and vaccines as an antimicrobial and antiseptic agent. The only safety test would be in 1931 on 22 people with meningitis, who all died. This Eli Lilly study would be considered a success, Merthiolate is claimed to be safe, and is licensed for use in childhood vaccines from 1935 to 2000.

1929 The Human Betterment Foundation is established in California, USA, as a nonprofit organization dedicated "to foster and aid constructive and educational forces for the protection and betterment of the human family in body, mind, character, and citizenship." In practice, the Foundation advocates and researches reproductive sterilization of the "socially and mentally unfit" according to eugenic principles.

1929–1930 In Lübeck, Germany, 256 newborn babies are vaccinated with an oral BCG vaccine resulting in 72–77 deaths and 130–173 cases of tuberculosis, leading to a highly publicized trial in court. Georg Deycke (head of the Lübeck general hospital) and Ernst Altstaedt are sent to prison for "bodily injury due to negligence" in 1932, for 24 months and 15 months respectively. Compulsory BCG vaccination is stopped, but is reinstated in East Germany in 1952.

1929 E. Gräfenberg adapts Richard Richter's silkworm gut ring intrauterine device (IUD) by wrapping it with a spiral of silver wire, producing a stiffer device with improved resistance to expulsion. Despite their availability, contraceptive IUDs would only become widely used from the 1960s.

1930–1939

1930s Influenza vaccines are tested on 800 "retarded" male subjects in a Pennsylvania state colony. Many vaccines are tested on mentally challenged minorities under state care. Several vaccines are also tested in colonies and developing nations in South America, Africa, and Asia.

1930s Vaccination with diphtheria toxoid becomes more widespread worldwide.

1930s Hungarian-born American immunologist Jules T. Freund develops Freund's Complete adjuvant, a mineral oil-based adjuvant containing heat-killed mycobacteria. It is potent but too toxic to include in vaccines.

1930 The Bank of International Settlements is founded in Basel, Switzerland, as an intergovernmental agreement or treaty, between the USA, UK, Germany, France, Belgium, Italy, Japan and Switzerland.

1930 The US Federal Drug Agency (FDA) is set up, with its headquarters based in Maryland.

1930 The US National Institute (later Institutes) of Health (NIH) is created out of the Public Health Service's Hygienic Laboratory as a result of the Ransdell Act. US Congress appropriates US$750,000 to build facilities and finance fellowships for biological and medical research.

1930 Polish biologist Rudolf Weigl invents the first typhus vaccine using 150 lice intestines to make a single dose of vaccine.

1930 The American Academy of Pediatrics (AAP) is founded by 35 pediatricians to address pediatric health standards.

1930 German physician Wolfgang Casper develops a gonococcal vaccine against gonorrhea, and conducts an unusual placebo-controlled clinical trial on 10 destitute volunteers, using a prostitute as the infectious "challenge."

1931 American researcher Margaret Pittman, PhD, classifies different types of *Haemophilus influenzae* bacteria, identifying the strain type b (called Hib, for *Haemophilus influenzae* type b) as being the cause of nearly all cases of *H. influenzae* meningitis.

1931 The LNHO holds a conference in London to establish the Permanent Commission on Biological Standardization, setting standards for gas-gangrene antitoxin, tuberculin, and diphtheria vaccine.

1931 American pathologist Ernest W. Goodpasture and American virologist Alice M. Woodruff publish their seminal paper *The Cultivation of Vaccine and Other Viruses in the Chorio-allantoic Membrane of Chick Embryos* using fowlpox. Their work facilitates the rapid advancement of virus study and egg culture virology, especially for influenza.

1932 Yellow fever vaccine prepared in mouse brain is developed, but requires an injection of human immune serum to prevent adverse events. A modified strain is subsequently developed that does not require immune serum; however it continues to be cultured in mouse brain and has neurotropic properties that result in a high incidence of encephalitis reactions.

1932 The first school-based diphtheria vaccine programs begin in Australia.

1932 Aluminum salts are first officially used in human vaccines as adjuvants, and are the only adjuvants in licensed vaccines for ~70 years.

1932 *The Journal of Pediatrics* is first published in July by the C.V. Mosby Company.

1932–1972 The Tuskegee Syphilis Study is conducted by the US Public Health Service on over 600 American-African men, leaving them untreated for syphilis to examine the natural course of the disease. The men are never given adequate treatment for their disease even though penicillin is available from 1947. This study ranks among one of the greatest medical ethics scandals in the history of the US. In 1973, a class-action lawsuit is filed and the government awards an out-of-court settlement of US$10 million to participants. A formal apology for this unethical act is issued by President Bill Clinton in 1997.

A Chronology of Vaccination

1932–1937

1932 American bacteriologist Herald R. Cox joins the Rockefeller Institute for Medical Research where he works on developing vaccines against eastern equine encephalomyelitis (EEE) and western equine encephalomyelitis.

1932 Methyl bromide or bromomethane, a colorless, odorless gas, is first used as a pesticide in France. It is used extensively worldwide until the early 2000s when it is phased out by most countries due to its highly neurotoxic properties.

1932 American physician Arthur W. Hedrich publishes the results of an observational study of measles data from 1900–1931. He concludes that when at least 55% of children under 15 years of age had natural measles, outbreaks stopped, thereby demonstrating the theory of herd immunity.

1933 Danish physician and director of the Danish State Serum Institute, Thorvald Madsen reports that two children die from adverse reactions to whole-cell pertussis vaccine.

1933 The first International Certificate of Inoculation and Vaccination is established by the International Sanitary Convention for Aerial Navigation in the Hague, Netherlands. It enters into force two years later.

1933 British scientists Wilson Smith, Christopher Andrewes, and Patrick Laidlaw of the UK Medical Research Council isolate influenza A virus from throat washings and infect ferrets. They will eventually produce influenza vaccines that are first tested on British army troops fighting in World War II.

1933 Merck & Co., Inc., establish the nonprofit, New Jersey-based corporation the Merck Institute for Therapeutic Research, for the "purpose of conducting investigations into the causes, nature and mode of prevention and cure of diseases in men and animals." The determination of the therapeutic value and safety of new drugs (including vaccines) is one of its main duties.

1933 The electron microscope is invented by German physicist Ernst Ruska, allowing scientists to view nanosized entities like viruses.

1933 The Medical Research Council in the UK announces that it will issue a supply of serum obtained from patients convalescent from acute poliomyelitis, to test it as a treatment of the preparalytic stage of the disease.

1934 Sulfonamide drugs, a group of synthetic antibiotics—a derivative of an old coal-tar compound—are released on to the global market, first by German company Bayer.

1934– 1944 Gerhard Schrader, a German chemist at IG Farben industries, and his coworkers synthesize about 2,000 organophosphate compounds, including parathion as a pesticide, and tabun, sarin, and soman as chemical warfare nerve agents.

1934 Mumps is identified as a transmissible virus by Johns Hopkins medical graduate and Vanderbilt professor Ernest W. Goodpasture.

1934 The first H1N1 flu virus that will be used in future influenza vaccine is isolated: A/PR8 (type A, Puerto Rico #8 strain). The viral strain is repeatedly passaged through various animals (mice, ferrets) and cell cultures to make it less virulent in humans.

1934 Italian Prime Minister Benito Mussolini opens the Rockefeller-endowed Institute of Public Health in Rome. The League of Nations would hold its course on malaria there.

1935 Polio vaccine trial disaster occurs in the US. American physician Maurice Brodie prepares a formaldehyde-killed polio vaccine while at the New York City Health Department. He tests it on chimpanzees, on himself, and finally on children, enrolling about 11,000 individuals in his trial. Meanwhile, physician John Kolmer in Philadelphia develops an attenuated poliovirus vaccine, which he tests on about 10,000 children. Several trial subjects die of poliomyelitis and many become paralyzed, sick, or suffer allergic reactions.

1935 ✪ Alum-precipitated toxoid (APT) diphtheria vaccine is commercialized in the UK by Burroughs Wellcome.

1935– 1941 The first clinical trials for live-attenuated influenza vaccines take place in the USA.

1935 In a letter from the Director of Biological Services of the Pitman-Moore Company to Dr. Jamieson of Eli Lilly & Company, thimerosal is declared as "unsatisfactory as a preservative for serum intended for use on dogs" due to its neurotoxic effects.

1936 ✪ Director of the Virus Laboratory at the Rockefeller Institute, Max Theiler, and his team develop a live-attenuated yellow fever vaccine (strain 17D) using tissue cultures prepared from embryonated chicken eggs. The vaccine is easily adapted for mass production and becomes the universal standard from the late 1930s.

1936 The Rockefeller Institute's researchers Peter Olitsky and Albert Sabin demonstrate a new cultivation method for poliovirus. They are able to grow virus in human embryonic brain tissue for the development of vaccine.

1936–1945 Japanese microbiologist and medical officer of the Imperial Japanese Army, Shirō Ishii, commands the newly established biological and chemical weapons facility in China, Unit 731—officially known as the *Epidemic Prevention and Water Purification Bureau.* The research involves using live human subjects for research on bubonic plague, cholera, anthrax, typhoid, gangrene, botulism, hemorrhagic fever and tuberculosis. Devices and vectors (including fleas) to spread germs and parasites are developed by Japan and used against China in the Second Sino-Japanese War (1937–1945). Numerous Japanese war criminals—Unit 731 veterans—will be employed to help develop the US biological weapons program, notably in labs outside the US to "carry out experiments on human subjects that could not be conducted legally in the US."

1936 British scientist Sir Harold Percival Himsworth publishes research dividing diabetes into type 1 and type 2 based on the degree of insulin sensitivity in patients.

1936 The International Health Division of the Rockefeller Foundation creates a Yellow Fever Research Institute at Entebbe, British East Africa, as a joint venture with the government of Uganda.

1936

After the death of pharmaceutical magnate Henry Wellcome, the Wellcome Trust is founded as a charitable foundation focused on health research. Based in London, it will become one of the world's leading medical institutions, holding an extensive library.

1936–1937 ✪ Connaught Laboratories launches a new pertussis vaccine based on fresh strain *B. pertussis.* National distribution of the vaccine begins through provincial public health departments across Canada.

1937 Due to the difficulty for virologists to satisfy any of Koch's Postulates to prove invisible particles assumed to be "viruses" exist and could cause disease, microbiologist Thomas M. Rivers of the Rockefeller Institute, waters down Koch's Postulates. The father of modern virology revises them to allow virologists more flexibility and expands the Postulates from four to six. However, even with this change, it remains challenging to prove viruses cause disease.

1937 ✪ Influenza vaccine is licensed in the USA, developed by Anatoli Smorodintsev.

1937 The Rockefeller 17D yellow fever vaccine is accidentally contaminated with hepatitis and given to two million people in Brazil during a national vaccination campaign. This vaccine is discontinued in 1943.

1937 ✪ Adsorbed tetanus toxoid vaccine is licensed and used extensively on military troops during World War II.

1937 One hundred and seven people are fatally poisoned by diethylene glycol—a solvent used in the production of an antibiotic syrup to treat strep throat. Also called "the Massengill disaster," this event hastens the signing of the Food, Drug and Cosmetic Act.

1937 The Division of Biologics Control is formed within the NIH.

Burroughs Wellcome & Co., Vaccine Pricelist 1938–1939
Chief offices and laboratories in London, UK
With branches worldwide in New York, Montreal, Sydney,
Milan, Shanghai, Cape Town, Bombay, Buenos Aires.

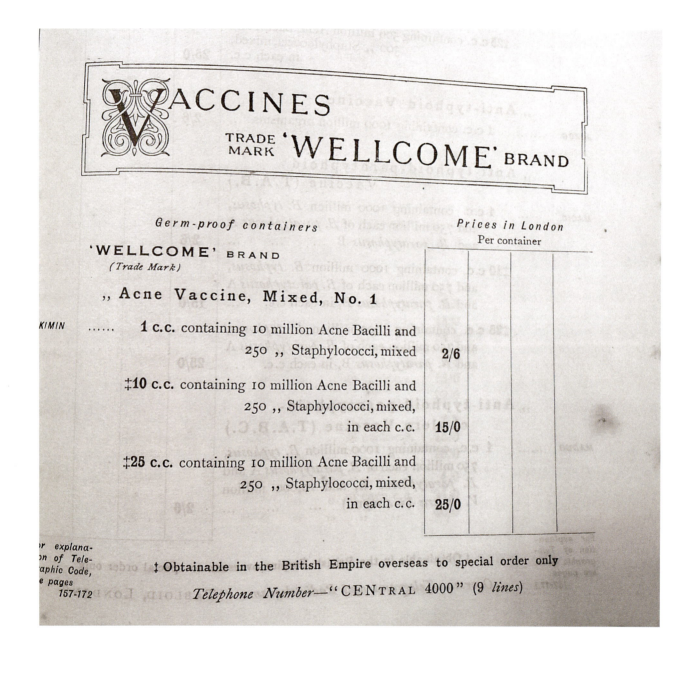

VACCINES
TRADE MARK 'WELLCOME' BRAND

Germ-proof containers	Prices in London Per container
'WELLCOME' BRAND (Trade Mark)	
,, Acne Vaccine, Mixed, No. 1	
...... **1 c.c.** containing 10 million Acne Bacilli and 250 ,, Staphylococci, mixed	**2/6**
‡**10 c.c.** containing 10 million Acne Bacilli and 250 ,, Staphylococci, mixed, in each c.c.	**15/0**
‡**25 c.c.** containing 10 million Acne Bacilli and 250 ,, Staphylococci, mixed, in each c.c.	**25/0**

‡ Obtainable in the British Empire overseas to special order only

Telephone Number—"CENTRAL 4000" (9 *lines*)

or explana-
on of Tele-
aphic Code,
e pages
157-172

Vaccines on the market in the late 1930s for humans and animals
Acne vaccines, typhoid vaccine, coryza (rhinitis or common cold) vaccine, *B.coli* vaccine, gonococcus vaccine, typhoid-paratyphoid-cholera (TABC) vaccine, influenza vaccine mixed, melitensis vaccine, rheumaticus vaccine, pneumococcal vaccine, staphylococcus vaccine, cholera vaccine, pertussis vaccine, and veterinary vaccines...

VACCINES—*continued*

Prices in
Per co

'WELLCOME' BRAND
(Trade Mark)

,, Acne Vaccine, Mixed, No. 2

...... 1 c.c. containing 125 million Acne Bacilli and 125 ,, Staphylococci, mixed | 2/6

‡10 c.c. containing 125 million Acne Bacilli and 125 ,, Staphylococci, mixed, in each c.c. | 15/0

‡25 c.c. containing 125 million Acne Bacilli and 125 ,, Staphylococci, mixed, in each c.c. | 25/0

,, Acne Vaccine, Mixed, No. 3

...... 1 c.c. containing 500 million Acne Bacilli and 500 ,, Staphylococci, mixed | 2/6

‡10 c.c. containing 500 million Acne Bacilli and 500 ,, Staphylococci, mixed, in each c.c. | 15/0

‡25 c.c. containing 500 million Acne Bacilli and 500 ,, Staphylococci, mixed, in each c.c. | 25/0

,, Anti-typhoid Vaccine

...... 1 c.c. containing 1000 million organisms ... | 2/6

,, Anti-typhoid-paratyphoid Vaccine (T.A.B.)

..... 1 c.c. containing 1000 million *B. typhosus*, and 750 million each of *B. paratyphosus* A and *B. paratyphosus* B ... ... | 2/6

‡10 c.c. containing 1000 million *B. typhosus*, and 750 million each of *B. paratyphosus* A and *B. paratyphosus* B, in each c.c. ... | 15/0

‡25 c.c. containing 1000 million *B. typhosus*, and 750 million each of *B. paratyphosus* A and *B. paratyphosus* B, in each c.c. ... | 25/0

,, Anti-typhoid-paratyphoid-cholera Vaccine (T.A.B.C.)

... 1 c.c. containing 1000 million *B. typhosus*, 750 million each of *B. paratyphosus* A and *B. paratyphosus* B, and 10,000 million *V. choleræ* (= 4 mgm.) ... ... | 2/6

‡ Obtainable in the British Empire overseas to special orde

Overseas Telegrams and Radiotelegrams— "TABLOID, Lo

W. & CO. PRODUCTS

REMEMBER THE
TRADE MARKS

VACCINES—*continued*

Prices in London
Per container

'WELLCOME' BRAND
(Trade Mark)

,, B. Coli Vaccine

...... 1 c.c. containing 10 million organisms ... | 2/6

‡10 c.c. ,, 10 ,, ,, in each c.c. | 15/0
‡25 c.c. ,, 10 ,, ,, ,, ,, | 25/0
1 c.c. ,, 50 ,, ,, ... | 2/6
‡10 c.c. ,, 50 ,, ,, in each c.c. | 15/0
‡25 c.c. ,, 50 ,, ,, ,, ,, | 25/0
...... 1 c.c. ,, 250 ,, ,, ... | 2/6
‡10 c.c. ,, 250 ,, ,, in each c.c. | 15/0
‡25 c.c. ,, 250 ,, ,, ,, ,, | 25/0
1 c.c. ,, 1000 ,, ,, ... | 2/6
‡10 c.c. ,, 1000 ,, ,, in each c.c. | 15/0
‡25 c.c. ,, 1000 ,, ,, ,, ,, | 25/0

,, Coryza Vaccine, No. 4

...... 1 c.c. containing 50 million each of *B. hofmanni*, *B. friedländer*, *M. catarrhalis*, and Staphylococci, mixed ; and 10 million each of Pneumococci and Streptococci, mixed ... ... | 2/6

‡10 c.c. containing 50 million each of *B. hofmanni*, *B. friedländer*, *M. catarrhalis* and Staphylococci, mixed ; and 10 million each of Pneumococci and Streptococci, mixed, in each c.c. ... ... | 15/0

‡25 c.c. containing 50 million each of *B. hofmanni*, *B. friedländer*, *M. catarrhalis* and Staphylococci, mixed ; and 10 million each of Pneumococci and Streptococci, mixed, in each c.c. ... ... | 25/0

,, Gonococcus Vaccine

...... 1 c.c. containing 5 million organisms ... | 2/6
‡10 c.c. ,, 5 ,, ,, in each c.c. | 15/0
‡25 c.c. ,, 5 ,, ,, ,, ,, | 25/0
...... 1 c.c. ,, 20 ,, ,, ... | 2/6
‡10 c.c. ,, 20 ,, ,, in each c.c. | 15/0
‡25 c.c. ,, 20 ,, ,, ,, ,, | 25/0
...... 1 c.c. ,, 200 ,, ,, ... | 2/6
‡10 c.c. ,, 200 ,, ,, in each c.c. | 15/0
‡25 c.c. ,, 200 ,, ,, ,, ,, | 25/0
...... 1 c.c. ,, 1000 ,, ,, ... | 4/0
‡10 c.c. ,, 1000 ,, ,, in each c.c. | 24/0
‡25 c.c. ,, 1000 ,, ,, ,, ,, | 40/0

‡ Obtainable in the British Empire overseas to special order o

Burroughs Wellcome & Co., Wellcome Brand Pricelist 1938–1939

REMEMBER THE
TRADE MARKS

PR

Telegraphic Code

VACCINES—*continued*

Prices
Per

'WELLCOME' BRAND
(Trade Mark)

,, Influenza Vaccine, Mixed

IAPIS	1 c.c. containing 400 million *B. influenzæ*, 200 million Pneumococci, and 80 million Streptococci	2/6
	‡10 c.c. containing 400 million *B. influenzæ*, 200 million Pneumococci, and 80 million Streptococci, in each c.c.	15/0
	‡25 c.c. containing 400 million *B. influenzæ*, 200 million Pneumococci, and 80 million Streptococci, in each c.c.	25/0

,, M. Melitensis Vaccine
(Mediterranean Fever)

EW	1 c.c. containing 100 million organisms ...	2/6
	‡10 c.c. ,, 100 ,, ,, in each c.c.	15/0
	‡25 c.c. ,, 100 ,, ,, ,, ,,	25/0

,, M. Rheumaticus Vaccine

IM	1 c.c. containing 10 million organisms	2/6
	‡10 c.c. ,, 10 ,, ,, in each c.c.	15/0
	‡25 c.c. ,, 10 ,, ,, ,, ,,	25/0
Y	1 c.c. ,, 50 ,, ,, ...	2/6
	‡10 c.c. ,, 50 ,, ,, in each c.c.	15/0
	‡25 c.c. ,, 50 ,, ,, ,, ,,	25/0

,, Pneumococcus Vaccine

V	1 c.c. containing 10 million organisms ...	2/6
	‡10 c.c. ,, 10 ,, ,, in each c.c.	15/0
	‡25 c.c. ,, 10 ,, ,, ,, ,,	25/0
......	1 c.c. ,, 50 ,, ,, ...	2/6
	‡10 c.c. ,, 50 ,, ,, in each c.c.	15/0
	‡25 c.c. ,, 50 ,, ,, ,, ,,	25/0

,, Staphylococcus Aureus Vaccine

......	1 c.c. containing 200 million organisms ...	2/6
	‡10 c.c. ,, 200 ,, ,, in each c.c.	15/0
	‡25 c.c. ,, 200 ,, ,, ,, ,,	25/0
	1 c.c. ,, 1000 ,, ,, ...	2/6
	‡10 c.c. ,, 1000 ,, ,, in each c.c.	15/0
	‡25 c.c. ,, 1000 ,, ,, ,, ,,	25/0

*ana-
Tele-
ode,* ‡Obtainable in the British E...

B. W. & CO. PRODUCTS

REMEMBER THE
TRADE MARKS

Telegraphic Code

VACCINES—*continued*

Prices
Pe

'WELLCOME' BRAND
(Trade Mark)

,, Staphylococcus Vaccine, Mixed

KEFOY	1 c.c. containing 200 million organisms ...	2/6
	‡10 c.c. ,, 200 ,, ,, in each c.c.	15/0
	‡25 c.c. ,, 200 ,, ,, ,, ,,	25/0
JOYAF	1 c.c. ,, 1000 ,, ,, ...	2/6
	‡10 c.c. ,, 1000 ,, ,, in each c.c.	15/0
	‡25 c.c. ,, 1000 ,, ,, ,, ,,	25/0

,, Streptococcus Vaccine,
Polyvalent

KEDOK	1 c.c. containing 10 million organisms ...	2/6
	‡10 c.c. ,, 10 ,, ,, in each c.c.	15/0
	‡25 c.c. ,, 10 ,, ,, ,, ,,	25/0
KEDUP	1 c.c. ,, 50 ,, ,, ...	2/6
	‡10 c.c. ,, 50 ,, ,, in each c.c.	15/0
	‡25 c.c. ,, 50 ,, ,, ,, ,,	25/0

,, V. Choleræ Vaccine

LUFEY	1 c.c. containing 4 mg m. (approximately 10,000 million organisms)	2/6
	‡10 c.c. containing 4 mg m. (approximately 10,000 million organisms) in each c.c.	15/0
	‡25 c.c. containing 4 mg m. (approximately 10,000 million organisms) in each c.c.	25/0

NUDAS	,, Whooping Cough Vaccine *(H. pertussis)* 10 c.c. containing 10,000 million organisms in each c.c.	12/6

,, Vaccines for Veterinary use
(see pages 155 *and* 156)

(For other methods of Laboratory Diagnosis, see
'WELLCOME' BRAND PRODUCTS FOR SERO-
LOGICAL DIAGNOSIS, *pages* 91–94)

FOR STANDARDISATION OF
VACCINES

'WELLCOME' BRAND
(Trade Mark)

MORIS	,, Opacity Tubes for the Standardisation of Bacterial Vaccines— Sets of 10 tubes (in cardboard outer), per set, 30/0	

*For explana-
tion of T...* ... British Empire overseas to s...

PRICE LIST OF

VETERINARY HYPODERMIC—*continued*

Prices in London
Per dozen tubes

'TABLOID' BRAND
(Trade Mark)
(VETERINARY HYPODERMIC)—

No.		Tubes of	
„ 207	Coniine Hydrobromide (S1 ¶), gr. 1 [0·065 gm.]	12	22/0
„ 210	Hyoscyamine Sulphate (S1 ¶), gr. 1/8 [0·008 gm.]	12	14/0
„ 211	Morphine Sulphate (DDA S1 ¶), gr. 1 [0·065 gm.]	12	24/0
„ 212	Morphine Sulphate (DDA S1 ¶), gr. 2 [0·13 gm.]	12	48/0
„ 213	Morphine Sulphate, gr. 2 [0·13 gm.] (DDA S1 ¶) Atropine Sulphate, gr. 1/2 [0·032 gm.]	12	63/0
„ 220	Physostigmine Hydrobromide (S1 ¶), gr. 1 [0·065 gm.]	6	86/0
„ 216	Sodium Arsenite (S1 ¶), gr. 1-1/2 [0·097 gm.]	12	12/0

Not only a leader in human vaccination products, British vaccine manufacturer Burroughs Wellcome is an important supplier of animal vaccines, serums, and antitoxins.

VETERINARY SERA AND SEROLOGICAL PRODUCTS, TRADE MARK **'WELLCOME'** BRAND

Germ-proof containers

Prices in London
Per container

'WELLCOME' BRAND
(Trade Mark)

„ Tetanus Antitoxin Serum	
........Containers of **2000** Units (= 1000 American Units) ...	2/6
.... „ „ **6000** Units (= 3000 American Units) ...	5/0

(DDA)—Within the scope of the **Dangerous Drugs** Acts
(SI)—Schedule 1 Poisons   ¶—Part 1 Poisons

Inland Telegrams—"TABLOID, CENT LONDON"

VETERINARY SERA—*continued*
Germ-proof containers

Prices i
Per c

'WELLCOME' BRAND
(Trade Mark)

„ Anti-leptospira Serum (Canine) Containers of **10** c.c.	3/6
„ Lamb Dysentery Serum Containers of **100** c.c.	18/0
„ Swine Erysipelas Serum ....Containers of **10** c.c.	1/0
.... „ „ **100** c.c.	6/0
„ Tetanus Prophylactic Alum Precipitated Toxoid ....Containers of **10** c.c.	2/6
.... „ „ **100** c.c.	20/0

TRADE MARK **'WELLCOME'** BRAND PRODU
FOR INJECTION

Prices
Pe

'WELLCOME' BRAND
(Trade Mark)
(VETERINARY)—

„ Calcium Borogluconate Containers of **2-1/2** oz.	21/9
„ Calcium Borogluconate Solution, 20 per cent. Rubber-capped containers of **400** c.c. ...	—

VETERINARY VACCINES
TRADE MARK **'WELLCOME'** BRAND

Price
Pe

Germ-proof containers

'WELLCOME' BRAND
(Trade Mark)

„ *†Bacillus Abortus Vaccine (Living) Containers of **25** c.c.	7/6

Orders must reach us <u>four clear working days</u> before this Vaccine is required

„ Bacillus Abortus Vaccine (Killed) Containers of **30** c.c. 100,000 million organisms in each c.c.	7/6

* *Supplied in Great Britain and Northern Ireland only*
† *For Sale to Veterinary Surgeons only*

Telephone Number—"CENTRAL 4000" (9 L

A Chronology of Vaccination

1937 Lederle Laboratories patents an acellular pertussis vaccine using soluble toxin—cells are removed and toxin inactivated with formaldehyde. It is actively marketed from 1944–1948, when it is then replaced with the trivalent DTwP vaccine Tri-Immunol®.

1937 The LNHO Standardization Commission announces that it has established international standards for eleven therapeutic sera, one bacterial extract, four vitamins, three sex hormones, and five other therapeutic agents. Thirty-one countries officially recognize and adopt the units and standards advocated by the LNHO.

1937 The National Cancer Institute Act is passed and establishes the National Cancer Institute as a government agency for cancer research and training in the USA.

1938 Japanese scientists S. Tasaka and Y. Hiro use nasal washings from children infected with rubella to then transmit the disease to healthy children, but they are unable to determine the causative agent of rubella.

1938 British Army Major J. Boyd undertakes tetanus vaccine trials on army troops. Tetanus toxoid vaccination is introduced routinely for all troops that year.

1938 ✪ Formaldehyde-inactivated *Rickettsia prowazekii* typhus vaccine is licensed in the US.

1938 US President Franklin D. Roosevelt signs the Federal Food, Drug, and Cosmetic Act into law. This gives the FDA responsibility to oversee and regulate the production, sale, and distribution of food, drugs, medical devices, and cosmetics. Drug companies must prove that their products are safe, but there is no obligation to demonstrate efficacy.

1938 The first synthetic organophosphate insecticide tetraethylpyrophosphate (TEPP) is introduced in the USA.

1938 US President Franklin Delano Roosevelt establishes the National Foundation for Infantile Paralysis (polio) with a March of Dimes campaign raising US$1.8 million (2024 value ~US$38 million).

1938 The American Academy of Pediatrics' (AAP) Committee on Infectious Diseases (COID)—called the Committee on Immunization Procedures, releases its first publication, an eight-page pamphlet whose red cover gives rise to the publication's official nickname the "Red Book." It provides information for doctors on the treatment and prevention of infectious diseases. The COID is responsible for making recommendations on vaccines in the US until the ACIP is born in 1964.

1938 Dorothy Andersen, MD, publishes the first characterization of a disease mysteriously taking the lives of children—cystic fibrosis.

1939 The Pasteur Institute in Dakar, Senegal, produces its first yellow fever vaccine given by scarification. This vaccine is prepared by desiccating mouse brains infected with neurotropic yellow fever virus that has had more than 250 passages through mice.

1939 Swiss chemist Paul Hermann Müller at Ciba-Geigy (later merged with Sandoz to become Novartis) invents DDT (dichloro-diphenyltrichloroethane), a powerful toxin used as a pesticide until the early 1970s.

1939– 1945 Under General Eli Lilly's leadership during World War II, the company manufactures encephalitis vaccine, antitoxin for gas poisoning, and vaccines for influenza and typhus. Encephalitis vaccine alone requires 15,000 white mice a week and proves both difficult and dangerous to produce.

1939 The UK begins its biological warfare program.

1939 Alhydrogel™ (aluminium oxyhydroxide) becomes commercially available and will be one of the most widely used vaccine adjuvants.

1939 World War II begins on September 1, ending in 1945 with an estimated 50 million deaths.

1939– 1945 During WWII, American troops are routinely given inactivated whole virus influenza vaccines, live smallpox vaccines, tetanus toxoid, whole-cell typhoid and paratyphoid A and B vaccines. In some situations whole-cell cholera, diphtheria toxoid, whole-cell plague, whole-cell typhus, live yellow fever, and whole-cell scarlet fever vaccines are administered.

1939 Pertussis vaccine is reported as being effective by American scientists Pearl Kendrick, PhD, and Grace Eldering, PhD, at the State of Michigan Health Department.

1939 Karl Landsteiner, Alexander Wiener, Philip Levine, and R.E. Stetson discover the Rh protein on the surface of red blood cells and identify positive and negative blood groups.

1939 The US Federal Security Agency (FSA) is established under the Reorganization Act with the aim to bring together in one agency all federal programs in health, education, and social security.

1940–1949

1940s British biologist Conrad Waddington coins the term epigenetics as "any process affecting the expression of genes that does not involve a change to the sequence of the gene itself."

1940s Freeze-drying (where a product is first frozen and then water is removed by sublimation of the ice) becomes an industrial process applied to blood plasma and penicillin. By the 1950s–1960s, freeze-drying becomes viewed as a multipurpose tool for pharmaceuticals and vaccines.

1940s ✪ DTP combination vaccine Triogen® by Parke, Davis & Co., is licensed in the USA, containing the mercury-derived preservative Merthiolate (thimerosal).

1940 The first influenza type B strain is isolated by American virologist and director of the Commission on Influenza of the Armed Forces Epidemiological Board (AFEB), Thomas Francis Jr., MD. Along with Jonas Salk, MD, he will be a key player in the development of the first inactivated influenza and polio vaccines.

1940–1947 The International Health Division of the Rockefeller Foundation ships 28 million doses of yellow fever vaccine worldwide. In the US, the yellow fever vaccine is given to virtually all recruits. Between January 1941 and April 10, 1942, an astounding 7 million doses are distributed for free, all manufactured by the International Health Division of the Rockefeller Institute.

1941 Polio virus is reported as entering the body via the mouth to the digestive system, then spreading to the nervous system.

1941–1947 The US Office of Scientific Research and Development is established by Executive Order 8807 to coordinate medical and scientific research for military purposes during the war. Through their Committee for Medical Research, they administer 450 contracts to universities, committing funds for nontherapeutic research on orphans, the mentally ill and challenged, and on prisoners. The results are the development of penicillin, cortisone, and gamma globulin, among others.

1941 Eli Lilly is charged as violating the Sherman Anti-Trust Act by combining with other insulin producers to "fix and maintain the prices of insulin" and therefore "prevented and restrained free and normal competition in the sale and distribution of insulin." Since 1924, insulin is sold at the same price of seven cents per day for the average diabetic (2024 value ~US$1.25).

1942–1945 The Chemical Warfare Services, a branch of the US Army, conduct "mustard gas experiments on around 60,000 servicemen, including 4,000 soldiers subjected to severe, full-body exposures."

1942–1944 The first series of pertussis vaccine field trials in the UK are organized by the Whooping Cough Immunization Committee of the Medical Research Council (MRC). The pertussis vaccines are supplied by Glaxo Laboratories, the Lister Institute of Preventive Medicine, and Lederle Laboratories. The trials are designed to assess the protective value of pertussis vaccine and to determine if this value can be assessed by a laboratory test. The results of these first trials show that there is no significant difference in the incidence or severity of the disease between the vaccinated and unvaccinated groups. Convulsions and paralytic polio are observed in some children following the injection.

1942 Johns Hopkins alumnae Pearl Kendrick, PhD, and Grace Eldering, PhD, combine diphtheria, tetanus, and pertussis vaccines into a single shot: the DTP vaccine, that is commercialized several years later.

1942 ✪ Diphtheria toxoid vaccines are routinely recommended in the UK.

A Chronology of Vaccination

1942	✪ Influenza vaccine using strains A/PR8 (H1N1) + B/Lee, produced in embryonated hen eggs, are introduced experimentally to the US Army.
1942	The Cox typhus vaccine consisting of killed rickettsia that are grown in egg yolk, is adopted for use in the US Army.
1942	✪ Rocky Mountain spotted fever vaccine is developed by AFEB-WRAIR and available to US Army troops until 1978.
1942	A hepatitis outbreak occurs following the Rockefeller Foundation's 17D yellow fever vaccine campaign in the US military. As a result, there are 81 deaths out of approximately 50,000 cases of "yellow atrophy of the liver"—vaccine-induced jaundice. This outbreak is later claimed to be due to infection with hepatitis B. *The Chicago Tribune* reacts with some outrage noting that "20 times as many soldiers have fallen victim to the vaccine as have been wounded thus far in the war." The paper calls for an investigation, claiming inadequate testing and questioning the need for all soldiers to receive the vaccine.
1942	American Herald R. Cox, who becomes the youngest principle bacteriologist for the US Public Health Service, joins Lederle Laboratories as Director of Viral Research. During his 26 years at the company, he helps develop live polio vaccine.
1942	The War Research Service, civilian agency of the US Army, is established. George W. Merck, president of Merck & Co., is appointed by President Franklin Roosevelt to direct the WRS. Its mission is to pursue research relating to biological warfare, including vaccines.
1942	The Office of Malaria Control in War Areas is established in Atlanta, USA, the predecessor of the CDC—to limit the impact of malaria and other vector-borne (such as typhus) diseases. It is located in Atlanta, Georgia, rather than Washington, DC, as the South is more affected by malaria transmission.
1942	✪ Formaldehyde-killed whole-cell bacterium *Yersinia pestis* plague vaccine by Cutter Laboratories is licensed in the USA.

1943	 The intensive indoor and outdoor use of the insecticide DDT begins in the USA and becomes popular worldwide during WWII to limit the spread of typhus and malaria.
1943	The US top secret military base at Fort Detrick opens in Frederick, Maryland, named after Frederick Louis Detrick, a distinguished surgeon at Johns Hopkins University Hospital. Named the US Army Biological Warfare Laboratories, the base is the center of biological warfare research and development for the military, becoming one of the world's largest users of laboratory animals. Detrick controls the procurement, testing, research, and development of all biological munitions and products, including vaccines and other defense (and offensive) measures. The three-part BW division consists of work on "antiplant, antianimal, and antiman" agents.
1943	 ✪ Connaught combines pertussis vaccine with diphtheria toxoid. After the introduction of Salk's polio vaccine in the mid-1950s, Connaught develops a new generation of combined vaccines, known as DTP-Polio, DT-Polio, and T-Polio.
1943–1944	The first placebo-controlled trial—for patulin, to treat the common cold—is conducted by the British Medical Research Council. Though the media is quick to report positive results, the published study in *The Lancet* concludes no effect compared to placebo.
1943	Diphtheria outbreaks occur during the war in Europe, with one million cases in Europe and 50,000 deaths (excluding the USSR).
1943–1944	✪ Toxoid tetanus vaccine products are licensed in the USA, made by Wyeth and Connaught.

1943	The Yellow Fever Research Institute is set up in Lagos, Nigeria, by the Rockefeller Foundation and the British West African Colonies. One of its purposes is to test and distribute yellow fever vaccine.
1943	Penicillin becomes mass produced and distributed worldwide.
1943	The antibiotic streptomycin is discovered by American biochemists Selman Waksman, Albert Schatz, and Elizabeth Bugie.
1943	The first mumps vaccine trials in 41 institutionalized children take place in the USA, using formol-inactivated virus obtained from an infected monkey. The results are reported three years later as "encouraging."

1944

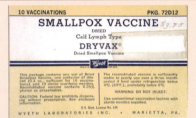

⭐ A US license is granted to Wyeth for its freeze-dried calf lymph smallpox vaccine, Dryvax®. By the 1960s, it is the only smallpox vaccine on the American market.

1944	American scientist and vaccine developer Maurice Hilleman, working at E.R. Squibb & Sons, helps develop a vaccine against Japanese encephalitis to protect US troops in the Pacific during WWII.
1944	In the USA, the Public Health Service Act is approved, establishing the federal government's quarantine authority. It also gives the USPHS responsibility for "preventing the introduction, transmission and spread of communicable diseases from foreign countries into the US" and for the regulation of biological products. The National Cancer Institute becomes a division of the NIH.
1944	The ability to grow viruses outside the body in cell culture (monkey kidney cells) is discovered by American scientists John F. Enders, Thomas H. Weller, and Frederick C. Robbins. The capacity to culture viruses from mucus is used to diagnose influenza. This process will become instrumental to the mass production of vaccines.

1944	The International Bank for Reconstruction and Development, soon called the World Bank, is founded following the Bretton Woods Conference (formally known as the United Nations Monetary and Financial Conference) in the USA.
1944	The US Public Health Service Act is enacted, providing a federal legislative basis for the provision of public health services.
1944	Canadian-born American bacteriologist Dr. Oswald T. Avery and his colleagues at the Rockefeller Institute publish evidence that genes are made of DNA.
1944	✪ E.R. Squibb & Sons launches two new quadrivalent vaccines for use against pneumococcal pneumonia. In the postwar years they are abandoned in favor of antibiotics.
1944	Pharmaceutical company E.R. Squibb & Sons opens the world's largest penicillin plant in New Brunswick, New Jersey.
1944	Glaxo owns four UK facilities dedicated to producing penicillin, including one split-funded with Wellcome that produces 7.5 billion units of penicillin annually. Around 80% of penicillin production in the UK was routed through Glaxo's Greenford site.
1944	Switzerland implements a federal mandate that smallpox vaccine be given to all children under the age of five, before entering school. The nationwide mandate is repealed in 1948.
1944	Switzerland experiences its worst ever polio epidemic affecting mainly children, with 224 deaths and 1,793 cases reported.
1944	✪ At the International Sanitary Convention for Aerial Navigation, Rockefeller's yellow fever vaccine (17D) is approved, followed by the French Pasteur Institute vaccine given by scarification.
1945	Sir Alexander Fleming, Ernst Boris Chain, and Sir Howard Walter Florey are awarded the Nobel Prize in Physiology or Medicine "for the discovery of penicillin and its curative effect in various infectious diseases."
1945	✪ A tetravalent pneumococcal vaccine with four strains is introduced, but largely ignored due to the success of penicillin.

A Chronology of Vaccination

1945	Dr. Thomas Francis writes in the July edition of *Bulletin of the New York Academy of Medicine* that "the presence of antibodies and immunity are not synonymous."
1945	American medical officer Karl Habel and American biomedical scientist John Franklin Enders successfully cultivate mumps virus in chick embryos and develop the first inactivated mumps vaccine a year later.
1945	The NIH send the mouse potency testing method (used to determine efficacy) to fourteen vaccine manufacturers of whole-cell pertussis. It becomes part of the NIH's First Minimum Requirements for Pertussis Vaccination in 1949.
1945–1992	The US conducts around 1,054 nuclear weapon tests, including 216 atmospheric, underwater, and space tests. Most of the tests take place at the Nevada Test Site and the Pacific Proving Grounds in the Marshall Islands and off Kiritimati Island in the Pacific. Ten other tests take place at varying locations, including Alaska, Nevada, Colorado, Mississippi, and New Mexico.
1945	The American military deploys two atomic bombs on the Japanese cities of Hiroshima and Nagasaki during World War II that marks the first use of a nuclear weapon in war. It is estimated that between 129,000 and 226,000 people, mostly civilians, are killed in the initial explosions and many more would later succumb to radiation poisoning.
1945–1956	The NIH funds a study that intentionally infects hundreds of impoverished Guatemalans, as young as ten years old, with syphilis without their consent. A US$1 billion lawsuit is filed in 2015 on behalf of the victims against the Johns Hopkins University, the Rockefeller Foundation, and Bristol-Myers Squibb for their role in unethical medical experiments. In April 2022, a federal judge dismisses the case, citing "insufficient evidence."
1945	✪ A toxoid tetanus vaccine is made available to civilians in Australia.
1945	Grand Rapids, USA, becomes the first city to add fluoride to public drinking water.
1945–1959	A top secret US government program of the Office of Strategic Services (predecessor to the CIA) named Operation Paperclip, sponsors the immigration of around 1,500 German and Austrian scientists, including Nazi war criminals, to exploit their expertise for military and industrial purposes. They are provided US citizenship and new identities in exchange for working with the government. Nazis are employed at NASA and within the biological and chemical weapons program.
1945	✪ The first formalin-inactivated viral influenza vaccine—A/PR8/34 and B/Lee/40 strains cultivated on embryonated chicken eggs according to the method of Australian virologist Frank Macfarlane Burnet—is approved in the US for use in the civilian population. One of the brands is Fluogen® by Parke, Davis & Co., and uses a calcium phosphate adjuvant.
1945	Between April 25 and June 26, delegates of 50 nations participate in the United Nations Conference on International Organization in San Francisco. The *Charter of the United Nations* is agreed upon, as well as the Statute of an International Court of Justice, which effectively creates the UN.
1945	Following the Bretton Woods Conference in 1944 (formally known as the UN Monetary and Financial Conference), the International Monetary Fund (IMF) is established.
1945	Rockefeller Foundation scientists test the toxicity of malaria drug, chloroquine, on Quechua Indians living and working on a Peruvian sugar plantation. The workers, including children from age six, are forced to take the weekly drug. Despite a high percentage of side effects in children due to an excessive dose, the conclusion is that chloroquine is minimally toxic. As a result of this and other experiments, chloroquine becomes widely used a cheap and effective prevention of, and therapy for, malaria.

1946 The AMA adopts the requirement of voluntary consent of human subjects for experimentation.

1946 The Communicable Disease Center (USA) is established, forerunner of the Centers for Disease Control and Prevention (CDC).

1946 In January, the US Army War Department releases a "slim, sanitized government monograph" called the *Official Report on Biological Warfare* a.k.a. the Merck Report, written by George W. Merck. While he does not reveal the biological agents investigated, Merck states that they "were made as virulent as possible" and admits to over 200 accidental exposures to these agents.

1946

The United Nations establishes the UN International Children's Emergency Fund (UNICEF) whose name is changed to the UN Children's Fund in 1953. Its mission is to provide relief to children devasted by the war. This mandate expands over the years to supplying humanitarian aid while supporting the long-term health and well-being of children and women worldwide—largely through vaccination campaigns.

1946 The Constitution of the World Health Organization (WHO) is adopted by the International Health Conference held in New York from June 19 to July 22, signed on July 22 by the representatives of 61 nations and enters into force on April 7, 1948. This date is commemorated as World Health Day. The Rockefeller Foundation is an observer to this first international health conference, renamed World Health Assembly (WHA). The WHA may adopt regulations on quarantine requirements and standardization of nomenclatures, diagnostic procedures, and biological and pharmaceuticals.

1946 US Congress passes the National Mental Health Act, allocating public funds to researching into the prevention, diagnosis, and treatment of mental illness. In 1949, Congress creates the National Institute of Mental Health (NIMH) to oversee the research.

1946 The first female to male sex reassignment surgery is performed by Sir Harold Gillies.

1946 The UK National Health Service Act is passed, creating the NHS two years later. The Vaccination Act will be repealed, making smallpox vaccination no longer compulsory.

1946 Financial compensation is awarded by the Swiss government to smallpox vaccine victims with neurological damage—cases of post-vaccinal encephalitis following the national smallpox vaccine mandate in 1944. An official document reports ~1 incident of injury per 2,220 vaccinations and warns to "expect 1 death due to post-vaccinal encephalitis per about 33,000 vaccinations."

1946–1948 A series of ten randomized field trials of different formulations of whole-cell, killed, pertussis vaccines is organized by the Whooping Cough Immunization Committee of the MRC. The pertussis vaccines are supplied by Glaxo Laboratories, Parke-Davis, and the Michigan Department of Health. One of Glaxo's vaccines is alum-precipitated and the "placebo" is a "catarrhal" (inflammation of the upper respiratory tract) vaccine provided by Burroughs Wellcome. Some 8,927 children aged 6–18 months are enrolled with parental consent. The US vaccine from the Michigan Department of Health is deemed the best performer. These trials and subsequent follow-up confirms the reliability of the mouse protection test as a potency assessment tool, enabling the MRC to set a minimum quality standard for evaluating whole-cell pertussis vaccine effectiveness. However, this test developed in the USA by Pearl Kendrick et al., at the Michigan Department of Health "could not adequately distinguish the differential effectiveness of different vaccines used in the field."

1947 In response to a single case of smallpox in New York City, the largest, fastest smallpox mass vaccination campaign is conducted in the USA, with over 6 million people injected. There are a total of 12 recorded smallpox cases, 2 smallpox deaths, and 46 cases of post-vaccinal encephalitis resulting in 8 reported smallpox vaccine-related deaths.

A Chronology of Vaccination

1947

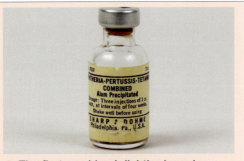

⭐ The first combined diphtheria and tetanus toxoids with whole-cell pertussis (DTwP or DTP) shot is commercialized and manufactured by Parke, Davis & Co., Lederle, Sharp & Dohme, National Drug Company, Cutter Laboratories and Pitman-Moore, according to NIH specifications. Some vaccines include toxoids adsorbed on aluminum hydroxide, like Cutter's DTP Alhydrox®.

1947 — Coxsackie A virus is first isolated during poliomyelitis outbreaks from feces of paralyzed children in Coxsackie, New York.

1947 — Johnson & Johnson launches an advertising campaign in the USA for its smallpox vaccine.

1947 — An outbreak of cholera in Egypt results in the country closing its borders.

1947 — Sweden introduces its first general vaccination program, shortly followed by the recommendation of tetanus and pertussis vaccines with the already available smallpox, tuberculosis, and diphtheria vaccines.

1947 — Toxaphene is created, and by the 1970s, it becomes the most heavily used insecticide in US history. In 1990, its use is banned in America due to its major negative effects on the central nervous system (CNS). It is banned globally by the 2001 Stockholm Convention on Persistent Organic Pollutants.

1947 — Pasteur Institute's live neurotropic yellow fever vaccine is used extensively in French African territories, with over 20 million vaccinations performed.

1947 — Zika virus, named after a forest in Uganda, is isolated by scientists at the Rockefeller Foundation from the blood of a forest sentinel rhesus monkey in Uganda.

1947 — The WHO creates the Expert Committee on Biological Standardization. It holds its first session in Geneva, and convenes annual meetings to establish standards—including detailed recommendations and guidelines for the manufacturing of biological products.

1947 — ⭐ In the USA, two hexavalent pneumococcal polysaccharide vaccines (PPVs) by E.R. Squibb & Sons are licensed, though with little success, as physicians prefer antibiotic therapy. Squibb ceases production in 1951 and withdraws both licences in 1954.

1947 — Dr. Frederick MacCallum defines two types of viral hepatitis; hepatitis A and hepatitis B. The first is also called "infectious hepatitis," while the latter is called "serum hepatitis."

1947 — American scientists Pearl Kendrick, Eldering, Dixon, and Misner publish their milestone scientific paper on the Intracerebral Mouse Protection Test (IMPT) for pertussis vaccines. They present this new method of measuring the virulence and protective properties of *Haemophilus pertussis*, the culture used for whooping cough vaccine production. The IMPT will become adopted as a minimal requirement by the WHO for assessing the safety and protection by whole-cell pertussis vaccines.

1947 — In the US Physicians' Desk Reference, the following vaccines are mentioned as being available: Acne vaccines by Lederle, National Drug Company, Parke-Davis, Wyeth, Cutter, Eli Lilly, Sharp & Dohme, Pitman-Moore; Arthritis vaccines by Cutter, Hollister-Stier, Pitman-Moore; Catarrhal vaccines by National Drug Company, Parke-Davis, Pitman-Moore, Lederle, Eli Lilly, Squibb, Wyeth; Cholera vaccines by Lederle, Eli Lilly, Wyeth; Cold vaccines by Merrell, Sharp & Dohme.

1947 — The first reported case of severe neurological damage after whole-cell pertussis vaccine is published in the *New York State Journal of Medicine* by M. Brody and R.G. Sorley.

1947 — An epidemic of Japanese encephalitis is reported in the Western Pacific Islands of Guam.

1947 The Microbiological Research Department at Porton Down is created by the UK government to study the impact of airborne germs. Ten years later, offensive biological warfare research is abandoned and stockpiles destroyed.

1947 The Nuremberg Military Tribunal's decision in the case of the *United States v. Karl Brandt, et al.,* includes what is now called the Nuremberg Code; a set of ethical research principles including a ten point statement delimiting permissible medical experimentation on human subjects. It highlights that voluntary consent of the human subject is absolutely essential.

1947 A,B,O blood-typing is performed on each unit of blood, becoming medical protocol.

1947 The first extensive poliomyelitis outbreak is recorded in Britain.

1948 The first official WHO bulletin features 30 pages (15% of the publication) dedicated to the issue of post-vaccinal encephalitis.

1948 Eighty-three children die in Kyoto, Japan, after an immunization campaign using an improperly manufactured diphtheria vaccine. Known as the "Kyoto-Shimane Diphtheria Tragedy," little information is reported to the public due to stringent censorship.

1948 Swiss scientist Paul H. Müller is awarded the Nobel Prize in Physiology or Medicine "for his discovery of the high efficiency of DDT as a contact poison against several arthropods."

1948 The first randomized trial using a placebo control is conducted by the MRC—a curative effect of streptomycin for tuberculosis is found.

1948 The National Health Service (NHS) is founded in the UK to provide free healthcare, financed through taxes.

1948 The WHO issues a report commending the freeze-dried vacuum-packed smallpox vaccine from the Vaccine Institute in Paris.

1948 DDT surpasses use of lead arsenate as a chemical insecticide. From 1947 to 1951, over 4.5 million US homes are sprayed with this pesticide, a potent insect nerve toxin.

1948 Coxsackie B virus is claimed to be isolated from cases of aseptic meningitis.

1948 The first World Health Assembly approves a modest budget of US$ 4.8 million, of which more than half goes to reimburse expenses incurred by the Interim Commission.

1948 Japan passes the Immunization Act, Preventive Vaccine Law, unprecedented in the world for its severity as it makes vaccination compulsory without exception and hefty fines are imposed on those who violate the law. The fines are lifted in 1977 though mandatory vaccination is maintained until 1994 when the law is revised, and mass vaccination in schools is discontinued. Japan's first routine schedule includes vaccines for smallpox, diphtheria, paratyphoid, pertussis, tuberculosis, typhus, plague, scarlet fever, influenza and Weil's disease (leptospirosis).

1948 Pioneering American vaccine developer Maurice Hilleman leaves E.R. Squibb & Sons to work for the Department of Respiratory Diseases at the Army Medical Center in Washington, DC—renamed in 1953 the Walter Reed Army Institute of Research.

1948 Compulsory immunization with smallpox vaccine is abolished in the UK.

1948 The WHO Influenza Center is established at the National Institute for Medical Research (NIMR) in London. Together with other centers in the USA, Australia, and Japan, they serve as the global influenza surveillance network "alert system for the identification of new influenza viruses, gathering information from 110 participating laboratories in 82 countries."

1948 ASHG

The American Society of Human Genetics is founded as a professional membership association for specialists in human genetics, including researchers, academics, clinicians, and nurses. Its annual meeting is the oldest and largest international human genetics conference worldwide, attracting thousands every year. One of the founders, German-born American psychiatrist Franz Josef Kallmann, MD, was a racial hygienist and member of the American Eugenics Movement at the beginning of the 20th century.

1947 Physicians' Desk Reference to Pharmaceutical Specialties and Biologicals

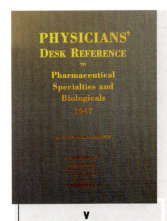

**PHYSICIANS'
DESK REFERENCE
TO
Pharmaceutical
Specialties
and Biologicals
1947**

V

VACCINE-PERTUSSIS (w/TOXOIDS, DIPHTHERIA & TETANUS)
D-P-T Alhydrox Diphtheria-Pertussis-Tetanus Combined (Cutter), page 291
D-P-T Plain (Diphtheria-Pertussis-Tetanus Combined (Cutter), page 291
Diphtheria-Tetanus-Pertussis Combined A.P. National (National), page 309
Pitman-Moore (Pertussis Vaccine—Diphtheria & Tetanus Toxoids), page 317
Sharp & Dohme (Diphtheria-Pertussis-Tetanus Combined), page 331

VACCINES-ACNE
Cutter Laboratories
Lilly, Eli & Company
Pitman-Moore Company

VACCINES-ACNE COMBINED
Lederle Laboratories, Inc.
National Drug Company, The
Parke, Davis & Company
Wyeth Incorporated

---

116   DRUG and PHARMACOLOGICAL INDEX   VA-VA

VACCINES—ACNE SEROBACTERIN
Sharp & Dohme, Inc.

VACCINES—ARTHRITIS
Cutter Laboratories
Hollister-Stier Laboratories
Pitman-Moore Company
Sherwood's Formula Arthritis Vaccine (Cutter)

VACCINES—CATARRHAL
National Drug Company, The
Parke, Davis & Company
Pitman-Moore Company

VACCINES—CATARRHAL COMBINED
Lederle Laboratories, Inc.
Lilly, Eli & Company
Squibb, E. R. & Sons
Wyeth Incorporated

VACCINES—CATARRHALIS INFLUENZA
Lederle Laboratories, Inc.

VACCINES—CHOLERA
Lederle Laboratories, Inc.
Lilly, Eli & Company
Wyeth Incorporated

VACCINES—"COLD" (ORAL)
Oravax (Merrell), page 309

VACCINES—"COLD" (PARENTERAL)
'Serobacterin' Catarrh (Cold) Vaccine Mixed (Sharp & Dohme), page 333

VACCINES—COLI COMBINED
Lilly, Eli & Company
National Drug Company, The

VACCINES—DUCERY (TEST)
Lederle Laboratories, Inc.

VACCINES—E. COLI MIXED
Pitman-Moore Company

VACCINES—FURUNCULOSIS
Pitman-Moore Company

VACCINES—GONOCOCCUS
Lilly, Eli & Company
Parke, Davis & Company
Wyeth Incorporated

VACCINES—GONORRHEA
Pitman-Moore Company

VACCINES—GONORRHEA MIXED
Hollister-Stier Laboratories

VACCINES—INFLUENZA
Parke, Davis & Company
Pitman-Moore Company

VACCINES—INFLUENZA (w/COLD)
'Serobacterin' Influenza (Cold) Vaccine Mixed (Sharp & Dohme), page 333

VACCINES—INFLUENZA COMBINED
Squibb, E. R. & Sons
Wyeth Incorporated

VACCINES—INFLUENZA MIXED
Hollister-Stier Laboratories
Lilly, Eli & Company
Upjohn Company, The

VACCINES—INFLUENZA—PNEUMONIA
Lilly, Eli & Company
Pitman-Moore Company

VACCINES—INFLUENZA VIRUS
Parke Davis (Influenza Virus Vaccine, Types A & B), page 313
Squibb (Influenza Virus Vaccine A & B), page 338

VACCINES—INFLUENZA VIRUS (TYPES A & B)
Lederle Laboratories, Inc.
Parke Davis (Influenza Virus Vaccine, Types A & B), page 313
Pitman-Moore (Influenza Virus Vaccine—Types A & B), page 317

VACCINES—MIXED RESPIRATORY
Cutter Laboratories
Hollister-Stier Laboratories
M.V.R.I. (Cutter)

VACCINES—ORAL
Immunovac Oral Vaccine Tablets (Parke Davis), page 313

VACCINES—PARENTERAL
Immunovac Parenteral (Parke Davis), page 313

VACCINES—PERTUSSIS
Ayerst, McKenna & Harrison (Pertussis Endotoxoid), page 271
Cutter Laboratories
Lederle Laboratories, Inc.
Lilly, Eli & Company
National Drug Company, The
Parke, Davis & Company
Pertussis Vaccine Phase I Super-Concentrate (Cutter), page 292
Pitman-Moore Company
Squibb, E. R. & Sons
Upjohn Company, The
Wyeth Incorporated

VACCINES—PERTUSSIS (w/DIPHTHERIA TOXOID)
Diph-Pertussis—Sauer—Alum Precipitated (Parke Davis), page 312
Upjohn Company, The

VACCINES—PERTUSSIS MIXED
Cutter Laboratories
Hollister-Stier Laboratories
Lilly, Eli & Company

VACCINES—PERTUSSIS SEROBACTERIN
Sharp & Dohme, Inc.

VACCINES—PLAGUE
Cutter Laboratories

VACCINES PNEUMOCOCCUS
Pneumococcus Antigen (Lilly)

VACCINES—PNEUMONIA COMBINED
Parke, Davis & Company

VACCINES—RABIES
Cutter Laboratories
Lederle Laboratories, Inc.

Lilly, Eli & Company
National Drug Company, The
Parke, Davis & Company
Pitman-Moore Company
Sharp & Dohme, Inc.
Squibb, E. R. & Sons
Wyeth Incorporated

VACCINES—RHEUMATOID ARTHRITIS
Lederle Laboratories, Inc.

VACCINES—ROCKY MT. SPOTTED FEVER
Lederle Laboratories (Rocky Mountain Spotted Fever Vaccine), page 302
Wyeth Incorporated

VACCINES—SMALLPOX
Cutter Laboratories
Lederle Laboratories, Inc.
Lilly, Eli & Company
National Drug Company, The
Parke, Davis & Company
Sharp & Dohme, Inc.
Squibb, E. R. & Sons
Wyeth Incorporated

VACCINES—STAPHYLOCOCCUS
Cutter Laboratories
Lilly, Eli & Company
National Drug Company, The

VACCINES—STREPTOCOCCUS
Cutter Laboratories
Lilly, Eli & Company
National Drug Company, The
Parke, Davis & Company
Pitman-Moore Company

VACCINES—STREPTOCOCCUS & STAPHYLOCOCCUS (COMBINED)
Parke, Davis & Company

VACCINES—TYPHOID
Cutter Laboratories
Lederle Laboratories, Inc.
Lilly, Eli & Company
National Drug Company, The
Parke, Davis & Company
Pitman-Moore Company
Squibb, E. R. & Sons
Wyeth Incorporated

VACCINES—TYPHOID—PARATYPHOID
National Drug Company, The
Parke, Davis & Company
Pitman-Moore Company
Wyeth Incorporated

VACCINES—TYPHUS
Parke, Davis & Company

VACCINES—ULCERATIVE COLITIS STREPTOCOCCUS
Parke, Davis & Company

VACCINES—UNDULANT FEVER
Lederle Laboratories, Inc.
National Drug Company, The
Parke, Davis & Company
Pitman-Moore (Undulant Fever Vaccines), page 317

VACCINES—WHOOPING COUGH
(see under Vaccines—Pertussis—this Section)

"PAGE NUMBERS INDICATE COMPLETE PRODUCT DESCRIPTION IN WHITE SECTION"

This commemorative 50th Anniversary reprint of the first edition of *Physicians' Desk Reference* by Medical Economics Inc., Rutherford, New Jersey, presents the inaugural issue of PDR that took two years to compile and produce.

It was distributed for free to over 100,000 practicing physicians in the USA through the courtesy of the 132 pharmaceutical companies whose professional products are fully described in the PDR's section four (pages 253-364).

The product descriptions are written by the companies themselves who pay US$115 per column-inch for the space.

View more pages online
www.TheUltimateVaccineTimeline.com

1947–1948 The WHO's First Official Bulletin

The purpose of the WHO is to develop, establish, and promote international standards on matters related to health. The WHO is officially established by constitution on April 7, a date that is commemorated as World Health Day.

The first publication features a section on the issue of post-vaccinal encephalitis. Since 1948, there have been no scientific vaccine safety studies to investigate the biological mechanisms behind the severe, long-term neurological and immunological complications seen after vaccination.

TABLE OF CONTENTS

(Authors are alone responsible for views expressed in signed articles.)

Link to view the publication online

https://web.archive.org/web/20140520155233/www.ncbi.nlm.nih.gov/pmc/issues/172498/

A Chronology of Vaccination

1948–1951

1948	In Canada, tetanus toxoid vaccine is routinely recommended and given to schoolchildren.
1948	US Congress approves a US$30 million (US$390 million equivalent in 2024) budget for antianimal weapons research. Due to the inherent danger, congress mandates that it takes place outside continental US, on an island that is not connected by a bridge to the mainland. Plum Island in the state of New York will be chosen for the location.
1949	American virologists, Dr. John F. Enders (often called the father of modern vaccines), Dr. Thomas H. Weller, and Dr. Frederick C. Robbins show that polioviruses could be grown in non-nervous tissue—namely human embryonic skin and muscle tissue. This would reduce reliance on live monkeys for growing and testing viruses, allowing research to forgo facilities that need large numbers of experimental and expensive animals. The cell culture methodology is born.
1949	The North Atlantic Treaty Organization (NATO) is founded as an intergovernmental military alliance between 31 member states—29 European and two North American.
1949	✪ The triple diphtheria, tetanus, and whole-cell pertussis (DTP) vaccine is officially licensed in the USA by the FDA—though it was already available on the market. It is recommended to babies and children.
1949	Portuguese ex-politician-turned-neurologist António Egas Moniz wins the Nobel Prize in Physiology or Medicine "for the discovery of the therapeutic value of leucotomy [lobotomy] in certain psychoses." Prefrontal lobe incisions are widespread in the 1940s and 1950s, but decline drastically when medication for mental illness is developed in the 1950s.
1949	An enclosed one million liter test sphere—used to experiment with the dispersal of aerosolized biological agents—is built at Fort Detrick, also known as the "8-ball."
1949	The last US case of smallpox is recorded.
1949–1970	The CIA and US Department of Defense project MKNAOMI, tied to Fort Detrick and MKULTRA (the CIA's mind control project 1953–1973), involves the unethical testing of incapacitating and lethal biochemical agents.
1949	The Council for International Organizations of Medical Sciences (CIOMS) is established by the WHO and UNESCO "as an international, non-governmental, nonprofit organization, [...] to provide a forum for biomedical research experts [...] to address the implications of advances in biomedical sciences upon diverse issues, ranging from health policy to medical education and bioethics."
1949–1969	During the 1977 Senate hearings of the Subcommittee on Health and Scientific Research, it is acknowledged that the US military and intelligence agencies executed field experiments with biological agents on at least 239 populated areas: New York City; San Francisco; Washington, DC; Alaska; Panama City—dropping and spraying pathogens from aircraft and ships.
1949	The Rockefeller Foundation and the British government provide US$500,000 in funding to the WHO for its first full year of operation.

1950–1959

1950s	American vaccine manufacturer Parke-Davis, the largest and oldest pharmaceutical company in the USA, hires Dr. Jonas Salk as a consultant on vaccine adjuvants.
1950s	Cell culture begins to be used widely in medical research and molecular biology. It is applied for the first time to mass-produce the polio vaccine developed by Dr. Jonas Salk, using monkey kidney cells. It is estimated that 1,500 monkeys are needed to produce 1,000,000 doses of vaccine.
1950s	Mass vaccination of smallpox, tetanus, diphtheria, and pertussis vaccines becomes widespread in the USA and Europe.
1950s	The US Army Chemical Corps develops an anthrax vaccine that is replaced by the BioThrax™ vaccine, licensed in 1970.
1950	DDT is sprayed directly on children to protect them from insect-borne diseases.

1950 Virulent myxoma virus is introduced via inoculation by Australian farmers in an effort to control wild rabbit populations. Initially highly effective, the virus and rabbit host adapt, with resistant rabbits bouncing back.

1950 The first reports of paralytic poliomyelitis following vaccination are published by Hill and Knowelden in *The British Medical Journal*.

1950 ✪ Generic inactivated killed mumps vaccine, made by Eli Lilly & Co., and Lederle Labs, is licensed in the USA, and then withdrawn in 1977–1978.

1950 ✪ Whole-cell pertussis vaccine is introduced in the UK, and is produced by Burroughs Wellcome, Glaxo, and the Lister Institute.

1950 The US Navy spray *Serratia marcescens* bacteria into the air near San Francisco over a period of six days. Dubbed "Operation Sea Spray," the project is intended to determine how susceptible the city was to a bioweapon attack. While the military initially think *Serratia marcescens* could not harm humans, an outbreak occurs, with some people developing urinary tract infections.

1950 ✪ BCG vaccine is introduced in the UK and added to the routine vaccination schedule. It becomes obligatory in France and is also added to the Swiss vaccine schedule, but the vaccine is never widely used in the US.

1950 Dr. Hilary Koprowksi, who would later direct the Wistar Institute in Philadephia, tests his experimental live oral polio vaccine made from cotton-rat brain and mouse spinal cord pulverized in a blender. His "volunteers" are twenty intellectually disabled children at Letworth Village in New York.

1950–1953 During the Korean War, the US expands their biological weapons arsenal with the creation of a new production facility in Pine Bluff, Arkansas. At the end of the war, "a defensive program is launched with the objective of developing countermeasures, including vaccines, and therapeutic agents, to protect troops from possible biological attacks."

1951 A third form of influenza, type C, is discovered, though it causes mild infections, and is not implicated in epidemics.

1951 Johns Hopkins alumni and public health professor Alexander Langmuir, MD, establishes the CDC Epidemic Intelligence Service (EIS) to protect against biological warfare and manmade epidemics. Langmuir also served as an advisor to the Defense Department's chemical and biological warfare programs, and was part of the eugenics movement that supported Planned Parenthood founder, Margaret Sanger.

1951 BCG vaccine is widely and routinely administered to newborns through the WHO's mass tuberculosis inoculation campaign in developing countries.

1951 Denmark introduces its childhood vaccination program including vaccines against diphtheria, tetanus, smallpox and tuberculosis. In 1952 and 1953 there is a national epidemic of infantile paralysis.

1951 The WHO issues its first infectious disease prevention regulations, the International Sanitary Regulations (ISR). The regulations focus on six notifiable and quarantinable diseases: cholera, plague, relapsing fever, smallpox, typhus, and yellow fever.

1951 HeLa cell line is sourced from an African-American woman called Henrietta Lacks, who died following radium treatments for her cervical cancer. She went to the Johns Hopkins Hospital in Baltimore, one of only a few hospitals that treated poor African-Americans at the time. Her cancer cells are the first immortalized human cell line and one of the most important used in medical research, notably in polio vaccine development. HeLa cells were shared worldwide and ended up contaminating many cell cultures used in research.

1951 Nutrient culture Medium 199 is developed by Connaught Laboratories in Ontario, Canada, enabling researchers to maintain live virus outside the human body, facilitating vaccine development. It will be used as the placebo for the Salk polio vaccine field trials.

1951 US Congress passes the Durham-Humphrey Amendment to the Harrison Narcotics Act, decreeing that most new drugs are available by prescription only, obtained via doctors.

A Chronology of Vaccination

1951 South African-American virologist Max Theiler is awarded the Nobel Prize in Physiology or Medicine "for his discoveries concerning yellow fever and how to combat it."

1952 57,628 polio cases are reported in the USA with more than 21,000 paralytic cases and 3,145 deaths. This is the worst epidemic recorded in the USA since 1916.

1952 In October, the UN publishes a 669-page report accusing the US of violating the Geneva Protocol—to be ratified by the US President Gerald Ford in 1975—by using biological weapons in the Korean War. Russian, Chinese, and Korean leaders accuse the US of "adopting Japanese technology to spread entomological weapons and hemorrhagic fever."

1952 Thalidomide is first synthesized by Swiss company Chemical Industry Basel, CIBA.

1952 American physican Howard Howe of Johns Hopkins University tests a formalin-inactivated trivalent polio virus vaccine on children at the Rosewood Training School in Maryland. He describes the children as "low grade idiots or imbeciles with congenital hydrocephalus, microcephaly, or cerebral palsy."

1952 Salk polio vaccine US clinical studies are conducted at the D.T. Watson Home for Crippled Children and Polk State School of "Mental Defectives" with no control groups.

1952 ✪ A heat-phenol inactivated typhoid vaccine is made by Wyeth and licensed in the US.

1952 American medical researcher Jonas Salk, MD, and Dr. Thomas Francis of the University of Michigan Medical School (Ann Arbor), with the support of the National Foundation for Infantile Paralysis, begin testing their first killed-virus polio vaccine. The first subjects are resident children in institutions for the physically and intellectually disabled. This campaign is to become one of the largest clinical trials (also called the Francis Field Trial) in medical history, vaccinating roughly 1.8 million children during the field test phase, which lasts until early 1955.

1952 The WHO Global Influenza Surveillance and Response System (GISRS) is established. Soon after, an Expert Committee on Influenza is created by the WHO, and it publishes its first Technical Report Series in 1953.

1952 Jewish Ukrainian-born American microbiologist, Selman A. Waksman, receives the Nobel Prize in Physiology or Medicine "for his discovery of streptomycin, the first antibiotic effective against tuberculosis."

1952 The first hydrogen bomb is detonated in the Pacific Marshall Islands—it is a thousand times more powerful than the atomic bomb.

1952 Pasteur Institute's live neurotropic yellow-fever vaccine is first employed on a large scale in Nigeria. The number and severity of reactions is high and use of this vaccine is discontinued. The vaccine is associated with multiple cases of encephalitis and for the first time, vaccine virus is recovered from affected brain tissue.

1952 For five days in December, London suffers from "The Great Smog," a lethal smog that covers the city caused by a combination of industrial pollution and high-pressure weather. This deadly mix of smoke and fog brings the city to a near standstill and results in thousands of deaths.

1952 Working at John Enders' laboratory at the Children's Hospital of Boston, American physician Thomas H. Weller, claims to isolate the varicella-zoster virus (VZV) from cases of chickenpox and shingles.

1952 ✪ Merck Sharp & Dohme markets a DTP vaccine Trinivac®/Trivivac® and from 1959, DTP-IPV Tetravax®. Merck will withdraw all whole-cell pertussis vaccines from the US in 1964, citing a fear of lawsuits.

1952 The *Diagnostic and Statistical Manual of Mental Disorders* (DSM) is created by the American Psychiatric Association.

1953 Zika virus (VR-84) is deposited into the American Type Culture Collection (ATCC) by Dr. Jordi Casals of the Rockefeller Foundation Virus Laboratory.

1953 The 3D structure of DNA is described by American scientist Dr. James Watson and English biophysicist Francis Crick.

1953 The US Army tests an experimental influenza vaccine on 18,000 troops, using Freund's Incomplete adjuvant—a less reactogenic version without heat-killed mycobacteria. It will also be used on army personnel in 1963–1964. This water-in-oil adjuvant is eventually abandoned due to persistent toxicity issues.

1953 ✪ Tetanus and diphteria toxoid vaccine—the adult formulation—is first licensed in the US, with a reduced concentration of diphtheria toxoid.

1953

The US Department of Health, Education, and Welfare (HEW) is created as a cabinet-level department of the US government until 1979, when it is renamed the Department of Health and Human Services (HHS). This government agency is responsible for the national vaccination programs.

1953 While working at the Walter Reed Army Institute of Research (WRAIR), Maurice Hilleman investigates an influenza outbreak at a military base. He isolates and discovers a new type of virus. Today that family of virus is known as the adenovirus and is claimed to cause up to 10% of all upper respiratory infections in children and adults.

1953 An initiative for global smallpox eradication is first proposed at the World Health Assembly, the WHO's governing body. It would take another thirteen years before the eradication program is approved and implemented.

1953 The US Biological Warfare Laboratories at Fort Detrick establish a program to study the use of anthropods (mainly ticks and mosquitos) for spreading anti-personnel biological warfare agents. They develop a "vast arsenal of weaponized insects as disease vectors."

1953– 1975 The US experiments with a diverse range of human, animal, and plant diseases as well as toxins on unsuspecting civilians and military volunteers. Mosquitos are bred to carry yellow fever, malaria, dengue, cholera and anthrax. Experiments testing biological agents on thousands of army volonteers, referred to collectively as Operation Whitecoat, take place at Fort Detrick's Chemical Corps Biological Lab. Many volunteers are Seventh-Day Adventists, who opt to participate in medical experiments instead of combat duty.

1953 ✪ Live yellow fever vaccine by Merrell-National Labs is licensed in the US.

1953 The Pfizer Foundation is established as an independent nonprofit to promote access to quality healthcare and nurture innovation.

1953 Under CIA director Allen Dulles, the clandestine project MKULTRA is created with the purpose of developing, testing, and stockpiling "chemical and biological materials capable of producing human behavioral and physiological changes."

1953 ✪ Combined diphtheria, tetanus, and pertussis vaccine is introduced in Australia.

1954 American physicians Drs. John F. Enders and Thomas C. Peebles collect fluids from 11-year-old David Edmonston. They claim to have isolated the measles virus (which would be passaged numerous times in animal and monkey kidney cells for attenuation) and develop a series of experimental vaccines.

1954 Dr. John Enders, Dr. Thomas H. Weller, and Dr. Frederick C. Robbins win the Nobel Prize in Physiology or Medicine for "their discovery of the ability of poliomyelitis viruses to grow in cultures of various types of tissue."

1954 The Castle Bravo thermonuclear weapon test at Bikini Atoll, Marshall Islands, is the largest nuclear weapon test ever conducted by the USA. The radioactive fallout and impact on health is the worst recorded in history, with neighboring populations exposed to high levels of radiation, consequently suffering from skin lesions, hair loss, and cancer.

A Chronology of Vaccination

1954–1956

1954

THORAZINE*

to stop intractable hiccups

'Thorazine' stopped hiccups (often after the first dose) in 56 out of 62 patients in seven different studies.

Thorazine® (chlorpromazine) is the first licensed psychiatric medication by Philadelphia-based pharmaceutical company, Smith, Kline & French (today GSK). The neuroleptic drug effectively acts like a chemical lobotomy. It is also marketed as a treatment for the hiccups. Dr. Arthur Sackler, father of Richard Sackler of Purdue Pharmaceuticals—who later would market the addictive substance Oxycontin®—is behind the marketing of Thorazine and would become a legend in medical advertising. Thorazine initiates a revolution in psychiatry, launching the beginning of the psychopharmacology era.

1954 The WHO's First Polio Technical Report, written by the Expert Committee on Poliomyelitis states "It must be strongly emphasized that paralysis is an infrequent complication of poliomyelitis infection, and that most persons who become infected either show no symptoms or else develop a transient abortive or 'minor' illness."

1954 ✪ The first Salk field trial of polio vaccine takes place in the USA, Switzerland, and other nations. Initial batches of vaccine are purified using a thick asbestos filter to efficiently separate the virus from the cellular debris. The US dual-protocol large field trials, led by Dr. Thomas Francis, included "placebos" of Medium 199 (in 84 regions) and unvaccinated controls (in 127 areas) with 1.8 million kids vaccinated with a 3-dose regimen. Pregnant women are also injected. Dr. Francis announces that the vaccine is "80–90% effective against paralytic poliomyelitis" which is the very rare complication of polio infection.

1954 DHS Science & Technology Office of National Laboratories (ONL) Plum Island Animal Disease Center (PIADC) is founded to serve as the nation's premier defense against accidental or intentional introduction of transboundary animal diseases (a.k.a. foreign animal diseases) including foot-and-mouth disease (FMD), rinderpest, and African swine fever (ASF). Plum Island's biowarfare ties date back to World War II and Operation Paperclip, a covert government program.

1954 The first successful human organ (kidney) transplant is performed on the Herrick twins in Boston, USA.

1954 ✪ An alum-adjuvanted and thimerosal-preserved anthrax vaccine is produced in the UK at Porton Down and used for the army.

1954 The Swiss Serum Institute imports Salk's experimental Merthiolate-containing inactived polio vaccine from American pharmaceutical company Eli Lilly to conduct field trials on Swiss schoolchildren. The vaccine will be officially introduced in Switzerland in 1957.

1954 The Ministry of Health in the UK states that the incidence of tetanus is insufficient to justify the introduction of routine vaccination—despite this, seven years later, it is added to the national childhood schedule.

1955 In April, the first Biological Safety Conference takes place at Camp Detrick in Frederick, Maryland. Fourteen representatives from the USDA, USPHS, and three principal US Army biological warfare (BW) laboratories—Camp Detrick; Pine Bluff Arsenal, Arkansas; and Dugway Proving Grounds, Utah—meet to share knowledge and experience on biosafety, chemical, and industrial safety issues common to BW operations.

1955 BURROUGHS WELLCOME FUND 𓂀

Burroughs Wellcome Fund is founded as an extension of the UK-based Wellcome Trust.

1955 The WHO launches a worldwide malaria eradication campaign.

1955

❂ The first inactived Salk polio vaccine is licensed with intense media coverage in the USA and worldwide. The vaccination campaign is launched with 500,000 children and pregnant women vaccinated with up to three doses of this new vaccine.

1955 The FDA approves Swiss company Ciba-Geigy's Ritalin® (methylphenidate—first synthesized in 1944 and patented in 1954) as a treatment for narcolepsy, depression, and to treat "organic brain syndrome." It would later be used most commonly to treat attention deficit disorder (ADD).

1955 The Cystic Fibrosis Foundation is established in the USA.

1955 The Division of Biologics Standards (DBS) is established as a specialized division with the NIH and becomes the gatekeeper of the US vaccine market.

1955 American anesthesiologist Henry K. Beecher publishes the article "The Powerful Placebo" in the *Journal of the American Medical Association*. He is credited as being the first scientist to quantify the placebo effect and to raise awareness of placebos as useful controls in clinical trial settings "to distinguish pharmacological effects from the effects of suggestion."

1955 The National Institute of Allergy and Infectious Diseases (NIAID) is founded and is part of the 27 institutes and centers that make up the National Institutes of Health (NIH), an agency of the United States Department of Health and Human Services.

1955 The definition of polio is modified to align with the diagnostic criteria applied during the 1952–1954 Salk vaccine field trials. Simply by changing the diagnostic criteria, the number of polio cases decrease.

1955 Cutter Laboratories produce an improperly inactivated Salk polio vaccine that is given to 200,000 people, mainly children. The vaccine campaign is temporarily suspended. It causes 40,000 polio cases, 10 deaths, and leaves 164 children severely paralyzed. Other companies (Eli Lilly, Parke-Davis, Pitman-Moore, Wyeth) are reported to also have live polio virus in their vaccines, but are never mentioned publicly. Two manufacturers settle lawsuits out of court.

1955 The Polio Vaccination Assistance Act is signed by President Dwight D. Eisenhower. It is the first federal involvement with vaccines, appropriating funds to the Communicable Disease Center (later the CDC) to help states and local communities acquire and administer polio vaccine.

1955– 1970 Dr. Saul Krugman, a respected New York pediatrician, experiments on residents of Willowbrook State School, a home for severely developmentally challenged children and adults. In order to study their immunity (but providing them zero benefits), children are injected with the hepatitis virus itself or made to drink chocolate milk mixed with feces from other infected children. Dr. Krugman is awarded for his unethical research—which was standard practice at the time—becoming president of the American Pediatric Society in 1972. Celebrated American vaccinologist Maurice Hilleman claims these studies "were the most unethical medical experiments ever performed on children in the US."

1956 The Wuhan Institute of Virology is founded as the Wuhan Microbiology Laboratory under the Chinese Academy of Sciences. In 1978, it is renamed the Wuhan Institute of Virology, which will become in 2018 the first fully operational BSL-4 laboratory in mainland China, working to develop vaccines and study SARS-coronaviruses.

1956 Despite warnings from the Department of Defense that live polio virus samples could be turned into biological weapons, Albert Sabin receives permission from the US State Department to ship them to the USSR.

A Chronology of Vaccination

1956–1959

1956 The risk of provocation poliomyelitis following inoculation with combined diphtheria, tetanus, and pertussis vaccine is noted in the MRC Committee on Inoculation Procedures and Neurological Lesions report. As a result, alum is removed from most DTP vaccines in the UK, only to be reinstated in 1961.

1956 ✪ The first inactivated Salk polio vaccine is introduced in Australia and added to the UK vaccine schedule (product made by Glaxo).

1956 An article in *The Lancet* features an analysis of 412 cases of poliomyelitis following vaccination and provides an estimate of polio incidence after alum diphtheria-pertussis vaccines as being 1 in 15,000.

1956 Vaccine passaged in human embryo lung marks the beginning of the transition from vaccines prepared from animal nerve tissue to embryonated eggs, followed by adaptation of virus to cultures of human diploid cells (like MRC-5) in the 1960s.

1956 The Walter Reed Army Institute of Research (WRAIR) creates an inactivated adenovirus types 4 and 7 oral vaccine. Another adenovirus vaccine against type 3 is developed by the NIH on monkey kidney cells, but due to manufacturing issues (and being widely contaminated with SV40), the license is revoked in 1963. A few years later two oral live-attenuated adenovirus viruses are available for the US military as tablets, grown using WI-38 cells, which are given regularly to all army recruits from 1971 until 1996, when the sole manufacturer, Wyeth, ceases production.

1956 The first large-scale human trial of the birth control pill is carried out in Puerto Rico, without informed consent. Around 1,500 women participate in the trial, informed only that the drug prevents pregnancy, not that it is experimental or may have side effects.

1957-1959 In the US, paralytic polio cases more than double to more than 6,000. This accelerates research into a more effective oral live polio vaccine that mimics the natural route of polio infection.

1957 Polish virologist and immunologist Dr. Hilary Koprowski becomes the director of the Wistar Institute and leads it for 34 years, ushering in the modern era of vaccine development and cancer research.

1957 ✪ The triple diphtheria-tetanus-pertussis (DTP) shot is introduced in the UK for infants from three months old. It is included in the first official childhood schedule four years later.

1957 **MERCK FOUNDATION**

The Merck Foundation is founded as a US-based, private charitable foundation, established by Merck & Co., Inc. It is entirely funded by the company and is its main source of funding to nonprofit charities.

1957 European scientists identify the first cytokine—a small protein important in cell signaling—and name it interferon, because it appears to interfere with the progress of viral infections and even tumor growth.

1957 Asian flu pandemic of influenza A virus subtype H2N2 originates in Guizhou, China. Over one million deaths globally are reported, with an estimated 116,000 deaths in the US.

1957 ✪ Burroughs Wellcome launches its inactivated polio vaccine Polimylex® in the UK, removing it from the market in 1960.

1957 Two American female NIH researchers Bernice Eddy and Sarah Stewart discover a mouse virus—they name "SE polyoma"—that causes multiple cancer tumors in a variety of small mammals. This watershed moment in cancer research will shift the focus and funding towards viruses.

1957 Maurice Hilleman joins Merck as director of its new department of virus and cell biology research. By 1954, under his leadership, Merck will have 37 product licences.

1957 The Treaty of Rome is signed by Belgium, France, Italy, Luxembourg, the Netherlands and West Germany, establishing the European Economic Community (EEC) and the European Atomic Energy Community (Euratom).

1957–1962

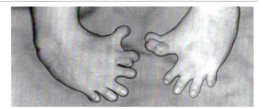

German company Chemie Grünenthal acquires thalidomide from Swiss company CIBA and launches it as an over-the-counter drug named Contergan®. It is sold as an effective treatment for morning sickness and proclaimed a "wonder drug" for insomnia. Under license from Chemie Grünenthal, 14 companies market thalidomide in 46 nations (including the UK from 1958–1962, Australia, and New Zealand) under at least 37 different brand names (Thalomid, Kevadon, Talidex, Neurosedyn, Distaval). Despite numerous requests for licensure in the US by William S. Merrell Company, Dr. Frances O. Kelsey of the FDA denies its approval. Even though it was never approved in the USA, Merrell distributes over 2.5 million doses "for investigational use" to more than 1,200 US doctors—its clinical trial evolves into an unauthorized marketing program. It turns out that thalidomide causes severe birth defects and becomes one of the worst medical disasters of all time. It is withdrawn from markets in 1961–1962 and banned worldwide by the end of the decade, after maiming an estimated 20,000 babies and killing 80,000.

1957 — Thomas Weller's laboratory is the first to cultivate cytomegaloviruses and describe the congenital transmission of the virus.

1958 — Measles vaccine, prepared in chick-embryo tissue culture and developed in the laboratory of American microbiologist Dr. John F. Enders, is first tested on 13 residents of a US state institution for the mentally deficient. By the time the measles vaccine is licensed in 1963, it has been tested on approximately 25,000 people from age four months to 32 years.

1958 — ✪ Bivalent (A)H2N2 and subtype B influenza vaccines are commercialized in the USA.

1958 — The Defense Advanced Research Projects Agency (DARPA) is established after the Soviet Union's launch of Sputnik.

1958–1960 — Dr. Hilary Koprowski and Dr. Stanley A. Plotkin launch an experimental oral live-polio vaccination campaign, using poliovirus type 1 Mahoney-derived CHAT strain, in Léopoldville, Belgian Congo. By the end of the campaign, over 75,000 children are vaccinated under the age of five (around 244,000 children in total), with clinical observations recorded for around 7,200 children, with a 13.5% adverse event rate. This field trial is funded by NIH and the Wistar Institute. This live vaccine is used on several continents, but not in the USA.

1958–1960 — Soviet Russian microbiologist and virologist Mikhail Petrovich Chumakov tests a vaccine made with Sabin's live polio seed virus on 20,000 children in the Soviet Union. He then gets the vaccine distributed to more than 15 million people in Soviet schools, hospitals, clinics, and nurseries, essentially conducting pivotal large-scale clinical trials—without control or placebo groups. By the end of 1960, the vaccine is given to 77 million children in the Soviet Union and 23 million in East Germany, Czechoslovakia, Hungary, Romania, and Bulgaria.

1958 — Dr. J.M. Berg notes in *The British Medical Journal* that there are 107 events in the medical literature of neurological adverse reactions associated with pertussis vaccine. "These sequelae have ranged from transient convulsions with complete recovery to gross crippling, mental retardation, and death."

1958 — American microbiologist Joshua Lederberg is awarded the Nobel Prize in Physiology or Medicine "for his discoveries concerning genetic recombination and the organization of the genetic material of bacteria."

1959 — Royal Army Medical Corps Captain P.D. Meers, from the Federal Laboratory Service Headquarters in Lagos, Nigeria, reports that a combined yellow fever vaccine (17D) and smallpox vaccine is tested in 1958 on 43 healthy male recruits of the Ikeja police force. It is concluded that yellow fever vaccine is more effective when taken alone.

1959 — The resolution to eradicate smallpox is passed at the WHO 12th World Assembly.

A Chronology of Vaccination

1959 Spanish biochemist Severo Ochoa and American biochemist Arthur Kornberg are awarded the Nobel Prize in Physiology or Medicine "for their discovery of the mechanisms in the biological synthesis of ribonucleic acid and deoxyribonucleic acid."

1959 A three-day symposium on "immunization in childhood" is held in the Wellcome Building, London. Attended by medical and school health officers, doctors, pediatricians, and epidemiologists, the subjects include the the neurological hazards of pertussis and smallpox vaccine, and paralytic poliomyelitis after aluminum-containing vaccines. The proceedings are published as a 139-page report, available in the public domain.

1959 The UK Medical Research Council (MRC) recommends that only pertussis vaccines showing adequate strength in the mouse protection test should be used to ensure potency and effectiveness.

1959 ✪ DTP and polio (IPV) combination vaccine Quadrigen® by Parke, Davis Co., is licensed in the USA and withdrawn in 1968. Instead of containing the preservative Merthiolate, used in Triogen® (DTP), benzethonium chloride (also called phemerol chloride) is used.

1960–1969

1960s Early in the decade, two DTP+IPV vaccines are available in the UK; Quadrilin® by Glaxo and Quadrivax® by Burroughs Wellcome.

1960s By the middle of the decade, most US states have mandated DTP vaccine for school.

1960s Madin-Darby canine kidney (MDCK) cell line is used for the cultivation of influenza virus.

1960s By the end of the decade, the US military develops a biological arsenal that includes numerous biological pathogens, toxins, and fungal plant pathogens that can attack crops to induce crop failure and famine.

1960s ✪ Influenza vaccines are introduced and routinely recommended in the UK.

1960 The AMA's revenues from drug advertising in its journals rise from US$2.5 million in 1950 to US$10 million.

1960 Australian virologist Sir Frank Macfarlane Burnet and Brazilian-British biologist Peter Brian Medawar are awarded the Nobel Prize in Physiology or Medicine "for discovery of acquired immunological tolerance."

1960 The US Surgeon General recommends annual influenza vaccines for people at high risk of flu complications, including pregnant women.

1960 The landmark legal case *Gottsdanker v. Cutter Laboratories* is judged in California. The plaintiffs, parents of two children who received Cutter's Salk polio vaccine and developed polio, are awarded US$139,000 and US$8,300 in special damages. The jury, "from a preponderance of the evidence concluded that the defendant, Cutter Laboratories, was not negligent either directly or by inference." However, Cutter is held liable for breach of "implied warranty," setting legal precedent for later lawsuits.

1960 Canada approves the drug thalidomide by prescription only. There are many different brands sold: Kevadon—produced by the US William S. Merrell Company—is released on the market a year later, even though some countries have already begun taking the product off the market. By 1962, the drug is withdrawn for the world market. In 1968, a criminal case in Germany is launched against Chemie Grünenthal, which it settles two years later, with no admission of guilt. By the mid-1990s, thalidomide is marketed by Celgene and approved in the USA for Hansen's disease (leprosy) and multiple myeloma cancer treatments.

1960 The fatal neurodegenerative disease Scrapie (native to sheep and goat) is adapted by scientists, from humans to mice.

1960 Dr. Bernice Eddy at the NIH discovers the simian virus 40 (SV40) in the Salk polio vaccine, potentially causing cancer. From 1955 to 1963, over 98 million Americans are injected with SV40 contaminated polio vaccines.

1960 ✪ In the US, Merck Sharp & Dohme markets an acellular pertussis vaccine until 1961.

1960–1962 American virologists Franklin A. Neva and Thomas H. Weller isolate the rubella virus in a series of experiments, paving the way for a vaccine against rubella.

1960 Enovid (nonethynodrel) is the first birth control pill to be approved by the FDA.

1960 ✪ A new modified DTP vaccine Trivax® by Burroughs Wellcome is available in the UK.

1961 The first national childhood vaccination schedule is introduced to the UK, including smallpox, tetanus, diphtheria, pertussis and polio vaccines.

1961 The American Academy of Pediatrics recommends a single injection of vitamin K1 (phytonadione) to be given to newborns, to prevent vitamin K deficiency bleeding (VKDB).

1961 The first ever program to provide no-fault compensation for an adverse event following vaccination is introduced in Germany.

1961 ✪ Trivalent inactivated split-virion influenza vaccine, Flushield® by Wyeth is approved in the US. It contains 1:10,000 thimerosal.

1961 The first International Conference on Measles Immunization takes place at the NIH in Bethesda, where the measles vaccine is declared to be 100% effective.

1961 Scientists at the California Institute of Technology extract RNA for the first time, later recognizing its molecular role as "middleman" between nuclear DNA and protein expression in the cell.

1961 Three experimental measles vaccines are tested on 56 mentally "subnormal" institutionalized children in England, using vaccines made by Parke-Davis and Burroughs Wellcome. One child dies eight days after vaccination—the death is labeled "coincidental." Nine children are reported as suffering a severe reaction.

1961 ✪ Albert Sabin's bivalent oral polio (types 1–2) vaccine is licensed in the USA and is soon used worldwide.

1961 ✪ In October, the Sabin oral polio vaccine available in Western Europe is made by Pfizer (monovalent type 2) and first administered via sugar cubes in Kingston-upon-Hull, England.

1961–1966 American pediatric virologist Dr. Harry Martin Meyer Jr. heads a team that carries out trials in Africa of the measles vaccine developed by Dr. John F. Enders. The CDC-USAID trial involves eight West African nations and more than a million children. The team demonstrates the practicality of giving vaccines in remote areas using the then-experimental jet injection gun.

1961 ✪ The tetanus vaccine is approved for routine childhood use in the UK.

1962 English molecular biologist Francis Crick, American molecular biologist James D. Watson, and New Zealand-born British biophysicist Maurice Wilkins win the Nobel Prize in Physiology or Medicine "for their discoveries concerning the molecular structure of nucleic acids and its significance for information transfer in living material." Despite the work of British chemist Rosalind E. Franklin's being central to the understanding of the molecular structures of DNA and RNA, her contributions to the discovery of the structure of DNA are largely unrecognized during her lifetime.

1962 Eli Lilly applies for a patent on a killed-virus measles vaccine with inactivated virus adsorbed upon hydrated aluminum phosphate. The patent US3132073A is granted in 1964 and expires in 1981.

1962 Sponsored by the National Institutes of Health, Leonard Hayflick, PhD, and Paul S. Moorhead, PhD, of The Wistar Institute in Philadelphia, establish the first diploid cell line of WI-38 female human fibroblasts (fetal lung cells) for viral vaccine development.

1962 ✪ Albert Sabin's trivalent oral polio vaccine (OPV) types 1–3 is licensed in the USA and soon replaces the bivalent version. OPV is commercialized in the UK.

1962 Avian leukosis virus is detected in a seed lot of yellow fever vaccine (17D strain) in the UK and in the US shortly after—a finding of potentially disastrous proportions. This virus displays oncogenic activity in animals. Thousands, if not millions, of people are inadvertently inoculated with this potentially cancer-causing virus.

1959   The Wellcome Symposium on Childhood Immunization

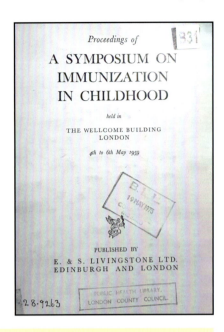

The very first session of the three-day Wellcome symposium is entitled "The Risks of Immunization: Provocation Poliomyelitis" and is presented by Dr. John Knowelden from the Department of Medical Statistics and Epidemiology, London School of Hygiene & Tropical Medicine and Medical Research Council (MRC) Statistical Research Unit.

"In 1949, three independent reports appeared which brought wide recognition to provocation poliomyelitis. [...] The common finding was that a number of patients with paralytic poliomyelitis had received inoculations of diphtheria and pertussis antigens in the previous weeks and that the paralysis was confined to, or concentrated mainly on, the limb last inoculated." Though initially believed to be coincidental, "an inquiry was conducted in the autumn of 1949 through the Medical Officers of Health of 33 areas in England and Wales. This study (Hill and Knowelden, 1950) confirmed that provocation poliomyelitis was a true entity. As suggested by the earlier reports, the association seemed particularly strong with the alum-precipitated mixed diphtheria-pertussis prophylactic, but alum-precipitated toxoid vaccine (APT) could not be exonerated."

As a result, the MRC "sponsored a large investigation in Britain. In the years 1951–1955, Medical Officers of Health provided reports of all notified cases of poliomyelitis stating whether there had been any prophylactic inoculations within twelve months prior to the onset of symptoms."

It is estimated that for every 15,000 injections with DP-alum, one case of paralytic polio is provoked.

PARALYTIC POLIOMYELITIS AFTER INOCULATION
C.B. Clinics England and Wales 1951-53

Prophylactics	Rates per 100,000 per month			Inoculations to provoke 1 Case
	1-28 Days	29-84 Days	Diff.	
Mixed with alum	8·0		6·7	15,000
Mixed without alum	6·5		5·2	19,000
P.T.A.P.	6·0	1·3	4·7	21,000
A.P.T.	3·4		2·1	48,000
Pertussis vaccine	1·9		0·6	170,000
F.T. and T.A.F.	1·4		0·1	1,000,000
All prophylactics	4·0	1·3	2·7	37,000

Mixed diphtheria-pertussis (DP) vaccine with / without alum

PTAP	Purified diphtheria toxoid aluminum phosphate precipitated	**FT**	Formol (diphtheria) toxoid
APT	Alum-precipitated (diphtheria) toxoid	**TAF**	Diphtheria toxoid-antitoxin floccules

In another table, it is shown that in all areas of England and Wales from 1951–1955, a total of 209 children were identified as having paralysis within 28 days of vaccination, most of whom received APT. Of the children paralyzed in the inoculated arm only, 80% received the mixed diphtheria-pertussis vaccine with alum. A further 163 children were identified as suffering from paralysis in the 29–84 days after vaccination, only 14 of whom were affected in the inoculated limb.

Proceedings of a Symposium on Immunization in Childhood **held in the Wellcome Building in London**
6th May 1959 / E & S Livingstone Ltd. – Obtained from the British Library

1960   The Present Status of Polio Vaccines, *Illinois Medical Journal*

Moderator: Herbert Ratner, MD, Director of Public Health, Oak Park, and Associate Clinical Professor of Preventive Medicine and Public Health, Stritch School of Medicine, Chicago.
Panelists: Herald Rea Cox, ScD, Pearl River, NY (Lederle)
Bernard G. Greenberg, PhD, Chapel Hill, NC
Herman Kleinman, MD, Minneapolis
Paul Meier, PhD, Chicago.

I pointed out that the discrepancy was purely a statistical one. There were two biases in the way the Public Health Service had calculated its rates of attack among the vaccinated and the unvaccinated.

First of all, the unvaccinated population figure for 5 to 9 year old children used in the Public Health Service report was the number given in the 1950 census minus the number of children vaccinated. The number of children aged 5 to 9 in 1955 was estimated, however, to be 101,000 more than it was in 1950. The Public Health Service did not take this increase into account. The omission of 101,000 children from the unvaccinated population would have increased the latter roughly from 236,000 to 337,000 children. Hence, the attack rate for unvaccinated children was overestimated by about 40 per cent.

In 1960 at a US polio symposium, Dr. Alexander Langmuir, in charge of the polio surveillance for the USPHS, states that the "resurgence of the disease, particularly in the paralytic form, provides 'cause for immediate concern' and that the upward polio trend in the US during the past two years 'has been a sobering experience for overenthusiastic health officers and epidemiologists alike.'"

In 1955, Dr. Langmuir predicted that "by 1957 there would be less than 100 cases of paralytic polio in the US." However in 1959, after 300 million doses of Salk vaccine, there were "approximately 6,000 cases of paralytic polio, 1,000 of which were in persons who had received three, four and more shots of Salk vaccine."

The complete article is available on
www.TheUltimateVaccineTimeline.com

Reasons for recent increase

If the vaccine was not as effective, one might wonder why the tremendous reduction occurred

in the 1955, 1956, and 1957 reported rates. Here, again, much of this reduction was a statistical artifact.

Prior to 1954 any physician who reported paralytic poliomyelitis was doing his patient a service by way of subsidizing the cost of hospitalization and was being community-minded in reporting a communicable disease. The criterion of diagnosis at that time in most health departments followed the World Health Organization definition: "Spinal paralytic poliomyelitis: Signs and symptoms of nonparalytic poliomyelitis with the addition of partial or complete paralysis of one or more muscle groups, detected on two examinations at least 24 hours apart."[11]

Note that "two examinations at least 24 hours apart" was all that was required. Laboratory confirmation and presence of residual paralysis was *not* required. In 1955 the criteria were changed to conform more closely to the definition used in the 1954 field trials: residual paralysis was determined 10 to 20 days after onset of illness and again 50 to 70 days after onset. The influence of the field trials is still evident in most health departments; unless there is residual involvement at least 60 days after onset, a case of poliomyelitis is not considered paralytic.

This change in definition meant that in 1955 we started reporting a new disease, namely, paralytic poliomyelitis with a longer lasting paralysis. Furthermore, diagnostic procedures have continued to be refined. Coxsackie virus infections and aseptic meningitis have been distinguished from paralytic poliomyelitis. Prior to 1954 large numbers of these cases undoubtedly were mislabeled as paralytic poliomyelitis. Thus, simply by changes in diagnostic criteria, the number of paralytic cases was predetermined to decrease in 1955-1957, whether or not any vaccine was used. At the same time, the number of nonparalytic cases was bound to increase because any case of poliomyelitis-like disease which could not be classified as paralytic poliomyelitis according to the new criteria was classified as nonparalytic poliomyelitis. Many of these cases, although reported as such, were not nonparalytic poliomyelitis. If this inaccurate number of cases of nonparalytic poliomyelitis reported in 1957 is accepted as accurate and considered as a base for subsequent comparisons, it is no wonder that we now say nonparalytic cases went down in 1958.

The criteria for the diagnosis of polio is modified.

A change in definition can itself create a reduction in cases.

Panel discussion edited from the transcript of the 120th Annual Meeting of Illinois State Medical Society in Chicago
26th May 1960 / Presented before the Section on Preventive Medicine and Public Health

1951   Cutter Laboratories, Alhydrox® advertisement

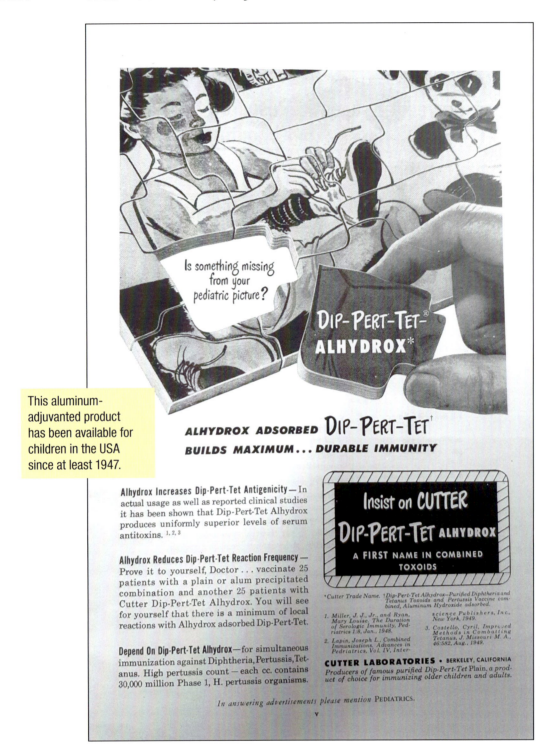

This aluminum-adjuvanted product has been available for children in the USA since at least 1947.

Cutter Laboratories – Alhydrox® advertisement in
Pediatrics – The Journal of the American Academy of Pediatrics Inc.
January 1951, Volume 7

1960   Merck Sharp & Dohme, Tetravax® advertisement

Merck's pioneering 4-in-1 inactivated vaccine is introduced to the US market in 1959. It is withdrawn in 1964 when Merck decides to remove all its whole-cell pertussis vaccines from the US market for fear of litigation.

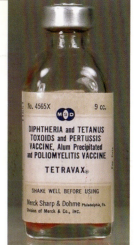

Merck Sharp & Dohme – Tetravax® advertisement in the
Illinois Medical Journal
1960, Volume 118

1960   Lederle advertisement

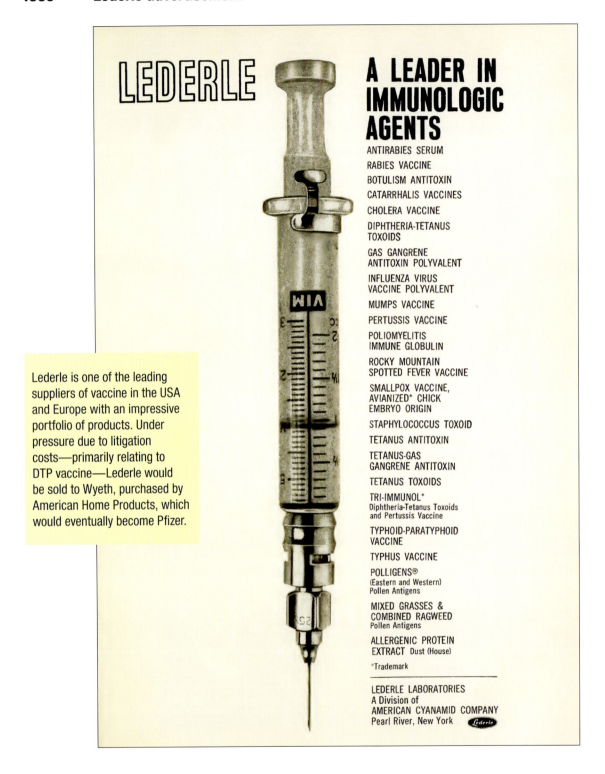

Lederle is one of the leading suppliers of vaccine in the USA and Europe with an impressive portfolio of products. Under pressure due to litigation costs—primarily relating to DTP vaccine—Lederle would be sold to Wyeth, purchased by American Home Products, which would eventually become Pfizer.

Lederle Immunologic Agents advertisement in the
Illinois Medical Journal
December 1960, Volume 118

1963   CDC Wellbee, oral polio advertisement

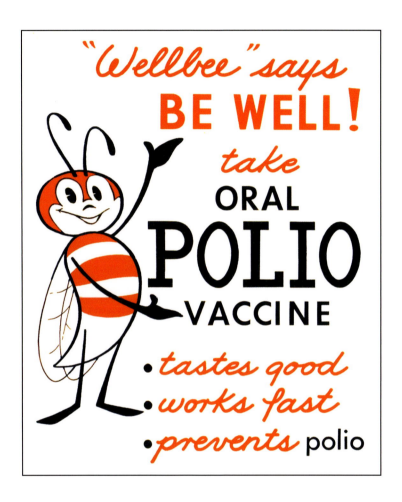

This 1963 poster featured CDC's national symbol of public health, the "Wellbee" (by Harold M. Walker) who was depicted here encouraging the public to receive an oral polio vaccine. CDC used the Wellbee in its comprehensive marketing campaign that used newspapers, posters, leaflets, radio, and television, as well as personal appearances at public health events. Wellbee's first assignment was to sponsor Sabin Type-II oral polio vaccine campaigns across the United States. Later, Wellbee's character was incorporated into other health promotion campaigns including diphtheria and tetanus immunizations, hand-washing, physical fitness, and injury prevention.

A Chronology of Vaccination

1962–1965

1962 Rubella virus is first isolated in tissue culture by two groups in the US, one based at the WRAIR and another based at Harvard School of Public Health.

1962 US President John F. Kennedy signs the Vaccine Assistance Act for Mass Immunization, making available federal funds (notably to the CDC) to ensure that all children under age five can receive vaccines.

1962 Following the tragedy of Thalidomide, President Kennedy signs the Kefauver-Harris Amendments to the Food, Drugs, and Cosmetic Act, also known as the Drug Efficacy Amendments. The bill requires that all new drug applications demonstrate substantial evidence of the drug's efficacy, in addition to demonstrating safety and adequate labeling for safe use. The amendments also require informed consent from subjects who participate in clinical trials.

1962 ✪ In the USA, Eli Lilly & Co. markets its first acellular or "fractionated" split cell pertussis vaccine under the name Tri-Solgen®. Despite capturing 65% of the market by 1977, it is withdrawn when Lilly halts all vaccine production. Lilly sells the rights for Tri-Solgen® to Wyeth, who abandons testing because it determines the vaccine would not be as profitable as the company wanted.

1962–1971 During the Vietnam War (1955–1975), the US Army uses defoliants and anti-crop agents, including Agent Orange and other herbicides. The US Air Force sprays nearly 19 million gallons of herbicides in Vietnam (supplied by Dow Chemical and Monsanto among others), of which at least 11 million gallons is Agent Orange. In total, 5.5 million acres are defoliated, exposing millions of Vietnamese to poison, causing hundreds of thousands of deaths in both Vietnamese and US military veterans, as well as millions of disabilities and other health issues, notably in children.

1962 According to Wikipedia, the most nuclear weapon tests worldwide occurs this year, with the USA detonating around a hundred bombs and the USSR forty.

1963 Twenty US states, the District of Columbia, and Puerto Rico have childhood vaccination mandates for enrollment into public schools. By 1970, this number rises to 29 states.

1963 ✪ Live-attenuated measles vaccine (Edmonston B strain) Rubeovax® is licensed in the USA by Merck—administered together with gamma globulin Gammagee®. Other live virus measles vaccines are eventually licensed: MVac® by Lederle, Pfizer-vax Measles-L® by Pfizer, and generic vaccines by Eli Lilly & Company, Parke-Davis, and Philips Roxane.

1963 The Joint Committee on Vaccination and Immunisation (JCVI) is founded and given the responsibility of recommending vaccines for the UK population.

1963 The US Federal Immunization Grant Program is established. Authorized under the Public Health Service Act, the grants are attributed to states to fund vaccine purchases.

1963 Following 27 smallpox cases declared in the Stockholm area, a voluntary mass smallpox vaccine campaign takes place in Sweden. Over 300,000 people are vaccinated, leading to adverse events and the set up of special clinics for patients suffering from complications.

1963 The CDC announces the first national measles eradication program. When the vaccine is launched, authorities declare that measles could be eradicated in the US by 1967 using the single-dose vaccine. The Ad Hoc Advisory Committee on Measles Control recommends the measles vaccine for all babies from nine months old. This expert committee includes CDC director Dr. James L. Goddard (chairman), Chief of the CDC virus disease surveillance programs Dr. Donald A. Henderson (secretary), measles vaccine developer Dr. John F. Enders, Dr. Frederick C. Robbins, and Dr. Saul Krugman.

1963 ✪ Sabin's live trivalent oral poliomyelitis vaccine, Orimune® by Lederle Laboratories, is marketed in the US.

1963 American president John F. Kennedy is assassinated in Dallas, Texas.

1963 Rachel Carson's pivotal book *Silent Spring* is released, exposing the hazards and harmful biological affects of DDT and highlighting the indiscriminate use of pesticides in agriculture.

1963 ✪ Formalin-inactived (killed) alum-adsorbed measles vaccines, Pfizer-vax Measles-K® and a generic version by Eli Lilly, are licensed in the USA. Pfizer's vaccine is recommended as a series of three doses at one month intervals—it is withdrawn in 1968 after concerns with "atypical measles."

1963 Belgian pharmaceutical company, Janssen, recently acquired by Johnson & Johnson, releases the drug fentanyl in Europe (the FDA approves it in 1968). It will soon become the world's most prescribed anesthetic drug.

1963 Swiss pharmaceutical giant Hoffman-La Roche launches Valium® (Diazepam) for the treatment of anxiety—primarily marketed for women. It becomes the bestselling drug in the West from 1968–1981.

1963 In the USA, good manufacturing practices (GMP) for production, processing, packing, and/or holding finished pharmaceuticals are first published.

1963 Dr. C.N. Christensen—Eli Lilly physician commissioned to study the whole-cell pertussis mouse toxicity test—states at the NIH International Symposium on Pertussis: "It is obvious that severe neurologic reactions have occurred in children after immunization with pertussis vaccines which have passed the toxicity and potency tests currently in use...It was clear that there was no correlation between the mouse toxicity test and the reaction rates in children."

1964 The World Medical Association develops the Declaration of Helsinki as a statement of ethical principles to provide guidance to physicians and other participants in medical research involving human subjects.

1964 The key ingredient of Merck's experimental killed influenza vaccine called Adjuvant 65, that contains peanut oil, is patented by Dr. Allen F. Woodhour and Dr. Thomas B. Stim.

1964 In the USA, the Civil Rights Act is signed into law by President Lyndon B. Johnson, intended to end all discrimination.

1964 The Surgeon General of the US Public Health Service (USPHS) establishes the Advisory Committee on Immunization Practices (ACIP), who is charged with the responsibility of recommending best practices to prevent communicable disease. The ACIP reviews the childhood vaccine schedule and makes recommendations for newly licensed vaccines.

1964 The International Standard for pertussis vaccine is established by the WHO, setting a minimum requirement for vaccine potency.

1964–1966 The first rubella (also called German measles or "three-day" measles) outbreak occurs in the USA, with 12.5 million cases reported—20,000 cases of congenital rubella, resulting in 11,250 abortions (spontaneous or surgical), 2,100 neonatal deaths, 11,600 cases of deafness, 3,580 cases of blindness, and 1,800 cases of mental retardation are reported. This epidemic leads to the rubella vaccine development campaign led by the NIH.

1964 AZT (azidothymidine), an antiretroviral compound, Is synthetically manufactured in the USA under a NIH grant.

1964 The Special Virus Cancer Program (SVCP) is launched with a budget of US$10 million and housed at Fort Detrick—the US Army's biological warfare facilities. Under the Viral Oncology program of the National Cancer Institute, the SVCP is funded to intensify virus-leukemia research, as scientists believe that identifying viruses and detecting them in tumors would contribute to finding the cause of cancer. A major goal of the SVCP is to control leukemia by the use of vaccines or other methods. Important contractors include Merck & Co., Inc., Bionetics Research Lab Inc. (Litton Industries), and Dow Chemical.

1965 In the USA, the recommended age for routine administration of measles vaccine is changed from 9 months to 12 months old.

1965 Medicare and Medicaid programs are created, making comprehensive healthcare available to millions of Americans.

1965 The first national childhood vaccine schedule is published in Switzerland by the Federal Office of Public Hygiene.

1963   Merck Sharp & Dohme, Rubeovax® advertisement

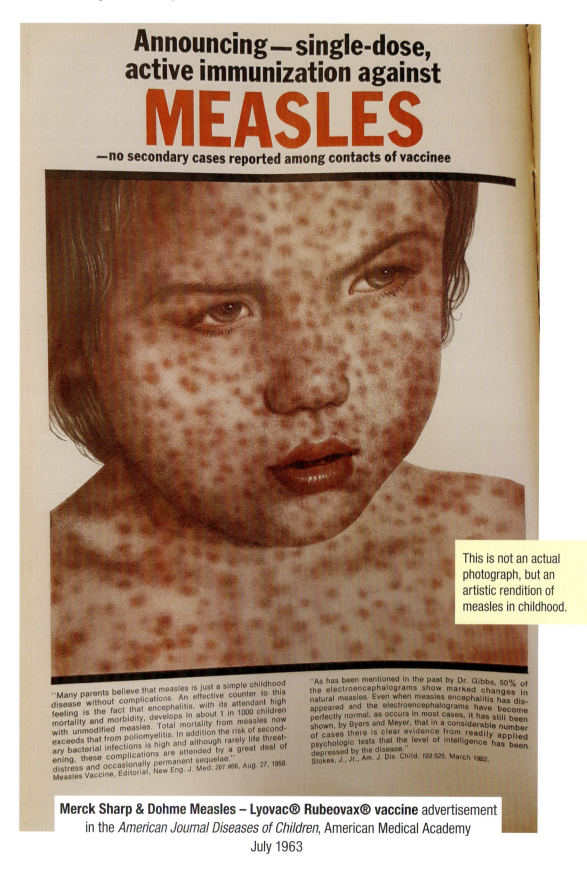

Announcing—single-dose, active immunization against

MEASLES

—no secondary cases reported among contacts of vaccinee

This is not an actual photograph, but an artistic rendition of measles in childhood.

"Many parents believe that measles is just a simple childhood disease without complications. An effective counter to this feeling is the fact that encephalitis, with its attendant high mortality and morbidity, develops in about 1 in 1000 children with unmodified measles. Total mortality from measles now exceeds that from poliomyelitis. In addition the risk of secondary bacterial infections is high and although rarely life threatening, these complications are attended by a great deal of distress and occasionally permanent sequelae."
Measles Vaccine, Editorial, New Eng. J. Med. 261:466, Aug. 27, 1959.

"As has been mentioned in the past by Dr. Gibbs, 50% of the electroencephalograms show marked changes in natural measles. Even when measles encephalitis has disappeared and the electroencephalograms have become perfectly normal, as occurs in most cases, it has still been shown, by Byers and Meyer, that in a considerable number of cases there is clear evidence from readily applied psychologic tests that the level of intelligence has been depressed by the disease."
Stokes, J., Jr., Am. J. Dis. Child. 103:525, March 1962.

Merck Sharp & Dohme Measles – Lyovac® Rubeovax® vaccine advertisement in the *American Journal Diseases of Children*, American Medical Academy July 1963

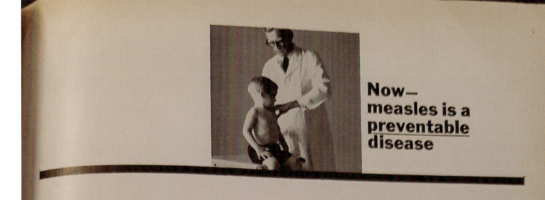

Now—
measles is a
preventable
disease

LYOVAC®

RUBEOVAX®

MEASLES VIRUS VACCINE, LIVE, ATTENUATED

0.5-cc. vial of lyophilized vaccine. (In an accompanying package, an 0.7-cc. ampul of Sterile Diluent for Reconstitution of LYOVAC RUBEOVAX and a sterile disposable syringe with needle are supplied.)

concurrent with

GAMMAGEE*

IMMUNE SERUM GLOBULIN (HUMAN)

2-cc. and 10-cc. vials.

- With a single dose — the magnitude of immunity is similar to that produced by natural measles — the duration is expected to be lifelong.
- CNS involvement has not been observed. Postinoculation electroencephalograms show no abnormality attributed to the vaccination procedure.
- Spread from vaccinees to susceptible contacts has not been reported.
- Clinical reactions caused by the vaccine are reduced by concurrent administration of GAMMAGEE. Fever is usually not accompanied by significant toxic reaction. If desired, aspirin can be used therapeutically or, if started on the fifth postvaccination day, prophylactically to reduce fever. A mild and transient rash occurs in approximately 16% of children.

This is the product launch advertisement of the first ever measles vaccine that initially was recommended to be administered concurrently with immune serum globulin.

It would be withdrawn in the USA in 1971, replaced with Merck's Attenuvax® and triple live measles, mumps, rubella vaccine MMR-I®.

BRIEF SUMMARY:
RUBEOVAX — *Indications:* Immunization of children against measles (rubeola). *Side Effects, Precautions, and Contraindications:* Co-administration of immune serum globulin (human) standardized for measles antibody content is recommended to minimize incidence and severity of fever and rash. Do not use in pregnancy, in persons with leukemia or untreated active tuberculosis, or in brain-damaged children under 1 year of age. Use with caution in adults, in children with history of febrile convulsions, or in persons under treatment with steroids. Defer use in presence of febrile respiratory illness, other active infection, or poliomyelitis epidemic. Severe reaction to egg protein possible. **GAMMAGEE** — *Indications:* Modification of mild measles induced by vaccination with live attenuated measles (rubeola) vaccine; prevention or modification of natural measles. *Side Effects, Precautions, and Contraindications:* Local tenderness and stiffness may occur after injection. Hypersensitivity, anaphylaxis possible. Before prescribing or administering, read product circular with package or available on request.

MERCK SHARP & DOHME, DIVISION OF MERCK & CO., Inc., WEST POINT, PENNSYLVANIA
*LYOVAC, RUBEOVAX, and GAMMAGEE are trademarks of Merck & Co., Inc.

1963   Lederle, Orimune® advertisement

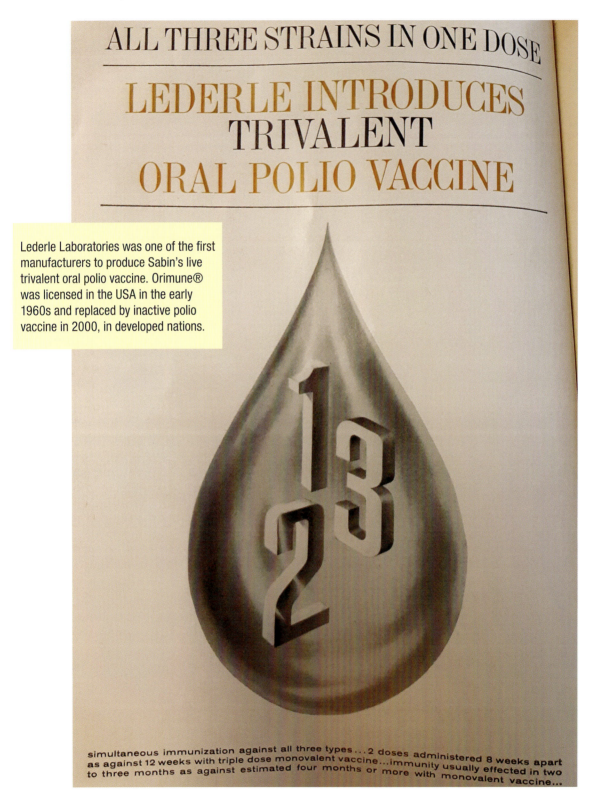

> Lederle Laboratories was one of the first manufacturers to produce Sabin's live trivalent oral polio vaccine. Orimune® was licensed in the USA in the early 1960s and replaced by inactive polio vaccine in 2000, in developed nations.

Trivalent polio Orimune® vaccine by Lederle advertisement in the
American Journal Diseases of Children, American Medical Academy
Volume 106, Nr. 1 – July 1963

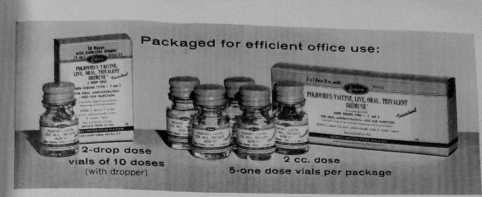

Packaged for efficient office use:

2-drop dose
vials of 10 doses
(with dropper)

2 cc. dose
5-one dose vials per package

ORIMUNE Poliovirus Vaccine, live, oral TRIVALENT presents, in a single vehicle, a means of simultaneously conferring immunity to poliomyelitis caused by types 1, 2 or 3 polioviruses. Each dose contains approximately 800,000 tissue culture immunizing doses ($TCID_{50}$) of type 1 attenuated live poliovirus, 100,000 of type 2, and 500,000 of type 3 (Sabin strains).

ORIMUNE Poliovirus Vaccine, live, oral TRIVALENT is available in two forms—a 2-drop dose and a 2-cc. dose. The 2-cc. dose contains sorbitol, to permit storage at the required low temperature without freezing. Both forms contain trace amounts (less than 5 mcgm. per 2-drop dose and less than 3 mcgm. per 2-cc. dose) of streptomycin, neomycin and nystatin, added to insure freedom from bacterial or fungal contamination. It contains no penicillin.

ADMINISTRATION — ORIMUNE Poliovirus Vaccine, live, oral TRIVALENT is to be administered orally only under the supervision of a physician. IT SHOULD NOT, UNDER ANY CIRCUMSTANCES, BE ADMINISTERED PARENTERALLY.

If the vaccine has become frozen in storage, let it stand until thawed, then agitate bottle or vial vigorously prior to use, to insure homogeneity of contents.

2-cc. dose—May be dispensed in a teaspoon or a cup. A small quantity of distilled water, or of tap water known to contain no free chlorine, may be added if given from a cup. Infants and small children may be fed the 2-cc. dose from a graduated dropper.

2-drop dose — May be dispensed directly from the dropper supplied with the vial, or from a teaspoon. Alternatively, the 2-drop dose may be administered mixed with liquids, such as distilled water, tap water known to contain no free chlorine, simple syrup U.S.P. or milk; or adsorbed on a small portion of bread or cake; or on a cube of sugar.

If the 2-drop dose is given directly from the dropper supplied with the vial, care should be taken to avoid contamination of the dropper by direct contact with the lips, teeth or tongue.

The color of the 2-drop dose may change when exposed to air or even, on occasion, before the bottle has been opened. Such color change may be disregarded.

DOSAGE—*Initial Vaccination*—Two doses given eight weeks apart. It is recommended that vaccination of infants not be commenced until the sixth week of life.

The two-dose, eight weeks schedule should be followed also when the history of previous vaccination is unknown or in doubt, or when there is reason to suppose that previous vaccination may not have conferred sufficient immunity.

Revaccination—Infants in whom initial vaccination was commenced between the sixth week and the end of the sixth month of life should receive a second course of two doses, eight weeks apart, commencing at 10-12 months.

This trivalent vaccine may be used as a substitute for the remaining portion of an incomplete series of other vaccines, only by administering the two doses, spaced eight weeks apart, recommended for initial vaccination.

The need for periodic booster doses of trivalent vaccine following completion of the 2-dose initial series, except in the case of infants at 10-12 months (see above) has not been definitely established. The current recommendation is that single booster doses of trivalent vaccine be considered at intervals of two to four years; or sooner, if an epidemic impends.

CONTRAINDICATIONS AND PRECAUTIONS — There are no known contraindications. Circumstances or conditions which, in the past, have been regarded as contraindications to polio vaccination, or as reasons for exercising special precautions, do not apply to ORIMUNE Poliovirus Vaccine, live, oral, TRIVALENT. These include: recent tonsillectomy, tooth extraction or vaccination against other diseases (diphtheria, tetanus, pertussis, smallpox, etc.); pregnancy, agammaglobulinemia, penicillin hypersensitivity, therapy with steroids. Known previous infection with poliovirus, or exposure to such infection, is not a contraindication.

Intercurrent infection with enteroviruses may impede the proliferation of the vaccine virus in the intestinal tract. If there is persistent vomiting or diarrhea suggesting such infection, it may be advisable to postpone vaccination.

The vaccine will not be effective in cases of incubating or established poliovirus infection. If there is reason to suspect such infection, it is suggested that vaccination be postponed.

STORAGE—The vaccine must be stored in the freezer compartment of a refrigerator. Because of its sorbitol content, the 2-cc. dose vaccine will remain fluid at temperatures above −14° C. (+7° F.).

Once the temperature of the vaccine rises above 0° C., or once the vial or bottle has been opened, it must be used within seven days. During this period it must be stored at a temperature no higher than 10° C. (50° F.).   ®

ORIMUNE

POLIOVIRUS VACCINE, LIVE, ORAL, TRIVALENT

LEDERLE LABORATORIES, A Division of AMERICAN CYANAMID COMPANY, Pearl River, New York

6666-3

A Chronology of Vaccination

1965

The bifurcated needle, invented by American microbiologist Dr. Benjamin Rubin while working at Wyeth, is licensed for the administration of smallpox vaccine.

1965 Doctors in the USA and Europe observe a new and severe measles-like disease in children previously vaccinated with killed measles vaccine, and subsequently exposed to measles infection or revaccination with live measles vaccine. Symptoms of what is called "atypical measles" include rash, swelling, prolonged fever, pneumonia, and the build up of fluid around the lungs.

1965 Parkman and Meyer launch their first human trial of HPV-77 rubella vaccine at the Arkansas Children's Colony, a state school for the intellectually disabled.

1965 ✪ Lirugen® single-shot measles vaccine, developed by Pitman-Moore (a division of Dow), is licensed in the USA for use on nine- to twelve-month-old infants. The PM measles strain is based on the Schwarz strain (created by researcher Anton Schwarz at American Home Products) derived from the original Edmonston strain. The manufacturer claims this vaccine will catch on quickly as it causes fewer reactions and does not require two shots like Rubeovax®—one shot of gamma globulin in one arm and the Rubeovax® vaccine in the other. In Canada, Lirugen® is given to all 9-month-old infants from February 1966 to September 1970, and administered to all one-year-old children from September 1970 to October 1972.

1965 Anatoxal Di Te Per® (Merthiolate-containing diphtheria, tetanus, whole-cell pertussis vaccine) by Berna is advertised in the *Revue Médicale de la Suisse Romande.*

1965 The International Agency for Research on Cancer (IARC) is founded in Lyon, France, as the specialized cancer agency of the WHO.

1965 ✪ The first live viral measles vaccines are introduced in the UK; Wellcovax® by Burroughs Wellcome and Mevilin-L® by Glaxo Laboratories. The first killed formalin-inactivated measles vaccine, Mesavac® by Pfizer, is also launched soon after.

1965 The first human coronavirus (B814) is isolated by British virologist David Tyrell and Malcolm M. Bynoe. Up until 2002, coronaviruses are considered to be the causative agent of the common cold.

1966

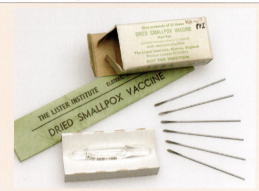

The WHO sets up the Smallpox Eradication Campaign over 100 years after smallpox vaccination began. The Lister Elstree strain of smallpox vaccine is mainly used. American medical doctor and epidemiologist at the CDC, Donald A. Henderson, directs the WHO's 10-year international intensified effort from 1967–1977, resulting in the declaration that smallpox is eradicated and launching the international childhood vaccination programs. He would later become Dean of the Johns Hopkins School of Public Health (1977–1990) and play a leading role in initiating federal biosecurity programs for public health preparedness and response following biological attacks.

1966 A landmark paper published in *The New England Journal of Medicine* exposes the scores of experiments and harm being done by top physicians and researchers operating with the support of the US government. This prompts the requirement of full informed consent for studies funded by US agencies.

1966 *Stromstodt v. Parke-Davis Company*—US$ 500,000 is awarded in damages to a vaccine-injured boy who received Quadrigen® (DTP-IPV) at 6 months old. A few hours after his injection, the infant suffered a series of convulsions that left him permanently brain damaged. The manufacturer is judged as having been negligent in failing "to test the drug adequately and to give adequate warning of the dangers inherent in its use." Several reactions were reported during the trials, yet "no effort was made to determine the cause of the high incidence of reactions."

1966 American pathologist at the Rockefeller University Peyton Rous wins the Nobel Prize in Physiology or Medicine "for his discovery of tumour-inducing viruses."

1966 US Congress signs into law the Laboratory Animal Welfare Act, being the first federal law to implement standards for the maintenance of some animals species. It fails to address the issue of lab research proccdurcs, however major revisions are made to the Act in 1970, 1976 and 1985.

1966 Serum Institute of India (SII) is founded by Cyrus S. Poonawalla and begins production of its anti-tetanus serum using horses.

1966 Initiated by J.P. Jacobs, the cell line MRC-5 (Medical Research Council, cell strain 5) is derived from the human lung tissue of a 14-week-old male fetus aborted from a 27-year-old woman. MRC-5 cell line is commonly utilized in vaccine development, as a transfection host in virology research and for in vitro cytotoxicity testing.

1966 The US government drops light bulbs filled with *Bacillus subtilis* variant niger into New York City subway stations during rush hour to see how far the bacili would spread. The army would acknowledge having conducted bacteriological tests on 239 populated areas in the US between 1949 and 1969 as part of a large-scale, secret program of germ warfare research and development.

1966 The first "internationally focused" and WHO-sponsored measles vaccine programs take place in Africa.

1966 An outbreak of 73 cases of smallpox *variola minor* occurs in Birmingham, UK. Patient zero is traced as being a 23-year-old photographer at the University of Birmingham Medical School, in the very building where a virology department are culturing smallpox viruses at the time.

1966 ✪ Sabin's live oral polio vaccine is introduced in Australia.

1966 The rubella virus is attenuated by Paul Parkman and American pediatric virologist Dr. Harry Martin Meyer Jr., leading to the production of a rubella vaccine and the first test to measure rubella immunity.

1966 ✪ In France, Pasteur commercializes adjuvant-free tetanus toxoid vaccines.

1966–1967 An intensive measles vaccine campaign is launched nationwide in the USA.

1966–1967 Projects MKSEARCH, MKOFTEN/CHICKWIT, Bluebird, and Artichoke are initiated by the CIA as a series of drug experimentation and mind control research programs.

1967–1977 Half of all US commercial vaccine producers stop making and distributing vaccines. During the late 1970s and early 1980s, the exodus from the vaccine business continues; Glaxo, Wellcome, Dow Chemical, Merrell-National Laboratories, Pfizer, and Eli Lilly are among the companies that discontinue their vaccine operations or sell off their vaccine components altogether—the primary reason being the high expense of liability exposure.

1967 A group of workers involved in polio vaccine development at a Behringwerke's Marburg Laboratory in Germany, at the Paul Ehrlich Institute in Frankfurt, and at the Institute of Virology, Vaccines and Sera in Belgrade, former-Yugoslavia/Serbia, suffer from various symptoms such as fever, diarrhea, vomiting and internal bleeding of organs—seven workers die. This is the first outbreak of Marburg hemorrhagic fever (same virus family as Ebola) most likely imported by African green monkeys from Uganda used for medical experimental lab research.

1967 The AMA is reported to receive 43% of its total income (US$ 13.6 million) from pharmaceutical advertisements.

A Chronology of Vaccination

1967 Japanese immunologist Kimishige Ishizaka, with his wife Teruko, discover the antibody class Immunoglobulin E (IgE).

1967 ✪ The first live mumps vaccine is developed by Maurice Hilleman at Merck, based on the Jeryl Lynn (his daughter) strain and licensed in the USA as Mumpsvax®. The vaccine is available in Europe and Canada soon after.

1967 ✪ The FDA licenses Eli Lilly's fractionated cell vaccine against pertussis, diphtheria, and tetanus, Tri-Solgen®, though it has been marketed since 1962.

1967 Fondation Mérieux is created in France as an independent family foundation, recognized for public utility, by pharmaceutical mogul Charles Mérieux. Their mission is to combat infectious diseases and improve access to healthcare in developing countries, notably through the use of vaccination.

1967 The first human heart transplant is performed by South African surgeon, Dr. Christiaan Barnard. The patient dies 18 days later due to septicemia and severe pneumonia.

1967 The first cases of multiple sclerosis following vaccination are reported as nine case studies and published in *The British Medical Journal*.

1968 ✪ Moraten-strain (further attenuated Edmonston strain) live-attenuated single-dose measles vaccine Attenuvax® by Merck, is brought to market in the USA and Switzerland. It will be included as the measles component of Merck's MMR.

1968 ✪ Rabies vaccine is developed by Tadeusz Wiktor, VMD, Dr. Hilary Koprowski, MD, and Mario V. Fernandes, DVM, at the Wistar Institute, causing fewer side effects than Pasteur's rabies vaccine. This vaccine is to become widely used in the USA and western Europe from the 1980s.

1968 ✪ Yugoslavia is the first country to license WI-38 propagated vaccines against polio and measles.

1968 The WHO Adverse Reaction Terminology (ART) is developed to describe side effects of medical procedures, drugs, and vaccines.

1968 American biochemists Robert W. Holley and Marshall W. Nirenberg, and Indian-American biochemist Har Gobind Khorana, win the Nobel Prize in Physiology or Medicine "for their interpretation of the genetic code and its function in protein synthesis."

1968 American physician and "godfather of vaccinology," Dr. Stanley A. Plotkin of The Wistar Institute (formerly of the Epidemic Intelligence Service), sends a patent request for his novel method of rubella vaccine development using WI-38 cells, derived from an aborted female fetus. Dr. Plotkin becomes an expert in the field of vaccinology and leading industry consultant, as well as the medical director for Pasteur Mérieux Connaught.

1968 Anti-Rh Antibody RhoGAM® or anti-Rh immunoglobulin is first licensed in the USA, developed by Dr. Vincent Freda and his team at Columbia University, and sold by Ortho Clinical Diagnostics, Inc. It contains 10.5 mcg of ethylmercury from thimerosal (until 2001), and is provided to all Rh-negative women during pregnancy and/or after delivery. In 1971, BayRho-D® is launched by Bayer with 35 mcg of ethylmercury, which is removed in 1996.

1968 The World Health Assembly adjusts its smallpox eradication strategy to focus on "surveillance and containment" with "ring vaccination" instead of "mass vaccination."

1968 While the last smallpox death in the US is reported in 1948, renowned American pediatrician C. Henry Kempe writes, "there have probably been 200 to 300 deaths from smallpox vaccination" in the last 20 years.

1968 The Hong Kong (A)H3N2 influenza pandemic kills an estimated 1 million people worldwide, with 100,000 deaths in the USA alone. Pandemic monovalent vaccines are launched at the end of the year in the US.

1969 ✪ Live-attenuated measles vaccine is introduced in Australia—the inactivated version is never licensed.

1969 ✪ An aluminium hydroxide adjuvanted DTP vaccine Trivax AD® by Burroughs Wellcome is commercialized in the UK.

1969 The UK Public Health Laboratory Service reports that the pertussis component of the triple DTP vaccine is "not very effective" in that it fails to control outbreaks or to protect fully vaccinated children.

1969 ✪ DTP vaccine is added to the mandatory childhood vaccination schedule in Japan.

1969 American biochemist and physician Dr. Baruch Blumberg independently discovers the hepatitis B virus. Together with American microbiologist Dr. Irving Millman submits the first patent for a hepatitis vaccine, that becomes available for general distribution in the 1980s. Blumberg is awarded the Nobel Prize in Medicine in 1976 for "discoveries concerning new mechanisms for the origin and dissemination of infectious diseases."

1969 ✪ Rubella live-attenuated vaccines, developed by Wistar Institute's Dr. Stanley A. Plotkin and Dr. Hilary Koprowski, are licensed in the USA, using HPV-77 cells grown on dog kidney cells and duck embryo culture. In total, three rubella virus vaccines are licensed in the US: HPV-77 strain (Rubelogen® by Philips Roxane/Parke-Davis); HPV-77 grown in duck-embryo culture (Meruvax® by Merck); and Cendehill strain grown in rabbit-kidney culture (Cendevax® by RIT-SKF, and Lirubel® and Lirutrin® by Dow Chemical). Rubella vaccine is also registered for use in Australia.

1969 The WHO revises the International Sanitary Regulations (ISR) and renames them the International Health Regulations (IHR). Cholera, smallpox, plague, yellow fever, relapsing fever, and typhoid are listed as diseases for which surveillance, quarantine, and port regulations are established. These IHR are revised in 1973, 1981, and in 2005.

1969 *Tinnerholm v. Parke-Davis Company*— US$651,783.52 in damages is awarded to parents of a three-month-old child, who was vaccinated in 1959 with Quadrigen® (DTP-polio) and subsequently brain damaged.

1969 Killed measles vaccine is withdrawn from the UK market following reports of death and brain damage post-vaccination. Wellcovax® is also taken off the market following cases of post-vaccinal encephalitis.

1969 German–American biophysicist Max Delbrück, American bacteriologist Alfred D. Hershey, and Italian microbiologist Salvador E. Luria are awarded the Nobel Prize in Physiology or Medicine "for their discoveries concerning the replication mechanism and the genetic structure of viruses."

1969 The subacute sclerosing panencephalitis (SSPE) national registry in the USA is set up at the University of Tennessee. It is the world's first disease registry on SSPE—a progressive and fatal illness of the nervous system—seen after both natural measles infection and measles vaccination.

1969 President Richard Nixon renounces all methods of offensive biological warfare and orders the destruction of existing stocks of biological agents. The US biological warfare program is confined strictly to research on defense measures such as vaccination. By this time, the military has already weaponized anthrax (Bacilus anthracis), Q fever (Coxiella burnetii), Venezuelan equine encephalitis virus, Brucellosis (Brucella suis) and tularemia (Francisella tularensis), among at least 20 other agents.

1969 By this date, the Pentagon is spending US$300 million per year on biological and chemical weapons, and delivery systems. It is reported that scientists at Fort Detrick are experimenting with "over 160 biological agents, including tularemia (rabbit fever), Q-fever (coxiella burnetii), viral equine encephalitis, pneumonic plague, pulmonary anthrax, biologically engineered anthrax, as well as virulent fungi, rusts, blights and rots weaponized to destroy food crops."

1970–1979

1970s ✪ A heat-inactivated herpes simplex virus vaccine, by German company Hermal-Chemie, called Lupidon (Lupidon H for HSV-1 and Lupidon G for HSV-2) is marketed in Europe until the early 2000s, and in Switzerland until 2010, for therapeutic and/or prophylactic use. As a treatment for herpes, 12 weekly injections over 1–3 months are required, and then 8 fortnightly injections in months 4–5.

A Chronology of Vaccination

1970–1972

1970s	By the beginning of the decade, there are only four DTP vaccine manufacturers in the USA: Eli Lilly, Wyeth, Lederle, and Connaught.
1970s	Natural measles infection is observed to reduce tumors in children with cancer.
1970s	George H.W. Bush sits on the Eli Lilly board of directors.
1970	✪ Cendehill strain rubella vaccine Cendevax®, grown in rabbit kidney cells, is licensed by Smith, Kline & French in the UK.
1970	The Environmental Protection Agency (EPA) is founded in the USA to set safety standards and to bring cohesion to the expanding federal environmental programs, notably pesticide registration and regulation.
1970	✪ Almevax® rubella vaccine, RA27/3 grown in WI-38 human fetal cells, is licensed by Burroughs Wellcome (now Evans Medical-GSK) in the UK. It will be discontinued in 1997.

✪ The first subunit protein-based anthrax vaccine, BioThrax™ (also known as AVA or Anthrax Vaccine Adsorbed) is approved in the US for the prevention of cutaneous anthrax, but not as a treatment for inhalation anthrax (post-exposure prophylaxis). It is the only vaccine licensed to prevent anthrax and contains aluminum. Initially produced by the Michigan Department of Public Health through its Michigan Biologics Products Institute, vaccine production will be taken over by BioPort Corporation in 1998—renamed Emergent BioSolutions in 2004.

1970	In the US Army's journal *Military Review*, the November article titled "Ethnic Weapons" is the first to publicly mention the possibility of using genetic information to target specific groups with chemical or biological weapons.
1970	✪ Adsorbed toxoid tetanus vaccine, Te Anatoxal Berna®, by Swiss Serum and Vaccine Institute Berne, is licensed in the US.

1970	NATO sponsors an *International Symposium on Uptake of Informative Molecules by Living Cells* in Mol, Belgium. Dr. Robert Gallo presents his work on the entry of nucleic acids into cells.
1970	The Department of Defense appropriates US$10 million over five years for "mutant" virus research to develop a "synthetic biological agent" from which there could be no natural immunity. One important defense contracter is Litton Bionetics, employer of American virologist Robert Gallo, future co-discoverer of the retrovirus HIV-1.
1970	✪ Combined live rubella-mumps vaccine Biavax® by Merck, is licensed in the USA.
1971	The European Management Forum—renamed the World Economic Forum in 1987—is founded in Switzerland by Henry Kissinger's protégé and German economist, Klaus Schwab.
1971	MSF, or Doctors Without Borders, is founded as a non-governmental organization by a small group of French doctors and journalists who seek to expand accessibility to medical care across national boundaries, irrespective of race, religion, creed or political affiliation.
1971	Routine smallpox vaccination for children is discontinued in the UK. The CDC recommends discontinuation of routine smallpox vaccination in the USA.
1971	American professor and virologist David Baltimore devises a system of classification for viruses based on symmetry and form, bringing order to the unfathomable magnitude and diversity of viral organisms.
1971	The National Cancer Act "The War on Cancer" is signed into US law by President Nixon. US$500 million is allocated to the National Cancer Institute, increasing to US$1 billion five years later. The Fort Detrick biological weapons facilities are converted into the world's foremost virus and cancer research laboratory.
1971	Litton Bionetics receives the largest single contract to date—US$6.8 million is awarded by the NIH for research on viruses and cancer.

1971 ✪ Combined MMR-I® measles, mumps, rubella live-attenuated vaccine is launched by Merck in the USA. It is a triple viral vaccine that combines the vaccine products Attenuvax® (measles), Meruvax® (rubella) and Mumpsvax®. No prelicensure clinical trials are done of the combined version.

1971 The Wildlife Preservation Trust International is founded. It is renamed The Wildlife Trust in 1999 and in 2010, EcoHealth Alliance.

1971 ✪ Serum Institute of India (SII) markets its first tetanus toxoid (TT) vaccine for active immunization, overcoming the shortage of this vaccine in India. SII begins production of DTP vaccine two years later.

1971 ✪ In Australia, rubella vaccine is introduced.

1971 A smallpox outbreak occurs on the Aral Sea in the USSR, as a result of an open-air field test at a Soviet biological weapons facility.

1972 ✪ The single live-attenuated measles vaccine Mevelin-L (Schwarz strain) by Evans Biologicals is licensed in the UK.

1972 The Special Programme of Research, Development and Research Training in Human Reproduction is established by the WHO with the World Bank, the Rockefeller Foundation, UNFPA and UNDP. Through its Task Force on Vaccines for Fertility Regulation, research is directed to developing "safe, effective and acceptable birth control vaccines" largely based on human chorionic gonadotropin (hCG). By 1980—in collaboration with the Population Council in New York and the National Institute of Immunology in New Delhi, India—the Task Force will produce a complex but potent prototype anti-hCG vaccine formulation consisting of the 109–145 CTP-beta-hCG peptide conjugated to diphtheria toxoid, mixed with a MDP (muramyl dipeptide) adjuvant, and suspended in a water-in-oil emulsion. Human phase I trials begin in 1986, and later tetanus toxoid is used as a carrier protein in the vaccine formulation.

1972 Agricultural use of DDT is banned in the US.

1972 The first recognized laboratory smallpox virus escape occurs with the infection of a laboratory assistant at the London School of Hygiene & Tropical Medicine.

1972 The UK National Institute for Biological Standards and Control is established as an agency responsible for biological standardization and independent regulatory testing of biologics, including vaccines.

1972 The first experiments on monkeys to develop a dental caries vaccine—targeting *Streptococcus mutans*—take place in the UK, led by British scientist Dr. William Bowen.

1972 ✪ The FDA approves Pfizer's live oral polio vaccine Diplovax® using Hayflick's WI-38 cells. It is withdrawn from the market in 1976 due to "supply shortages and a campaign by Lederle to sow distrust in the vaccine."

1972 The Biological and Toxin Weapons Convention is agreed in Geneva, Switzerland, prohibiting defensive biological warfare programs and the stockpiling of bacteriological and toxin weapons. Entered into force in 1975, the convention is signed by 140 nations and ratified only by 22 countries, however it contains no provisions for the inspection of biological weapon facilities. Another loophole is that the convention explicitly permits working with biological agents and toxins that may be used as warfare agents, but that are justified "for prophylactic, protective or other peaceful purposes."

1972 US biochemists Herbert Boyer and Stanley Cohen develop recombinant DNA (rDNA) in the *Escherichia coli (E. coli)* bacterium, ushering in the era of modern biotechnology. A pioneer in gene splicing and genetic modification using plasmids, Herbert Boyer establishes the company Genentech in 1976 to apply this groundbreaking science to the development of medical products.

1972 American biochemist Paul Berg inserts SV40 into *E. coli* for transfection into mammalian cells.

1972 American biologist Gerald M. Edelman and British biochemist Rodney R. Porter are awarded the Nobel Prize in Physiology or Medicine "for their discoveries concerning the chemical structure of antibodies."

1972 Eli Lilly & Company find thimerosal to be "toxic to tissue cells" in concentrations as low as one part per million (PPM), 100 times more diluted than in a typical vaccine.

A Chronology of Vaccination

1972 The Consumer Product Safety Act (CPSA) is enacted by the 92nd US Congress, creating the Consumer Product Safety Commission (CPSC) as an independent federal regulatory agency. The CPSC is "tasked with identifying consumer products that pose an unreasonable risk of injury and creating standards to remove or lessen that risk." Senate hearing minutes from the second session reveal the details of the 32 "worthless vaccines" manufactured by Eli Lilly & Co., Merck Sharp & Dohme, and Parke-Davis.

1972 The Division of Biologic Standards is transferred from the NIH to the FDA and renamed the Bureau of Biologics. The DBS establishes and maintains standards of quality and safety of all biologics, including serums, vaccines, antitoxins, and blood products, that come under the jurisdiction of the US Public Health Service. In the early days, some 180,000 mice, 3,500 rabbits, 2,000 monkeys, and 8,600 guinea pigs are used annually for control tests.

1972 In March, the US Government Accounting Office releases a report calling out the DBS for allowing 32 ineffective vaccines to remain on the market "for at least 10 years while drug regulators quietly exchanged memos." The DBS also released flu vaccines on the market even though its own potency tests showed less than 1% of standards. The GAO report inspires the FDA to propose a "sweeping review of the effectiveness, safety and labeling of all 1,100 licensed vaccines and biological products." Eight years later in the published review, it is deemed that none of the products assessed were safe or effective.

1972 ✪ Beecham launches Pollinex® hay fever vaccine in the UK, with grass pollen allergens.

1972 Merrell-National develops a safer acellular pertussis vaccine that passes the FDA potency and toxicity tests. Due to low yields in production, the company decides not to pursue commercialization, and later sells the vaccine to Connaught Laboratories.

1972 Dr. Stanley A. Plotkin's patent on his rubella vaccine is granted by the US Patent and Trademark Office, with all proceeds flowing to The Wistar Institute who in turn direct 15% to Plotkin as the inventor of the vaccine. This is despite the fact that research for the vaccine is funded by the NIH, who issue a waiver allowing the research to be patented—like with rabies vaccine.

1972 Smallpox virus escapes from a lab in the London School of Hygiene & Tropical Medicine, causing the death of two people. Guidelines for laboratories handling dangerous pathogens are established as a result and are the direct precursor to current BSL protocols.

1972 In the USA, there are five manufacturers of DTwP vaccine: Lederle, Wyeth, Parke-Davis, Pitman-Moore, and National Drug Co. Eli Lilly is the only maker of a "safer" split-cell DTaP.

1973 The UK Association of Parents of Vaccine Damaged Children is set up.

1973 Monsanto launches the genetically engineered and patented glyphosphate pesticide Roundup® for use on soybean crops.

1973 The American Academy of Pediatrics begins to release public service announcements on the need for complete childhood vaccination.

1973 The WHO begins to issue recommendations on the composition of annual influenza vaccines.

1973 The US Federal Register features the first regulations on "General Biological Products Standards." This includes section 610.15(a), constituent materials: "In no event shall the recommended individual dose of a biological product contain more than 0.85 mg of aluminum, determined by assay, or more than 1.14 mg of aluminum, determined by calculation on the basis of the amount of aluminum compound added." This amount was selected empirically from data that demonstrated it enhanced the effectiveness of the vaccine, and was not based on safety.

1973 The WHO removes cholera vaccination from international travel requirements.

1973 ✪ Live M-M-vax® by Merck is licensed in the US against measles and mumps.

1973 ✪ In Egypt, tetanus toxoid vaccination is introduced for pregnant women.

1973 The CIA orders the destruction of documents on its MKULTRA program, along with the records of human experimentation.

1974 ✪ In the US, Dow Chemical is granted a license, until 1978, to manufacture a live mumps vaccine.

1974 ✪ In France, the Pasteur Institute launches IPAD® DT, a reduced proteinic diphtheria-tetanus vaccine formula adsorbed on calcium phosphate. IPAD stands for "Institut Pasteur ADsorbed on calcium phosphate."

1974 The WHO's Special Program of Research, Development and Research Training in Human Reproduction, convenes an international symposium on immunological approaches to fertility control in partnership with the Karolinska Institutet in Stockholm, Sweden.

1974 ✪ MMR® by MSD is first licensed in Switzerland and advertised in government-issued public health bulletins.

1974 The World Health Assembly of the WHO establishes the Expanded Program on Immunization (EPI) in response to low vaccination levels in developing countries. The diseases targeted are tuberculosis, polio, diphtheria, tetanus, pertussis, and measles.

1974 ✪ Oka strain of varicella-zoster virus (VZV) is discovered by Dr. M. Takahashi and a live attenuated chickenpox vaccine is developed by him and his team at Osaka University in Japan, licensed in 1976. In 1981, Merck will license the Oka strain for its vaccine.

1974 The National Research Act is passed in the USA in an effort to create regulations that protect subjects in human research trials. It implements the creation of an ethics task force, and the National Commission for the Protection of Human Subjects of Biomedical and Behavioral Research. The commission publishes the Belmont Report in 1979—a comprehensive guideline of basic ethical principles for modern clinical trials and scientific research.

1974 The US National Security Council completes a top-secret document *The National Security Study Memorandum* or NSSM-200—also named the Kissinger Report after Henry Kissinger, the US Secretary of State. Declassified in 1990, this document tackles the subject of population growth, national security, and foreign policy. Published shortly after the first major international population conference in Bucharest, Romania, NSSM-200 is the result of a collaboration among the CIA, USAID, and the US Departments of State, Defense, and Agriculture.

1974 ✪ The first meningococcal polysaccharide monovalent group C vaccine is licensed in the USA, manufactured by Merck.

1974 In the US case of *Reyes v. Wyeth Laboratories,* Judge John M. Wisdom suggests that compensation for unavoidable vaccine-related injuries should be borne primarily by price increases passed to the public, without the possibility of finding negligence on the part of the manufacturers. The court concedes that vaccines are "unavoidably unsafe."

1974 Merck's patented peanut-oil-based Adjuvant 65 is licensed for general use in the UK. Having failed to obtain approval in the US, Maurice Hilleman develops several other patented formulations of the same adjuvant to satisfy government requirements. FDA policy is to "approve adjuvants as they appear within a complete vaccine formula. Vaccine formulae are approved and patented with peanut-oil adjuvants, but not with Adjuvant 65 as a stand-alone product."

1974 The first transgenic mouse is created by the Massachusetts Institute of Technology (MIT) Professor of Biology, Rudolf Jaenish, by introducing foreign DNA into its embryo.

1974 Eli Lilly stops manufacturing thimerosal, but maintains the patent and licensing rights.

1974 Belgian-American cell biologist Albert Claude, Belgian biochemist Christian de Duve, and Romanian-American cell biologist George E. Palade win the Nobel Prize in Physiology or Medicine "for their discoveries concerning the structural and functional organization of the cell."

A Chronology of Vaccination

1974 Wolfgang Ehrengut, the influential director of the State Vaccine Institute of Hamburg in Germany, suggests that around one in 20,000 vaccinated children may suffer from neurological damage after the whole-cell pertussis vaccine.

1975 In Switzerland, the Federal Office for Public Hygiene changes its name to become the Federal Office for Public Health.

1975 ✪ Beecham launches the allergy vaccine Pollinex® R (Ragweed) in Canada, to prevent or reduce allergic hay fever symptoms caused by exposure to ragweed pollen.

1975 Lyme disease is first recognized in the USA when a cluster of 51 juvenile rheumatoid arthritis cases occur in Lyme, Connecticut. The syndrome is named Lyme disease in 1979, with symptoms including neurological problems and severe fatigue.

1975 Some 140 scientists involved in the field of molecular biology representing 17 countries—including Stanford Professor of Genetics Joshua Lederberg—meet at the first Asilomar Conference on Recombinant DNA in California. They discuss the potential biohazards and the regulation of biotechnology and genetic engineering.

1975 American virologist David Baltimore, Italian–American virologist Renato Dulbecco, and American geneticist Howard M. Temin win the Nobel Prize in Physiology or Medicine "for their discoveries concerning the interaction between tumour viruses and the genetic material of the cell."

1975 The US ratifies the Geneva Protocol of 1925, the Biological Weapons Treaty of 1972, and completes the destruction of all biological weapons. The Pentagon transfers the virus section of Fort Detrick's Center for Biological Warfare Research laboratory to the National Institutes of Health (NIH) and the National Cancer Institute (NCI). It is documented that the bioweapons program remains intact and continues covertly under the "Special Virus Cancer Program" run by the NCI since 1964.

1975 The UK National Biological Standards Board is formed at the National Institute for Medical Research (in 2016 renamed the Francis Crick Institute) in Mill Hill, UK.

1976 The Swiss Federal Office for Public Health officially recommends the measles vaccine for all babies at 12 months of age.

1976 The CDC first produces "Vaccine Information Statements" which are given to patients and describe the risks and benefits of vaccines.

1975–1976 Eli Lilly & Company gets out of the business of making vaccines, ceasing production of fourteen vaccines, selling some off to Wyeth.

1976 ✪ In the Netherlands, a mass measles vaccination campaign is launched using Attenuvax® by Merck Sharp & Dohme (MSD).

1976 ✪ Following a small flu outbreak among army troops at Fort Dix, Congress passes the Public Law 94–266 which provides US$135 million of taxpayers' funds to pay for a national swine flu inoculation campaign. Gerald Ford appeals to the US population via CBS TV station to get vaccinated. Vaccine makers negotiate immunity from liability for their never field-tested experimental H1N1 swine flu vaccines, which are primarily supplied by Merrell-National and Merck. Over 40 million Americans are vaccinated resulting in 538 cases of paralysis and 58 deaths. The vaccine campaign is halted nationwide after 12 reported deaths. US taxpayers foot the US$220 million payout for injury compensation, notably neurological issues.

1976 The NIH releases the first ever guidelines for the safe laboratory use and development of recombinant DNA research, published on July 7 in the *Federal Register*.

1976 Biotechnology becomes commercialized, allowing private companies to experiment with inserting genes from one species to another for medicinal or food purposes.

1976 ✪ The first TBE vaccine by Immuno AG (Austria) is licensed in Europe. It is formalin-inactivated with an aluminum hydroxide adjuvant and will be replaced with a "highly purified" version called FSME-Immun® in 1980.

1976 American physician and geneticist Baruch S. Blumberg and American physician Daniel Carleton Gajdusek—known for his discovery of prion disease—are awarded the Nobel Prize in Physiology or Medicine "for their discoveries concerning new mechanisms for the origin and dissemination of infectious diseases."

1976 In the USA, the age recommended for MMR vaccine is raised from 12 to 15 months.

1976 Ebola virus is isolated in Zaire (now the Democratic Republic of the Congo), named after the Ebola River. Outbreaks occur in Sudan—with 284 reported cases and 151 fatalities allegedly at a cotton factory—and in Zaire with 318 cases and 280 deaths.

1976 An investigation by the US Government Accountability Office uncovers the unethical forced sterilization of over 25% of Native Americans in the early 1970s.

1976 ✪ WHO introduces a conjugate tetanus toxoid with human chorionic gonadotropin (hCG), producing a birth control vaccine.

1977 Dow Chemical (Pitman-Moore) gets out of the animal and human vaccine business and ceases the production of twelve vaccines.

1977–1978 The re-emergence of the H1N1 strain (found to be closely related to viruses circulating in 1949-1950) triggers the Russian Flu influenza outbreak. It is determined that this emergence may likely be the result of a laboratory leak (in Russia or China) while working on an experimental H1N1 influenza vaccine.

1977 The Royal Commission on Civil Liability, in vaccine injury related correspondence to the UK Ministry of Health, describes "screaming to fits" as a sign of brain damage.

1977 The CDC launches a nationwide MMR vaccination campaign in the USA, coinciding with the beginning of an initiative to increase the number of vaccinated children in the country to 90% within two years.

1977 The antibiotic tetracycline, approved for prescription since 1954 in the USA, is added to the WHO's *List of Essential Medicines.*

1977 It is estimated that about 50–60 million women worldwide have been fitted with contraceptive intrauterine devices (IUDs).

1977–1993 In the USA, over a 15-year period the cumulative increases (in 1993 dollars and excluding excise tax) in the contract and catalog prices for DTP vaccine, are US$1.55 (1,033% increase) and US$5.22 (2,847% increase) respectively.

TABLE 4-7 Vaccine Prices (in dollars) in the United States from 1977–February 1993

Year	DTP		OPV		MMR		Hib-CV	
	CP	FC	CP	FC	CP	FC	CP	FC
1977	0.19	0.15[a]	1.00	0.30	6.01	2.42	NA	NA
1978	0.22	0.15[a]	1.15	0.31	6.16	2.35	NA	NA
1979	0.25	0.15[a]	1.27	0.33	6.81	2.62	NA	NA
1980	0.30	0.15[a]	1.60	0.35	7.24	2.71	NA	NA
1981	0.33	0.15[a]	2.10	0.40	9.32	3.12	NA	NA
1982	0.37	0.15[a]	2.75	0.48	10.44	4.02	NA	NA
1983	0.45	0.42[a]	3.56	0.58	11.30	4.70	NA	NA
1984	0.99	0.65[a]	4.60	0.73	12.08	5.40	NA	NA
1985	2.80	2.21	6.15	0.80	13.53	6.85	NA	NA
1986	11.40	3.01	8.67	1.56	15.15	8.47	NA	NA
1987	8.92	7.69	8.07	1.36	17.88	10.67	NA	NA
1988	11.03	8.46[b]	8.07	1.36	24.11	16.18	13.75	11.00
1989	10.65	7.96	9.45	1.92	24.11	16.18	13.75	6.00
1990	10.65	6.91	9.74	1.92	24.07	14.71	14.55	5.20
1991	9.97	6.25	9.91	2.09	25.29	15.33	14.55	5.16[c]
1992	9.97	6.25	9.91	2.09	25.29	15.30	14.55	5.16[c]
1993	10.04	5.99	10.43	2.16	25.29	15.33	15.13	5.37

NOTE: CP, Catalog Price; FC, Federal contract Price; NA, vaccine not licensed. From 1988 to 1992, prices include federal excise tax for the Vaccine Injury Compensation Program. Excise taxes are set at $4.56 per dose of DTP, $4.44 per dose of MMR, and $0.29 per dose of OPV.
[a] No federal contract. The price represents the average price charged to the states.
[b] Federal contract price was $9.62 for a portion of 1988.
[c] Merck federal contract price was $8.25 for use among Native American populations.
SOURCE: Division of Immunization, Centers for Disease Control and Prevention, 1993.

1977-2000 Lederle's live oral polio vaccine made using African green monkey kidneys is the only available live polio vaccine on the US market.

1977 ✪ The first smallpox vaccine based on the Modified vaccinia Ankara virus (MVA), a highly attenuated strain of vaccinia virus developed in Germany between 1953–1968, is approved in Germany and used until 1980. It is produced by over "500 serial passages of vaccinia virus (from a wild strain discovered by the Turkish vaccine Institute of Ankara) in chicken embryo fibroblasts."

1977 In the March 4 issue of *Science*, Darrell and Jonas Salk warn that "live virus vaccines against influenza or poliomyelitis may in each instance produce the disease it is intended to prevent....the live virus against measles and mumps may produce such side effects as encephalitis (brain damage)."

1977 Following the death of two babies in Japan after their DTP vaccines in 1976, the Japanese government support Dr. Yugi Sato to develop an acellular pertussis vaccine.

1977 The last known case of naturally acquired smallpox infection is recorded in Somalia.

1977 https://nap.nationalacademies.org/catalog/2224/the-childrens-vaccine-initiative-achieving-the-vision

1977   Royal Commission on Civil Liability – Vaccine Damage

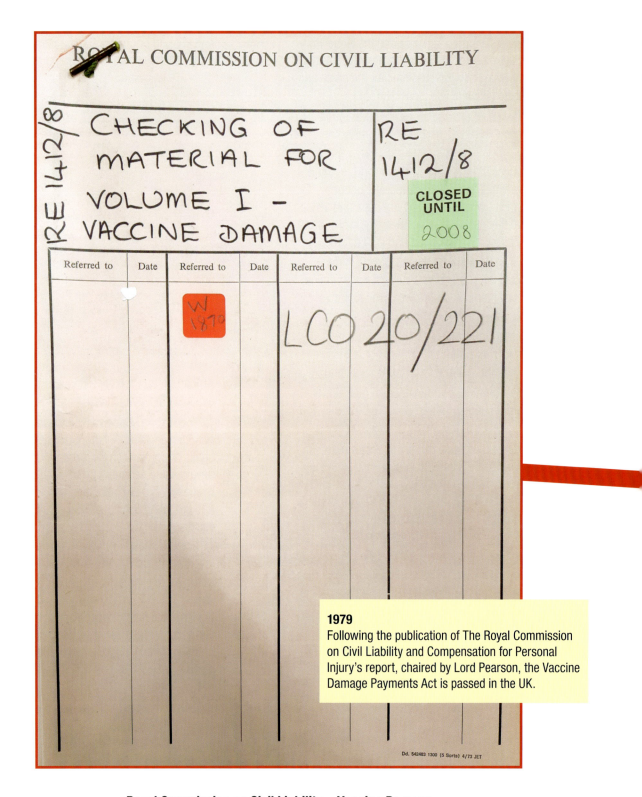

Royal Commission on Civil Liability – Vaccine Damage
1977 – Cover of dossier obtained from the UK National Archives in 2018 – Documents in the public domain

Government documents describe "screaming to fits" as a sign of <u>brain damage!</u>

"The effects of this brain damage include disturbed nights with crying and screaming to fits, loss of consciousness; all forms of physical and mental handicap and, sometimes, death."

This important admission is edited down to instead read:

"This brain damage may lead to forms of physical and mental handicap and, sometimes, death."

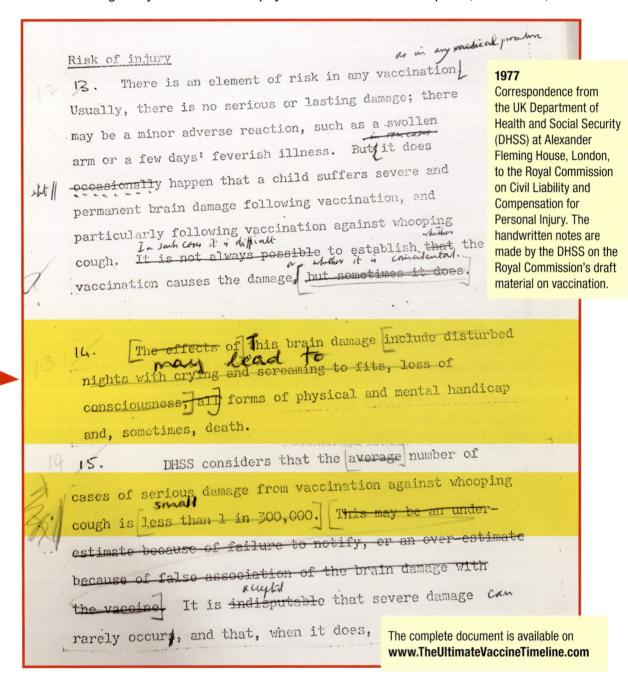

1977
Correspondence from the UK Department of Health and Social Security (DHSS) at Alexander Fleming House, London, to the Royal Commission on Civil Liability and Compensation for Personal Injury. The handwritten notes are made by the DHSS on the Royal Commission's draft material on vaccination.

The complete document is available on **www.TheUltimateVaccineTimeline.com**

UK Department of Health and Social Security correspondence with Royal Commission on Civil Liability

A Chronology of Vaccination

1977 PATH is founded in Seattle as the Program for the Introduction and Adaptation of Contraceptive Technology with a focus on family planning. PATH soon widens its scope to work on global health issues like health technology, maternal and child health, reproductive health, vaccines, and emerging and epidemic diseases. In 1981, it is renamed Program for Appropriate Technology in Health.

1977 ✪ A capsular polysaccharide pneumococcal vaccine (PPV) Pneumovax® is licensed in the USA, for people 50 years and older, and children two years and older with certain underlying health conditions. This phenol-preserved 14-serotype vaccine is the result of the collaboration between Merck and Dr. Robert Austrian of the University of Pennsylvania, whose vaccine research had been supported by NIAID since 1967.

1977 In alignment with the WHO's theme of the year "Immunize and Protect All Children," US President Jimmy Carter announces the Childhood Immunization Initiative to increase vaccination rates from 65% to 90% for seven "vaccine-preventable" diseases.

1977 According to an internal document, scientists at Wyeth analyze the Eli Lilly fractionated DTP vaccine formula (that they had bought the rights to after Lilly left the vaccine market) and find that the purification process yields 80% less of the pertussis-fighting component than the more reactogenic whole-cell formula. Changing the formula would result in "a very large increase in the cost of manufacture."

1978 India bans the export of Rhesus monkeys, first used in biomedical research for the development of polio vaccines, where over 100,000 monkeys were sacrificed.

1978 ✪ Meningococcal polysaccharide monovalent vaccine groups A and C (Menomune-A® and Menomune-C® by Connaught) and bivalent Menomune A/C, are licensed in the USA for people over 55 years old, against the bacterium *Neisseria meningitidis.*

1978 Killed mumps vaccine is withdrawn in the US.

1978 The International Conference on Primary Health Care (PHC) takes places in Alma-Ata, USSR (present-day Kazakhstan), sponsored by the WHO and UNICEF. The Alma-Ata Declaration identifies primary health care as "the key to the attainment of the goal of Health for All." The conference reaffirms that health is a fundamental human right and that it is a "state of complete physical, mental, and social well-being, and not merely the absence of disease or infirmity."

1978 ✪ The first trivalent influenza vaccines (A) H1N1, (A)H3N2 + B are marketed in the USA.

1978– 1990 The CDC conducts the Monitoring System for Adverse Events Following Immunization (MSAEFI) in the public sector, collecting data on adverse events occurring 4 weeks after vaccination. From 1979 to 1984, 6,483 reports are received, including 74 SIDS deaths and 573 febrile convulsions. The system is replaced by VAERS in 1990.

1978– 1984 According to reports from the three manufacturers of DTP vaccine—Wyeth, Connaught, and Lederle—during the seven-year period from 1978–1984, 140 lawsuits are filed in the USA, with the average damages claimed per case rising from US$10 million to US$46.5 million, posing an existential threat to DTP vaccines. Only one injury lawsuit is filed in 1978 compared to 73 in 1984.

1978 Biotech company Genentech announces the production of the first genetically engineered human insulin expressed in *E. coli*, using recombinant DNA techniques.

1978 Data from the first year of the UK National Childhood Encephalopathy Study (established in 1976) are reviewed to see whether any relation is apparent between pertussis vaccination and brain disease. After 1,000 cases are analyzed, a significant association is shown between serious neurological illness and pertussis vaccine, estimating 1 in 100,000 children is left permanently brain-damaged by the DTP shot.

1978 ✪ Yellow fever vaccine YF-Vax® by Connaught is licensed in the USA.

1978–1981 Experimental trials of a formalin-inactivated hepatitis B vaccine, prepared from HBsAg positive plasma, take place in the US. Targeted at young homosexual men, the first "placebo-controlled," randomized, and double-blind trial begins in November in Manhattan—led by Polish Professor of epidemiology at the Columbia University School of Public Health, Dr. Wolf Szmuness. Largely supported by grants from the CDC, NIH, and NIAID, trials continue in Denver, St. Louis, Chicago, Los Angeles, and San Francisco under the supervision of CDC epidemiologist Donald Francis. Over 5,000 young, mainly white homosexual men are given the vaccine produced by Merck Sharp & Dohme, Heptavax-B®. The placebo consists of alum alone in the vaccine diluent.

1978 ✪ In France, Pasteur Institute launches its monovalent tetanus vaccine IPAD® Tétanique Pasteur, adsorbed on calcium phosphate.

1978 The US City of Hope National Medical Center successfully inserts cloned DNA fragments coding for insulin into genes of *E. coli* bacteria.

1978 The WHO Program for International Drug Monitoring is set up in Uppsala, Sweden—beginning with just three pharmacists. The Uppsala Monitoring Center (UMC) develops, maintains, and analyzes VigiBase, the single largest drug and vaccine safety data repository of individual case safety reports in the world—using WHO's ART language.

1978 The first successful birth using in vitro fertilization (IVF) is achieved in the UK.

1978 The Indian government begins its Expanded Program on Immunization, renamed the Universal Immunization Program (UIP) in 1985—the largest of its kind. It includes the tuberculosis BCG vaccine, Sabin's oral polio vaccine (OPV), DTP, tetanus toxoid and measles vaccines. Serum Institute of India (SII) is the major vaccine supplier for this program.

1978 The last death from smallpox variola major is recorded. British forty-year old medical photographer Janet Parker is infected as a result of an accidental release of lab-kept virus at the University of Birmingham Medical School in England.

1978 The Tennessee SIDS Cluster: Eleven infants die suddenly within eight days of their DTP vaccination, nine of whom had received the same lot of DTP vaccine from Wyeth Laboratories: lot 64201. Four of the eleven infants die within 24 hours. Wyeth's solution to clustering is resolved by distributing a single lot to various geographic areas, thereby making it harder to identify "hot lots."

1978 ✪ The updated MMR-II® measles, mumps, rubella live-attenuated vaccine is approved in the USA. The new formula uses rubella strain RA 27/3 cultivated on WI-38 human diploid cells (developed by Dr. Stanley A. Plotkin) instead of HPV-77 (High Passage Virus—passed 77 times through monkey kidney cells—developed at the NIH Division of Biologics Standards) duck embryonic strain. No placebo-controlled studies are required for licensure.

1978 Scientists at American company Biogen are the "first to announce synthesis in bacteria (expression) of hepatitis B protein antigens." This patented process becomes the basis for producing hepatitis B vaccine, the first ever recombinant, genetically modified vaccine—first marketed in 1981 in the USA. Biogen licenses its hepatitis B technology to SmithKline Beecham and Merck in 1988.

1978–1984 The worst epidemic of anthrax occurs in Rhodesia (former British colony, now called Zimbabwe) affecting both livestock and humans, with 10,000 cases and ~200 human deaths. Evidence suggests Rhodesian and South African forces deliberately released anthrax spores in cattle herds.

1979 The US Department of Health, Education and Welfare (HEW) is renamed the Department of Health and Human Services (HHS).

1979 The whole-cell pertussis vaccine is discontinued in Sweden and Germany following increasing safety concerns. A new acellular pertussis vaccine will be introduced in Sweden in 1996, and in Germany in 1995.

1979 ✪ Merck's Meruvax-II® replaces Meruvax® worldwide to include rubella strain RA 27/3.

1979 ✪ MMR-II® replaces MMR-I® and is licensed for use in Switzerland.

1979   Letter from Prof. Gordon T. Stewart, MD, to CSM Chairman Sir Eric Scowen

Miss Higgs   Sir Eric to reply

UNIVERSITY OF GLASGOW

MECHAN PROFESSOR

GORDON T. STEWART, M.D.

GTS/LS

RUCHILL HOSPITAL
GLASGOW G20 9NB
TEL: 041-946 7120

1st November, 1979

DEPARTMENT OF COMMUNITY MEDICINE

Sir Eric Scowen,
Chairman,
Committee on Safety of Medicines,
Finsbury Square House,
33/37A Finsbury Square,
LONDON, EC2A 1PP

Dear Sir Eric,

        Thank you for yours of 24th October. I am well aware of the importance of confidentiality and protocol in our deliberations. This does not relieve us entirely of our wider responsibility which I interpret as being to help in protecting people against those dangers which come to our attention.

        Even acknowledging the imperfections of some of our data, we seem agreed that triple vaccine, probably on account of the pertussis component, can cause severe reactions associated occasionally with brain damage and even, very occasionally, with sudden death. The comparative rarity of these disasters is offset by the fact that when the vaccine is given on a national and international scale, a large number of children are placed at risk. Inspection of our records leaves little doubt that many lives of children and their families have been gravely damaged by adverse reactions following vaccination and that prior knowledge of this possibility, of contra-indications and of early warning signs might well have prevented, and could still prevent, such disasters.

        I am not asking you to disregard protocol but I do suggest that you should echo the note of urgency in our recommendations and see that the facts are published. Through its other agencies, which apparently are not inhibited by the same constraints, the DHSS or its authorised spokesmen are saying that the vaccine is perfectly safe. This statement should not be allowed to pass uncontradicted.

                                        Yours sincerely,

Copy to: Dr. T.W. Meade

Committee on Safety of Medicines (CSM) – Acknowledgment of severe reactions following DTP vaccine
November 1979 – Letter from Professor Gordon T. Stewart, MD, to CSM Chairman Sir Eric Scowen
obtained from the UK National Archives in 2018

1977–1982   Vaccine Complications and Underreporting

III. THE COMPLICATIONS OF IMMUNIZATION

The Problem of the Assessment of Complications

35. It is important here to recognise the limitations of the data available from the central organizations established to monitor the safety of drugs. The original Committee on Safety of Drugs (CSD) was set up as a direct consequence of the thalidomide disaster in the 1960s with the task of providing an early warning of adverse reactions to a drug. Later, under the Medicines Act of 1968, the Committee on Safety of Medicines (CSM) was given the statutory duty of monitoring adverse reactions to all drugs, including vaccines, but again its role was to provide an early warning system; it was never anticipated that such a reporting system would be sufficient to provide figures on the incidence of reactions. It is known that there is considerable under-reporting of all drug reactions, even of the most severe type. It has been argued that reactions to vaccines, particularly the severe ones, are reported less frequently than other drug reactions because the prescribing practitioner may be apprehensive that his action may have been responsible for the child's injury. Whatever the reason, apprehension, lack of understanding of the need to report vaccine reactions, or occasional forgetfulness, the fact remains that there is under-reporting and there is little reason to suppose that there will be any material change in reporting procedures in the future. This procedure cannot provide valid data on the incidence of complications to vaccines and other methods of obtaining information have therefore to be devised.

Joint Committee on Vaccination and Immunization – III. The Complications of Immunization
May 1977 – UK Department of Health and Social Security "Whooping Cough Vaccination"

Data from the UK National Childhood Encephalopathy Study 1976–1979

Only 7 days observation!

TABLE 6

Attributable risks of serious neurologic disorders occurring among children in the United Kingdom within 7 days after immunization with diphtheria-tetanus-pertussis (DTP) vaccine*

	Risk ratio	95% confidence limits
All children, all outcomes	1:110,000 immunizations	1:44,000, 1:360,000
Assuming 25% underreporting	1:85,000 immunizations	1:33,000, 1:270,000
Previously normal children with evidence of subsequent neurologic damage	1:310,000 immunizations	1:54,000, 1:5,310,000
Assuming 25% underreporting	1:230,000 immunizations	1:41,000, 1:3,970,000

Underreporting is acknowledged currently as being closer to 90-99%

* Source of data: The National Childhood Encephalopathy Study (reported in references 70 and 71).

"Assuming 25% underreporting" is likely a substantial underestimate!

**1982 *Epidemiologic Reviews* – Vol. 4 "Whooping Cough and Whooping Cough Vaccine:
the Risks & Benefits Debate" – D.L. Miller, R. Alderslade and E.M. Ross**
Extract of study published by The Johns Hopkins University School of Hygiene & Public Health,
found within the dossier of the Ministry of Health's CSM / JCVI.

A Chronology of Vaccination

1979–1982

1979 The film *A Gift, An Obligation* is produced by the American Academy of Pediatrics with a grant from the Division of Health Information Services, Merck Sharp & Dohme. It reaches more than six million people nationwide as part of a campaign to increase vaccination levels in children.

1979 In the US lawsuit *Brown v. Stone*, the Mississippi Supreme Court removes "unconstitutional" religious exemptions to school vaccine mandates. The court rules that the state has an "overriding and compelling public interest" to protect children from harm, "even when such rights conflicted with the religious rights of the parents."

1979 The Vaccine Damage Payments Act in the UK takes effect, paying out £10,000 to victims able to prove at least 80% disability. This would be increased to £20,000 in 1985, £30,000 in 1991, and then £40,000 to the year 2000. By 2020, the compensation has risen to £140,000.

1979 PATH begins collaboration with the WHO to develop time-temperature sensitive labels called a "vaccine vial monitor." The small sticker is applied to a vaccine vial and changes color as it is exposed to heat over time. It takes over thirty years for the widespread adoption of this sticker by vaccine makers, which becomes mandatory for all UNICEF vaccine purchases.

1979 In April, an accidental laboratory leak of anthrax spores from a military biological warfare facility triggers an outbreak of inhalation anthrax in Sverdlovsk, USSR (now Yekaterinburg, Russia). In response to the outbreak, thousands of civilians are vaccinated with the Soviet STI anthrax vaccine, manufactured by the Scientific-Research Institute of Vaccines and Sera in Tbilisi, Georgia. Initially the laboratory origin is denied by Soviet authorities, but in 1992 Russian President Boris Yeltsin admits the source of the outbreak was a laboratory leak.

1979 The WHO declares smallpox eradicated. Countries worldwide suspend vaccination.

1979 ✪ Lederle receives approval in the USA for its new 14-valent polysaccharide pneumococcal vaccine, Pnu-Imune®.

1979 ✪ The UK Health Protection Agency at Porton Down receives licensing approval for its aluminum and thimerosal containing anthrax vaccine in the UK. The cell-free formula had been used since 1954 and is today produced by Porton Biopharma Ltd, a Company owned by the UK Dept. of Health.

1980–1989

1980s ✪ In the early 1980s, Smith Kline-RIT markets its 17-valent pneumococcal polysaccharide vaccine Moniarix™ in Europe.

1980s The USSR bans the use of thimerosal in childhood vaccines, stating that this ethylmercury-based derivative is highly toxic to cells and even capable of changing cell properties, therefore inadmissible in biological products intended for children.

1980 The FDA announces that over 3,000 drugs will be removed from the market whose efficacy has not been proven.

1980 The Bayh-Dole Act or Patent and Trademark Law Amendments Act enables federally funded scientists to patent their discoveries. This enables nonprofit institutions to commercialize discoveries and make profits off projects funded by US taxpayer money.

1980 A landmark legal decision affirms the patentability of microbial life forms and "anything under the sun that is made by the hand of man." The US Supreme Court rules in *Diamond v. Chakrabarty* that genetically modified organisms can be patented, allowing General Electric Co., to patent its artificially constructed bacterium *Pseudomonas putida,* capable of breaking down crude oil. This spurs the expansion and profitability of the biotech industry.

1980 ✪ Pneumovax® by Merck is available in Switzerland, marketed by Chibret.

1980 ✪ In Australia, mumps vaccine is licensed.

1980 At the World Health Assembly, the WHO certifies the world to be smallpox-free.

1980 All fifty US states have some form of vaccine-related law regarding school entry—medical exemptions are allowed in all states.

1980 Venezuelan-American immunologist Baruj Benacerraf, French immunologist Jean Dausset, and American mouse geneticist George D. Snell win the Nobel Prize in Physiology or Medicine "for their discoveries concerning genetically determined structures on the cell surface that regulate immunological reactions."

1980 ✪ Human diploid-cell (WI-38) vaccines against rabies, Imovax® by Institut Mérieux and Wyvac® by Wyeth, are licensed in the USA. Wyvac® is removed from the US market in 1985 due to vaccine failure, leaving Imovax® and Verorab® (licensed in 1985) as the only available US rabies vaccines.

1980 The third edition of the *Diagnostic and Statistical Manual* (DSM-III) identifies attention deficit disorder (ADD) for the first time. Seven years later it is renamed ADHD, attention deficit hyperactivity disorder.

1981 Xanax® by Upjohn is launched as an anti-anxiety medication for panic disorder.

1981 ✪ Meningococcal polysaccharide quadrivalent vaccine, Menomune® A, C, Y, W-135 by Connaught, is licensed in the USA for people over 55 years old.

1981 Mice are cloned by scientists in Switzerland.

1981 NIAID begins its Program for Accelerated Development of Vaccines to focus and enhance research activities leading to new vaccines for important diseases. Its journal *The Jordan Report* is launched to update "researchers and policy makers" on the "accomplishments and future directions of vaccine research."

1981 The UK National Childhood Encephalopathy Study (NCES) preliminary report is first published. The UK High Court, represented by Justice Sir Murray Stuart-Smith, hears the case and in 1988 rules that the overall evidence is insufficient to prove a causal relation between the injection of DTP vaccine and chronic encephalopathies.

1981 In October, Merck runs the first ever direct-to-consumer (DTC) print advertisement in *Reader's Digest* for the vaccine Pneumovax®.

1981 ✪ In Japan, Takeda's acellular pertussis vaccine replaces the highly reactogenic whole-cell component in the triple DTP shot. This safer vaccine contains significantly less endotoxin and less active pertussis toxin. Only one dose is recommended before age two.

1981–1985 According to the CDC, DTP manufacturers distribute just under 20 million doses of vaccine and face 219 lawsuits filed for damages—with an average compensation demand of US$ 26 million. Insurance costs are a factor in the impressive vaccine price increase (from US$ 0.11 a dose in 1984 to US$ 11.40 a dose in 1986) with US$ 8 for insurance costs alone. The total compensation sought by these cases is over US$ 3 billion, more than 30 times greater than the 1985 market value of all DTP vaccines sold (at the private sector price of US$ 4.29). The CDC coins the term "litigation epidemic" and calls for an urgent resolution to ensure a continued supply of vaccines in the US. Vaccines are one of the first medical products to be almost completely eliminated by lawsuits.

1981–1995 Project Coast is established to "develop a range of chemical and biological agents designed to control, poison and kill people within and outside South Africa." About 18% of all projects are said to be on fertility and fertility control studies, with the objective to develop anti-fertility vaccines that could be administered to black South African women without their knowledge.

1981 ✪ Institut Mérieux commercializes its 14-valent pneumococcal vaccine, Imovax Pneumo 14® in France.

1981 ✪ The first viral formalin-inactivated alum-absorbed hepatitis B surface antigen (HBsAg) plasma-derived vaccine by Merck Sharp & Dohme, Heptavax-B®—developed by Dr. Baruch Blumberg and Dr. Irving Millman—is licensed in the USA. Pasteur Institute in France launches its brand of hepatitis B vaccine called Hevac-B®.

1982 Recombinant human insulin Humulin® by Eli Lilly—who licenses the technology from Genentech—is the first genetically engineered drug approved by the FDA.

1982   Merck Sharp & Dohme – Heptavax-B® advertising

1982 – Heptavax-B® advertising material provided by Merck Sharp & Dohme

This plasma-derived hepatitis B vaccine, against liver cancer, contains 20 mcg of HBsAg formulated in 620 mcg of aluminum hydroxide, 1:20,000 Merthiolate (thimerosal—mercury derivative) preservative and trace amounts of formaldehyde from the viral inactivation process. Residual amounts of genetic and protein materials are also included.
The pediatric dose contains 10 mcg HBsAg and half the amount of aluminum and Merthiolate.

The same technology and manufacturing process is used to formulate the hepatitis B plasma-based vaccines by SmithKline French, Glaxo, Institut Mérieux, Chibret—all of whom purchase licensing rights or make distribution agreements with the Institute of Cancer Research. It is one of the first biological products resulting from US taxpayer funded research that is patented and commercialized by private pharmaceutical companies. It is the most expensive vaccine on the market at US$ 90–100 for three doses, with the longest manufacturing process of 65 weeks from collection of plasma to the release of purified vaccine product.

Military Medicine is the journal of the Association of Military Surgeons of the United States (AMSUS).

Merck Sharp & Dohme Hepatitis B – Heptavax-B® vaccine advertisement in
Military Medicine, Volume 147, Issue 11, November 1982 © AMSUS
https://academic.oup.com/milmed/issue/147/11

NEW **Heptavax·B**® Vials, 3 ml
(Hepatitis B Vaccine|MSD)

References

1. Denes, A. E. et al.: Hepatitis B infection in physicians: Results of a nationwide seroepidemiologic survey, JAMA 239(3):210-212, January 16, 1978.
2. Dienstag, J. L. and Ryan, D. M.: Occupational exposure to hepatitis B virus in hospital personnel: Infection or immunization?, Am. J. Epidemiol. 115(1):26-39, 1982.
3. Perrillo, R. P. and Aach, R. D.: The clinical course and chronic sequelae of hepatitis B virus infection, Seminars in Liver Disease I(1):15-25, February 1981.
4. Wright, R.: Type B hepatitis: Progression to chronic hepatitis, Clin. Gastroenterol. 9(1):97-115, January 1980.
5. Dienstag, J. L., Wands, J. R., and Koff, R. S.: Acute hepatitis, in Harrison's Principles of Internal Medicine, ed. K. J. Isselbacher et al., 9th ed., New York, McGraw-Hill Book Company, 1980, pp. 1459-1470.
6. Zuckerman, A. J.: Hepatitis B: Its prevention by vaccine, J. Infect. Dis. 143(2):301-304, February 1981.
7. Szmuness, W. et al.: Hepatitis B vaccine: Demonstration of efficacy in a controlled clinical trial in a high-risk population in the United States, N. Engl. J. Med. 303(15):833-841, October 9, 1980.
8. Francis, D. P. et al.: The prevention of hepatitis B with vaccine: Report of the CDC multi-center efficacy trial, in press.
9. Krugman, S. et al.: Immunogenic effect of inactivated hepatitis B vaccine: Comparison of 20 μg and 40 μg doses, J. Med. Virol. 8:119-121, 1981.
10. Szmuness, W. et al.: A controlled clinical trial of the efficacy of the hepatitis B vaccine (Heptavax B): A final report, Hepatology 1(5):377-385, September-October 1981.

Indications and Usage: Immunization against infection caused by all known subtypes of hepatitis B virus in persons 3 months of age or older, especially those who are at increased risk of infection with this virus; will not prevent hepatitis caused by other agents such as hepatitis A virus, non-A, non-B hepatitis viruses, or other viruses known to infect the liver.
NOTE: Further study is required to determine the effectiveness of HEPTAVAX-B in preventing hepatitis B when the vaccine regimen is begun after an exposure to the hepatitis B virus has already occurred (i.e., use for postexposure prophylaxis). However, it has been demonstrated that doses up to 3 ml of Hepatitis B Immune Globulin, when administered simultaneously with HEPTAVAX-B at separate body sites, do not interfere with the induction of neutralizing antibodies against hepatitis B virus.

Contraindications: Hypersensitivity to any component of the vaccine.

Warnings: Persons with immunodeficiency or those receiving immunosuppressive therapy may require larger vaccine doses to develop adequate circulating antibody levels.
Because of the long incubation period for hepatitis B, it is possible for unrecognized infection to be present at the time HEPTAVAX-B is given. HEPTAVAX-B may not prevent hepatitis B in such patients.

Precautions: General: As with any parenteral vaccine, epinephrine should be available for immediate use should an anaphylactoid reaction occur. Any serious active infection is reason for delaying use of this vaccine except when, in the opinion of the physician, withholding the vaccine entails a greater risk. Caution and appropriate care should be exercised in administering this vaccine to individuals with severely compromised cardiopulmonary status or to others in whom a febrile or systemic reaction could pose a significant risk.

Pregnancy: Pregnancy Category C—It is not known whether the vaccine can cause fetal harm when administered to pregnant women or can affect reproductive capacity; not recommended for use in pregnant women.

Nursing Mothers: It is not known whether this drug is secreted in human milk. Caution should be exercised when this vaccine is administered to a nursing woman.

Pediatric Use: Safety and effectiveness in children below the age of 3 months have not been established. Use of this vaccine in children below the age of 3 months is not recommended at present.

Adverse Reactions: HEPTAVAX-B is generally well tolerated. No serious adverse reactions attributable to vaccination have been reported during the course of clinical trials involving administration of HEPTAVAX-B to over 6,000 individuals. In two double-blind placebo-controlled studies among 2,485 persons, the overall rates of adverse reactions reported by vaccine recipients (24.3% and 21.5%) did not differ significantly from those of placebo recipients (21.4% and 18.7%). Approximately half of all reported reactions were injection-site soreness, which occurred somewhat more frequently among vaccine recipients. Other less common local reactions have included erythema, swelling, warmth, or induration. These signs and symptoms of local inflammation are generally well tolerated and usually subside within 2 days of vaccination.
Low-grade fever (less than 101 F) occurs occasionally and is usually confined to the 48-hour period following vaccination. Although uncommon, fever over 102 F has been reported.
Systemic complaints, including malaise, fatigue, headache, nausea, dizziness, myalgia, and arthralgia, are infrequent and have been limited to the first few days following vaccination. Rash has been reported rarely.

Administration: The immunization regimen consists of 3 doses of vaccine given intramuscularly. Do not inject intravenously, subcutaneously, or intradermally. Store at 2-8 C (35.6-46.4 F). Do not freeze, since freezing destroys potency. The vaccine is used as supplied. No dilution or reconstitution is necessary.

How Supplied: Vials containing 3.0 ml of liquid vaccine for use with syringes only, each ml containing 20 mcg hepatitis B surface antigen formulated in an alum adjuvant and thimerosal (mercury derivative) 1:20,000 added as a preservative.
For more detailed information, consult your MSD Representative or see Prescribing Information. Merck, Sharp & Dohme, Division of Merck & Co., INC., West Point, PA 19486.

A Chronology of Vaccination

1982 Wyeth and E.R. Squibb / Connaught interrupt vaccine production due to product liability exposure and rising insurance costs.

1982 The DTP documentary *Vaccine Roulette* is released by WRC-TV in Washington and stirs up controversy in the medical community and in the general public. On April 20 segments of the program are aired nationally on the NBC *Today Show*.

1982 The FDA proposes a ban of over-the-counter (OTC) products containing thimerosal, concluding that thimerosal is not generally recognized as safe "for OTC topical use because of its potential for cell damage if applied to broken skin and its allergy potential. It is not effective as a topical antimicrobial because its bacteriostatic action can be reversed." It takes 16 years before this ban actually goes into effect.

1982 *Malek v. Lederle Laboratories* is the first court case concerning Lederle's DTP vaccine Tri-Immunol® that is brought to jury trial. Sherry Malek, mother of Edward Malek III, brings this personal injury action against defendant Lederle Laboratories following the permanent injury of her four-month-old son, who received Tri-Immunol® in December 1978. Within 5–7 hours after receiving the vaccine, Edward Malek experienced convulsions and seizures, leaving him permanently brain damaged. The complaint alleges that the vaccine was unreasonably dangerous and not accompanied by adequate warnings. This court case is won by Lederle Laboratories.

1982 Swiss-born American medical entomologist and biological weapons research scientist Willy Burgdorfer attributes Lyme disease to a vector-borne infection transported by ticks—caused by a previously unrecognized spirochetal bacteria causing multi-system and multi-stage inflammation. The bacteria is named *Borrelia burgdorferi,* in his honor.

1982 ✪ Merck's pneumococcal 14-strain Pneumovax® vaccine is licensed in Australia.

1982 J.P. Morgan launches its annual healthcare conference, which is to become the biggest biotech conference where industry leaders, pharmaceutical companies, biotech start-ups, and the healthcare service come together to network and make deals.

1982 Educational organization "The Dissatisfied Parents Together" (DPT) is founded by Barbara Loe Fisher, and other parents whose children are injured or dead following DPT/DTP vaccine. DPT evolves to become the National Vaccine Information Center (NVIC). The co-founders work with Congress on the National Childhood Vaccine Injury Act of 1986, which acknowledges that vaccine injuries and deaths are real and that the vaccine injured and their families should be financially supported.

1982 In the US, the cost to vaccinate every child is US$ 6.69 (public sector) and US$ 23.29 (private). Ten years later, this increases significantly—due to higher costs and more vaccines on the schedule—to US$ 128.79 (public) and US$ 243.90 in the private sector.

1983 ✪ A live-attenuated oral bacterial vaccine against typhoid *Salmonella typhi* Ty21a, Vivotif® by Swiss Serum and Vaccine Institute Berne (BERNA), is first licensed in Switzerland and Europe.

1983 ✪ Pluserix® MMR and Rimparix® MM by SmithKline Beecham (using Urabe Am9 mumps strain) are licensed in Switzerland and withdrawn in 1992 due to safety concerns over the mumps component.

1983 Before any scientific paper is peer-reviewed, Margaret Heckler, Secretary of Health and Human Services (HHS), announces at a press conference that Dr. Robert Gallo of the National Cancer Institute (NCI) has discovered the human immunodeficiency virus (HIV), the "probable cause" of AIDS. She also states that a vaccine will likely be available in two years.

1983 ✪ Merck's 23-strain pneumococcal vaccine Pneumovax® is licensed in the USA. Lederle also receives approval for its version of this PPV23 vaccine, called Pnu-Imune 23®.

1983 ✪ Merck's combined measles-mumps vaccine and plasma-containing hepatitis B vaccines are introduced in Australia.

1983 The US Congress passes the Orphan Drug Law Act (ODA), amending the Federal Food, Drug, and Cosmetic Act. ODA incentivizes pharmaceutical manufacturers to develop drugs targeted at treating rare diseases. About 20–25 million patients suffer from around 5,000 "rare" diseases. By 1990, the FDA designates 370 "orphan" drugs.

1984 The Task Force for Child Survival is founded at the Rockefeller Foundation's Bellagio Conference, by former CDC director Dr. William H. Foege and two former colleagues, Carol Walters and Bill Watson. Created as a global health coalition supported by the WHO, UNICEF, the World Bank, and the United Nations Development Program, its original mission focuses on increasing global childhood vaccination rates. By 1990, the percentage of vaccinated children worldwide rises from 20% to 80% as a result of its work. In 1991, the name is changed to The Task Force for Child Survival and Development to reflect its broadened objectives. In 2009, the name is modified again to become The Task Force for Global Health.

1984 The first genomic sequencer method and first genome sequence is developed by American molecular engineer Dr. George M. Church.

1984 Danish immunologist Niels K. Jerne, German biologist Georges J. F. Köhler, and Argentine biochemist César Milstein are awarded the Nobel Prize in Physiology or Medicine "for theories concerning the specificity in development and control of the immune system and the discovery of the principle for production of monoclonal antibodies."

1984 The Rules and Regulations of the *US Federal Register,* Vol. 49, No. 107 dated June 1, states that "Any possible doubts, whether or not well founded, about the safety of the [polio] vaccine cannot be allowed to exist in view of the need to assure that the vaccine will continue to be used to the maximum extent consistent with the nation's public health objectives."

1984 By this date, there are only two remaining whole-cell pertussis vaccine manufacturers in the USA: Lederle and Connaught.

1984 Dr. Anthony Fauci, American physician-scientist, is appointed director of NIAID at the NIH—a position he holds until 2022 when he will retire at the age of 81.

1984 ✪ Pasteur's inactivated polio vaccine IPOL® is licensed in the UK. It is withdrawn in 2004, replaced by combination DTP-IPV vaccines.

1985 The remaining smallpox vaccines are removed from markets worldwide.

1985 The Swiss Federal Office for Public Health officially recommends the MMR-II® vaccine for all infants, with a first dose at 12 months old and a second dose between 15–24 months.

1985 ✪ A pure polysaccharide *Haemophilus influenzae* type B (Hib) unconjugated vaccine by Praxis Biologics b-CAPSA 1® against meningitis (for babies 24–29 months old), is first licensed in the USA and then withdrawn in 1988. Two other Hib vaccines, Hib-VAX® by Connaught (for babies 18–24 months old) and Hib-IMUNE® by Lederle (for babies from 15 months old) are also licensed in the USA. They are both withdrawn in 1988–89, replaced with conjugate Hib vaccines.

1985 ✪ Pasteur's next generation purified Vero cell rabies vaccine Verorab® is licensed in the USA for pre- and post-exposure, for all ages.

1985 American scientist and founder of biotech company Vical Inc., Dr. Philip Felgner, develops the first lipid nanoparticle "Lipofectin" DNA transfection technology and commercializes it as a reagent in 1987. It becomes the most widely used approach for introducing nucleic acid into cultured cells and key to mRNA vaccine development.

1985 Based on the scientific and commercial potentials of NIAID inventions on hepatitis, SmithKline Beecham Biologicals takes a nonexclusive license on the related patents and, a year later, establishes a cooperative research and development agreement (CRADA) to develop Havrix®—the world's first commercially available hepatitis A vaccine, licensed in 1991 in Europe, and in 1995 in the USA.

A Chronology of Vaccination

Year	
1985	Novo Nordisk introduces the first insulin pen delivery system called the NovoPe.
1985	The Vaccine Damage Payments Act in the UK increases compensation to £20,000.
1985	In the case of *Shackil v. Lederle Laboratories*, the mother of a DTP vaccine injured child, Clara Morgan Shackil, sues Lederle Laboratories. Within 24 hours of her 1972 inoculation, Clara's daughter Deanna displayed symptoms of extreme pain followed by the rapid deterioration of her condition, resulting in the loss of her then-acquired verbal, motor, and mental capacities. Deanna is diagnosed as having chronic encephalopathy and severe retardation requiring institutionalization and constant care. During discovery for this trial, corporate documents show that 1 in 300 infants under six months suffer from generalized seizures following DTP, with persistent crying in 1 in every 13 children.
1985	British pharmaceutical giant Burroughs Wellcome files for a patent for AZT. In the USA, the FDA recommends the approval of the drug for use against HIV, AIDS, and pre-AIDS illness via the newly implemented FDA accelerated approval system.
1985	The first transgenic livestock (pigs, sheep, rabbits) are created.
1985	The FDA, WHO, and European Medicines Agency (EMA) set an upper limit for the amount of residual DNA permitted in each vaccine dose as 100 pg/dose (picogram).
1986	Pablo DT Valenzuela, research director of Chiron Corporation, succeeds in making the world's first recombinant vaccine by modifying the yeast *Saccharomyces cerevisiae* to make hepatitis B antigen, replacing the previous human-blood plasma derived hepatitis B vaccine.
1986	✪ In the UK, intradermal BCG vaccine by Evans Vaccine Ltd., is approved for use.
1986	The nuclear reactor explosion and radioactive disaster of Chernobyl takes place near Pripyat, USSR (today Ukraine).
1986	The National Childhood Vaccine Injury Act (NCVIA) is signed into law by President Ronald Reagan, providing partial liability for manufacturers and creating a "no-fault" compensation fund for the vaccine-injured. HHS establishes the Vaccine Adverse Event Reporting System (VAERS) to be co-administered by the FDA and the CDC. To comply with Section 2105 of the Public Health Service Act, Congress enacts the National Vaccine Program (NVP) and sets up the National Vaccine Advisory Committee (in 1987) under HHS, to coordinate vaccine research programs of the NIH, CDC, FDA, NIAID, and the Department of Defense.
1986	In the USA, American vaccine manufacturer Lederle lowers the DTP vaccine price down to US$8.92 from US$11.40 as a result of the reduced product liability exposure awarded by the NCVIA and VICP.
1986	✪ The first genetically engineered recombinant DNA hepatitis B vaccine (pioneered by Chiron Vaccines) as Recombivax-HB® by Merck, is licensed in the USA and West Germany—by Merck Sharp & Dohme—replacing Heptavax-B®. This surface antigen subunit vaccine is known as H-B-Vax II® in Europe and other countries outside the USA. The vaccine contains synthetic hepatitis B surface antigens (HBsAg) produced in yeast cells, less than 15 mcg/mL of residual formaldehyde and 0.5 mg/mL aluminum hydroxide, which is later replaced by Merck's proprietary adjuvant amorphous aluminum hydroxyphosphate sulfate (AAHS).
1986	Cattle in Britain begin to suffer from a condition similar to scrapie in sheep, nicknamed "mad cow disease" due to the behavior of the sick cows. While the cause is unknown, some suspect the feeding of rendered scrapie-infected sheep to cattle. By 1993, 120,000 cattle are diagnosed with BSE in Britain—a total of 4.4 million cattle are slaughtered during the outbreak.

1986 ✪ Stamaril® yellow fever vaccine (17D strain) is launched in France by Pasteur. It is prequalified by the WHO in 1987 and will be approved in over 70 countries, though never in the USA.

1986 A review of the vaccines licensed for use in the US shows that approximately half are new, which is markedly different from the situation in the 1970s, where the majority of vaccines licensed were improvements of old versions.

1986 ✪ SmithKline Beecham's mumps Urabe Am9 strain MMR Trivirix® vaccine is licensed for use in children in Canada. A year later, cases of mumps meningitis after this MMR vaccine are reported.

1986 In the USA, the ACIP recommends that all healthcare workers receive an annual influenza vaccine.

1986 In November, the WHO holds the first International Conference on Health Promotion in Ottawa, Canada. An international agreement *The Ottawa Charter for Health Promotion* is signed, providing an outline for action to achieve Health for All by the year 2000.

1986– 1988 The WHO Task Force on Vaccine for Fertility Regulation launches the first human phase I clinical trial of a synthetic peptide anti-human chorionic gonadotropin (hCG) vaccine—109–145 CTP-beta-hCG:diphtheria toxoid (DT)—on thirty women aged 26–43 years who had all been surgically sterilized. The anti-hCG vaccine is composed of a portion of the beta hCG subunit complexed to an antigenic carrier protein, diphtheria toxoid, in a water-in-oil emulsion with an adjuvant.

1987 Switzerland launches a national measles vaccine campaign targeting children using MMR-II®. Over 200 medical doctors sign a petition against it, highlighting the potential dangers of such a widespread campaign for a new triple live viral vaccine. All their concerns are systematically dismissed by the public health authorities.

1987 ✪ Pasteur Institute (acquired by Institut Mérieux in 1985) no longer commercializes its calcium phosphate adsorbed IPAD® vaccines.

1987 ✪ Inactivated polio vaccine, Poliovax® (grown in MRC-5 cell line) by Connaught, is approved in the US, though never marketed.

1987 ✪ The first protein conjugate vaccine (diphtheria toxoid conjugate PRP-D), ProHIBIT® by Connaught, is brought to market in the USA against *Haemophilus influenzae* type b. It is withdrawn in 2000. This adjuvant-free vaccine (including trace amounts of thimerosal) is initially licensed for babies from 18 months old, but in December 1989 is changed to 15 months.

1987 The US Department of Defense admits to operating 127 bioweapons research projects at universities and government laboratories around the country, despite the treaty banning research and development of weaponized biological agents.

1987 In 1979, Kevin Toner, then a three-month-old infant, is vaccinated with Tri-Immunol® and suffers a rare condition of the spine known as transverse myelitis, leaving him permanently paralyzed from the waist down. The jury in this *Toner v. Lederle Laboratories* case finds that the pertussis component of the vaccine caused Kevin's paralysis; although in a special verdict the jury rejects the strict liability and breach of warranty claims, it finds Lederle negligent and awards damages of US$1,131,200. Lederle appeals, but the decision is upheld all the way to the Supreme Court.

1987 CRISPR repeats are first discovered in *E. coli* by Japanese biologist Yoshizumi Ishino.

1987 In 1980, Michelle Graham's daughter received Wyeth's DTP vaccine at three months old and suffered from encephalopathy resulting in permanent brain damage. She sues Wyeth in the lawsuit *Graham v. Wyeth* and after a seven-week jury trial, US$15 million is awarded for damages. Wyeth appeals the decision, requesting a retrial based on "evidentiary error at trial" which is granted by the 10th District Court of Appeals in 1990.

1987 Under the NCVIA, the "Mandate for Safer Childhood Vaccines" becomes effective in December. It requires a report on the progress of the development of safer childhood vaccines and calls for the establishment of the Task Force on Safer Childhood Vaccines (enacted in December 1989) under NIAID. Its first and only report is published in 1998.

1987   Merck Sharp & Dohme – Recombivax-HB® advertising

1987 – Recombivax-HB® advertising material provided by Merck Sharp & Dohme

This first ever recombinant subunit vaccine contained 20 mcg of HBsAg formulated in 620 mcg of aluminum hydroxide, 1:20,000 Merthiolate (thimerosal—mercury derivative) preservative and trace amounts of formaldehyde from the viral inactivation process. Residual amounts of genetic and protein materials would have also been included. The pediatric dose contains 10 mcg HBsAg and half the amount of aluminum and Merthiolate. From 1991, this vaccine will be given to all newborns within 24 hours of birth.

The same technology and manufacturing process are used to formulate the hepatitis B recombinant-based vaccines by SmithKline Beecham, Pasteur Mérieux, and Berna Biotech—all of whom have purchased licensing rights or made distribution agreements with patent-owning companies Genentech and Biogen.

Merck Sharp & Dohme Hepatitis B – Recombivax-HB® vaccine advertisement in
Infection Control & Hospital Epidemiology, Volume 8, Issue 4, April 1987
© The Society for Healthcare Epidemiology of America
https://www.cambridge.org/core/journals/infection-control-and-hospital-epidemiology/article/dr-charles-bryan-responds-to-joan-ottermans-letter/A226DD510ED27438C59A680C4C223C75

Recombivax-HB®

(Hepatitis B Vaccine [Recombinant] | MSD)

INDICATIONS AND USAGE

RECOMBIVAX-HB is indicated for immunization against infection caused by all known subtypes of hepatitis B virus.

RECOMBIVAX-HB will not prevent hepatitis caused by other agents, such as hepatitis A virus, non-A, non-B hepatitis viruses, or other viruses known to infect the liver.

Vaccination is recommended in persons of all ages who are or will be at increased risk of infection with hepatitis B virus. In areas with high prevalence of infection, most of the population are at risk of acquiring hepatitis B infection at a young age. Therefore, vaccination should be targeted to prevent such transmission. In areas of low prevalence, vaccination should be limited to those who are in groups identified as being at increased risk of infection.

CONTRAINDICATIONS

Hypersensitivity to yeast or any component of the vaccine.

WARNINGS

Patients who develop symptoms suggestive of hypersensitivity after an injection should not receive further injections of RECOMBIVAX-HB (see CONTRAINDICATIONS).

Because of the long incubation period for hepatitis B, it is possible for unrecognized infection to be present at the time RECOMBIVAX-HB is given. RECOMBIVAX-HB may not prevent hepatitis B in such patients.

PRECAUTIONS

General

As with any percutaneous vaccine, epinephrine should be available for immediate use should an anaphylactoid reaction occur.

Any serious active infection is reason for delaying use of RECOMBIVAX-HB except when, in the opinion of the physician, withholding the vaccine entails a greater risk.

Caution and appropriate care should be exercised in administering RECOMBIVAX-HB to individuals with severely compromised cardiopulmonary status or to others in whom a febrile or systemic reaction could pose a significant risk.

Pregnancy

Pregnancy Category C. Animal reproduction studies have not been conducted with RECOMBIVAX-HB. It is also not known whether RECOMBIVAX-HB can cause fetal harm when administered to a pregnant woman or can affect reproduction capacity. RECOMBIVAX-HB should be given to a pregnant woman only if clearly needed.

Nursing Mothers

It is not known whether RECOMBIVAX-HB is excreted in human milk. Because many drugs are excreted in human milk, caution should be exercised when RECOMBIVAX-HB is administered to a nursing woman.

Pediatric Use

RECOMBIVAX-HB has been shown to be usually well tolerated and highly immuno-

RECOMBIVAX-HB®
(Hepatitis B Vaccine [Recombinant], MSD)

genic in infants and children of all ages. Newborns also respond well; maternally transferred antibodies do not interfere with the active immune response to the vaccine. See DOSAGE AND ADMINISTRATION for recommended pediatric dosage and for recommended dosage for infants born to HBsAg positive mothers.

ADVERSE REACTIONS

RECOMBIVAX-HB is generally well tolerated. No serious adverse reactions attributable to the vaccine have been reported during the course of clinical trials. No serious hypersensitivity reactions have been reported. No adverse experiences were reported during clinical trials which could be related to changes in the titers of antibodies to yeast. As with any vaccine, there is the possibility that broad use of the vaccine could reveal adverse reactions not observed in clinical trials.

In a group of studies, 3258 doses of vaccine were administered to 1252 healthy adults who were monitored for 5 days after each dose. Injection site and systemic complaints were reported following 17% and 15% of the injections, respectively.

The following adverse reactions were reported:

Incidence Equal to or Greater than 1% of Injections

LOCAL REACTION (INJECTION SITE)
Injection site reactions consisting principally of soreness and including pain, tenderness, pruritus, erythema, ecchymosis, swelling, warmth, and nodule formation.

BODY AS A WHOLE
The most frequent systemic complaints include fatigue/weakness; headache; fever ($\geq 100°F$); malaise.

DIGESTIVE SYSTEM
Nausea; diarrhea.

RESPIRATORY SYSTEM
Pharyngitis; upper respiratory infection.

Incidence Less than 1% of Injections

BODY AS A WHOLE
Sweating; achiness; sensation of warmth; lightheadedness; chills; flushing.

DIGESTIVE SYSTEM
Vomiting; abdominal pains/cramps; dyspepsia; diminished appetite.

RESPIRATORY SYSTEM
Rhinitis; influenza; cough.

NERVOUS SYSTEM
Vertigo/dizziness; paresthesia.

INTEGUMENTARY SYSTEM
Pruritus; rash (non-specified); angioedema; urticaria.

MUSCULOSKELETAL SYSTEM
Arthralgia including monoarticular; myalgia; back pain; neck pain; shoulder pain; neck stiffness.

HEMIC/LYMPHATIC SYSTEM
Lymphadenopathy.

PSYCHIATRIC/BEHAVIORAL
Insomnia/disturbed sleep.

SPECIAL SENSES
Earache.

UROGENITAL SYSTEM
Dysuria.

CARDIOVASCULAR SYSTEM
Hypotension.

Potential ADVERSE EFFECTS
In addition, a variety of adverse effects, not observed in clinical trials with RECOMBIVAX-HB, have been reported with HEPTAVAX-B® (Hepatitis B Vaccine, MSD) (plasma-derived hepatitis B vaccine). Those listed below are to serve as alerting information to physicians:
Hypersensitivity: An apparent hypersensitivity syndrome of delayed onset has been

RECOMBIVAX-HB®
(Hepatitis B Vaccine [Recombinant], MSD)

reported days to weeks after vaccination. This has included the following findings: arthritis (usually transient), fever, and dermatologic reactions such as urticaria, erythema multiforme, or ecchymoses.
Nervous System: Neurological disorders such as optic neuritis, myelitis including transverse myelitis; acute radiculoneuropathy including Guillain-Barré syndrome; peripheral neuropathy including Bell's palsy and herpes zoster.
Hematologic: Thrombocytopenia.
Special Senses: Tinnitus; visual disturbances.

DOSAGE AND ADMINISTRATION

Do not inject intravenously or intradermally.

RECOMBIVAX-HB is for intramuscular injection. The deltoid muscle is the preferred site for intramuscular injection in adults. Data suggest that injections given in the buttocks frequently are given into fatty tissue instead of into muscle. Such injections have resulted in a lower seroconversion rate than was expected. The anterolateral thigh is the recommended site for intramuscular injection in infants and young children.

RECOMBIVAX-HB may be administered subcutaneously to persons at risk of hemorrhage following intramuscular injections. However, when other aluminum-adsorbed vaccines have been administered subcutaneously, an increased incidence of local reactions including subcutaneous nodules has been observed. Therefore, subcutaneous administration should be used only in persons (e.g., hemophiliacs) at risk of hemorrhage following intramuscular injections.

The immunization regimen consists of 3 doses of vaccine. The volume of vaccine to be given on each occasion is as follows:

Group	Formulation	Initial	1 month	6 months
Younger Children (Birth to 10 years of age)	Pediatric 5 mcg/0.5 mL	0.5 mL	0.5 mL	0.5 mL
Adults and Older Children	Adult 10 mcg/1.0 mL	1.0 mL	1.0 mL	1.0 mL

Since there have been no clinical studies in which a vaccine series was initiated with HEPTAVAX-B® (Hepatitis B Vaccine, MSD) and completed with RECOMBIVAX-HB, or vice versa, it is recommended that the 3-dose series be completed with the same vaccine that was used for the initial dose.

Whenever revaccination or administration of a booster dose is appropriate, RECOMBIVAX-HB may be used.

For dosage for infants born of HBsAg positive mothers and for dosage for known or presumed exposure to HBsAg, see the Prescribing Information.

The vaccine should be used as supplied; no dilution or reconstitution is necessary. The full recommended dose of the vaccine should be used.

Storage
Store vials at 2–8°C (35.6–46.4°F). Storage above or below the recommended temperature may reduce potency.

Do not freeze since freezing destroys potency.

For more detailed information, consult your MSD Representative or see Prescribing Information. Merck Sharp & Dohme, Division of Merck & Co., Inc., West Point, PA 19486.   (200)

MSD
MERCK SHARP & DOHME

Recombivax-HB® is the first ever genetically modified vaccine, and will be recommended to all newborns from 1991, along with SmithKline Beecham's Engerix-B®. Adult clinical trial participants were monitored for just five days after each dose.

A Chronology of Vaccination

1987–1989

1987 The WHO establishes the Vaccines Prequalification Program, a service provided to UN procurement agencies—mainly UNICEF—to ensure that the vaccines supplied are "consistently safe and effective under conditions of use in national immunization programs." Prequalification is a key process in developing nations' ability to access, distribute and use vaccines.

1987 The British newspaper *The Times* publishes a provocative front page article called "Smallpox vaccine 'triggered AIDS virus.'" The newspaper reports that the WHO is "studying new scientific evidence suggesting immunization with the smallpox vaccine *Vaccinia* awakened the unsuspected, dormant human immuno defence virus infection (HIV). The discovery of how people with subclinical HIV infection are at risk of rapid development of AIDS as a vaccine-induced disease was made by a medical team working with Dr. Robert Redfield at Walter Reed." The WHO strongly denies the claims made in the article.

1987 In a recorded interview with Dr. Edward Shorter, author of *The Health Century*, American microbiologist, and inventor of over forty vaccines, Maurice Hilleman, states, "I think that vaccines have to be considered the bargain basement technology for the 20th century." He also admits to importing African green monkeys for use in vaccine development: "I brought African greens in. I didn't know we were importing AIDS virus at the time."

1987 Merck & Co., Inc., announces plans to donate Mectizan (ivermectin), a new medicine designed to combat onchocerciasis "river blindness," for as long as it might be needed. This is a result of Merck working in collaboration with international experts in parasitology, the WHO, the World Bank and UNICEF to reach those affected by the illness. This unusual decision came twelve years after the discovery of ivermectin by Merck scientists and nearly seven years after human clinical trials in Dakar, Senegal.

1987 After eleven years of litigation, the American Medical Association, the American College of Surgeons, and the American College of Radiologists are found guilty of conspiring to destroy the profession of chiropractic.

1987 ✪ Varicella® chickenpox vaccine by Smith, Kline & French is licensed in Switzerland.

1988 ✪ MMR-II® by Merck is licensed in the UK.

1988 The WHO, UNICEF, Rotary International, and CDC launch the Global Polio Eradication Initiative, following the passing of a World Health Assembly resolution to eradicate polio by 2000, with vaccination being the key tool.

1988 ✪ An adsorbed rabies vaccine by the Michigan Department of Public Health is licensed by the FDA in the US.

1988 ✪ Connaught's highly purified vero-cell derived "enhanced potency" inactivated, adjuvant-free, polio vaccine IPOL® is licensed in the US. It contains traces of preservatives 2-phenoxyethanol and formaldehyde.

1988 Fluoxetine, sold under the brand name Prozac®, is launched in the US by Eli Lilly. An antidepressant drug, Prozac is the first of the selective serotonin reuptake inhibitor (SSRI) class, and becomes one of the bestselling drugs of all time.

1988 ✪ SmithKline Beecham's mumps Urabe Am9 strain MMR Pluserix/Immravax® vaccine is recommended to children in the UK, even though it is taken off the market in Canada (under the brand name Trivirix®) due to reported cases of aseptic meningitis. It is eventually withdrawn in the UK in 1992 after four million children are vaccinated. The same vaccine will be used for a mass vaccination campaign in Brazil in 1997.

1988 ✪ In South Korea, Korean pharmaceutical company Green Cross develops (and Rhein Biotech produces) the world's first epidemic hemorrhagic fever vaccine, Hantavax®. With viral strains harvested from suckling mouse brains, the vaccine contains aluminum hydroxide adjuvant, thimerosal preservative, and gelatin as stabilizer.

1988 On April 12, OncoMouse becomes the first transgenic animal to be patented in the US (Patent 4,736,866). OncoMice are genetically engineered to have an active oncogene, rendering them highly likely to develop cancer and making them useful test subjects in cancer research.

1988 During his employment at the Salk Institute and collaborating with Vical Inc., American microbiologist Dr. Robert Malone invents the system of using mRNA inside lipid nanoparticles (cationic liposomes) for transfection into various cell types. This method lays the foundation for future mRNA-based gene therapies and vaccines.

1988 The National Vaccine Injury Compensation Program (NVICP) is created in the USA to provide compensation for injury or death following vaccination, as an alternative to civil litigation. The law establishes a Vaccine Injury Table with a list of recognized events and associated time period requirements for vaccine injuries that can be compensated. It becomes fully operational from 1989.

1988 ✪ Conjugate vaccine *Haemophilus influenzae* type B, HibTITER® by Lederle-Praxis is licensed in the USA, and is recommended as a single dose to all babies at 18 months old. This is modified to 15 months old in 1989, and then to four doses at 2, 4, 6, and 12–15 months from October 1990. HibTITER® is the first oligosaccharide conjugate (HbOC) vaccine (developed by Porter Anderson at University of Rochester, manufactured and licensed by Praxis and distributed by Lederle—later purchased by Wyeth) and the first to contain mutated diphtheria protein carrier CRM197, developed by Italian vaccinologist Rino Rappuoli. HibTITER® would be removed from the US market in 2007.

1988 The reporting of Adverse Events Following Immunization (AEFI) is implemented in India. In Switzerland it becomes mandatory by law, however, there is little oversight and doctors who do not comply are not penalized.

1988 NATO holds its first *Advanced Study Institute on Immunological Adjuvants and Vaccines* conference in Greece.

1988 In an effort to compensate for adverse events from US government-mandated vaccines, as well as to offset vaccine manufacturers' liability concerns, an excise tax is added to the price of each of the government-mandated childhood vaccines. The taxes—US$4.56 per dose of DTP, US$4.44 per dose of MMR, and US$0.29 per dose of OPV—are paid into a special trust fund that is used to pay the claims of those with vaccine-related injuries.

1988 The Center for Biologics Evaluation and Research (CBER) is created within the FDA to regulate blood products, vaccines, and other biological products.

1988 The UK Joint Committee on Vaccination and Immunisation (JCVI) notes anaphylaxis to be a severe and potentially fatal complication of measles-containing vaccines.

1988 Director of infectious diseases and senior physician at Children's Hospital of Philadelphia, Dr. Stanley A. Plotkin (who becomes medical and scientific director at Pasteur Mérieux Connaught from 1991 to 1998) and Edward Mortimer author the very first edition of *Vaccines,* published by W.B. Saunders Co, Philadelphia. The 630+ page book becomes the definitive reference on the subject, with updates every several years.

1988 The HHS National Vaccine Advisory Committee—a lobby-free think tank evolving into a federal advisory committee who provides vaccine policy recommendations—holds its first meeting.

1988 ✪ Recombinant hepatitis B vaccines Engerix-B® by Smith, Kline & French and GenHevac-B® by Pasteur Mérieux are approved for use in Switzerland. GenHevac-B® is also licensed in France as a 4-dose regimen.

1988 ✪ MMR-II® by Merck is licensed in the UK as a single-dose regimen. A second dose is added to the schedule in 1996.

1989 ✪ Following vaccine failures in schoolchildren, the ACIP and AAP recommend adding a second dose of live-attenuated measles vaccine to the routine childhood vaccination schedule in the USA.

A Chronology of Vaccination

1989 Vaccine manufacturer Merck signs an agreement with the People's Republic of China, under which the company will provide the technology needed to produce its recombinant hepatitis B vaccine. This vaccine will become mandated at birth in China.

1989 The US HHS CDC-USDA program of "Select Agents" is established, identifying around sixty biological and chemical agents that could affect humans, plants, and animals. The CDC opens its first high-containment BSL-4 laboratory, the second in the USA.

1989 The British Eugenics Society is renamed the Galton Institute.

1989 ✪ Immravax® by Pasteur Merieux is licensed as the second Urabe-containing MMR vaccine in the UK, for protection against measles, mumps, and rubella.

1989 The WHO recommends that infants in developing countries receive the high-titer, live-attenuated measles vaccine at six months old. Long-term studies show that "girls who received the high-titer vaccine were more likely to die before the age of three years than those who received the standard vaccine." By 1992, the WHO no longer recommends high-titer measles vaccine for children.

1989 The US Biological Weapons Anti-Terrorism Act (written by American bioweapons expert Dr. Francis Boyle) is signed, with the purpose of implementing the Biological Weapons Convention, providing protection from biological terrorism. The Act amends the federal criminal code to "impose criminal penalties upon anyone who knowingly develops, produces, stockpiles, transfers, acquires, retains, or possesses any biological agent, toxin, or delivery system for use as a weapon or assists a foreign state or any organization to do so." Under the treaty loophole of "dual-use" technologies, "defensive" weapons are developed under the cover of vaccine research.

1989 The Medicines Control Agency is set up to make sure that medicine and biologicals licensed in the UK are safe and effective.

1989 novo nordisk **fonden**

The Novo Nordisk Foundation (NNF) is set up with its main focus on awarding grants in biomedicine and biotechnology, with significant funding of vaccine research. By 2023, NNF is the wealthiest charitable fund by endowment value—even richer than the BMGF—with US$167 billion.

1989 ✪ The thimerosal and aluminum-containing recombinant (genetically engineered, containing <5% yeast protein) hepatitis B vaccine, Engerix-B® by SmithKline Beecham is licensed by the FDA in the USA, and in the UK. Thimerosal is phased out of this vaccine by the year 2000. There are two formulations, one for babies and children Engerix-B®10 with 10 mcg of HBsAg and 0.25 mg of aluminum hydroxide, and an adult version Engerix-B®20 with 20 mcg of HBsAg and 0.5 mg of aluminum hydroxide.

1989 ✪ Japan introduces its first national MMR vaccine using Kitasato's measles, BIKEN's mumps, and Takeda's rubella vaccines. Adverse events such as aseptic meningitis following vaccination are soon reported, with an incidence ratio of around 1:1000. BIKEN's mumps vaccine strain is identified as the culprit, and the company pays out a settlement to the victims' families. This triggers the closing of the country to all new vaccines (except hepatitis A in 1995) until 2007. The MMR vaccine is withdrawn from Japan in 1993, and replaced in 2005 with a measles-rubella vaccine. This MR vaccine and DTaP would be the only combination vaccines available in Japan until 2012.

1989 ✪ Merck's single-dose PedvaxHIB® meningococcal conjugate Hib vaccine is licensed in the USA for children from 15 months old. A year later it is approved as a 3-dose regimen for babies at 2, 4, and 12–15 months old.

1989 The Environmental Protection Agency (EPA) later identifies this year as the start of the autism epidemic, according to Robert F. Kennedy Jr.

1989 The first humanized mouse is created to simulate humans at a cellular and molecular level. Dr. Mel Bosma and his team at the US Fox Chase Cancer Center discovered severe combined immunodeficiency (SCID) mutated mice in 1983. They are the first type of genetically engineered, "humanized" mice (that have functioning human genes, cells, tissues and/or organs) used for biological and medical research for human therapeutics.

1989 The first genetically modified mice with genes removed called "knockout mice" are created by Mario Capecchi and Oliver Smithies. Their success is enabled by Dr. Martin J. Evans' work in the UK.

1989 ✪ Oral live-attenuated bacterial typhoid vaccine (Ty21a) Vivotif®, produced by the Swiss Serum and Vaccine Institute Berne (BERNA), is licensed in the US.

1989 ✪ MMR vaccine as a single-dose regimen is introduced in Australia and is recommended at twelve months of age. In 1992, a second dose is added to the national schedule.

1990–1999

1990 By this date, the polio vaccine, IPOL® by Pasteur Mérieux Sérums et Vaccins (who acquired Connaught) is the only standalone inactivated polio vaccine available in the US.

1990 Gene therapy is administered for the first time in September, to a four-year-old girl named Ashanthi De Silva suffering from adenosine deaminase (ADA) deficiency—a severe metabolic and enzymatic disorder that causes immunodeficiency. In the context of an NIH clinical trial under geneticist Dr. French Anderson, De Silva receives an infusion of genetically engineered T-lymphocytes developed by the National Cancer Institute and the National Heart, Lung and Blood Institute. She remains reliant on medication to keep her disorder in control, and is a staunch advocate of gene therapy.

1990

The Foundation for the National Institutes of Health (FNIH) is established by US Congress as a charitable organization with the mission to raise private-sector funds, and create and manage strategic alliances with public and private institutions in support of the NIH. The largest donor will be the BMGF, donating over US$10 million from 2001–2020. Other large donors include Eli Lilly, GSK, Johnson & Johnson, Pfizer, and the Wellcome Trust, who each contribute between US$5,000,000 and US$9,999,999.

1990 The license for MMR vaccine Trivirix® is revoked in Canada, followed by Malaysia, Philippines, Singapore and Australia.

1990 Secretary of Defense Dick Cheney orders all US troops deployed during the Iraq Gulf War to be inoculated—without informed consent—with experimental non-FDA-approved vaccines against anthrax and botulinum toxin. Many thousands of troops, including those who never went to Iraq, suffer from a debilitating illness named "Gulf War Syndrome."

1990 The routine childhood vaccine schedule is expanded in the USA and worldwide, with the introduction of meningitis and Hib vaccines.

1990 The first recombinant plant-derived pharmaceutical protein is created: human serum albumin, initially produced in transgenic tobacco and potato plants.

1990 In the USA, the CDC recommends a second dose of rubella vaccine to all children before entering school.

1990 Robert Malone and Jon A. Wolff et al., demonstrate the concept of nucleic acid-encoded drugs by directly injecting in vitro transcribed (IVT) mRNA or plasmid DNA into the skeletal muscle of mice that express the encoded protein in the injected muscle. This is the first evidence that IVT mRNA could deliver the genetic information necessary to produce proteins within living cells.

A Chronology of Vaccination

1990 From October, the ACIP recommends use of any of the licensed Hib conjugate vaccines HibTITER® or PedvaxHIB® for children as young as 2 months of age, in a three or four-dose regimen instead of one.

1990 In the USA, the National Coalition for Adult Immunization (NCAI) develops the first *Standards for Adult Immunization Practices,* which are endorsed by more than 60 professional organizations from both public and private sectors.

1990– 1991

The Children's Vaccine Initiative (a predecessor to Gavi) is launched at the World Summit for Children and is co-sponsored by UNICEF, the United Nations Development Program (UNDP), the WHO, the World Bank, and the Rockefeller Foundation. Both a concept and an organization, its purpose is to harness new technologies to advance the vaccination of children—aiming to reduce deaths from preventable infectious diseases. CVI is based out of the WHO headquarters, and funded by an annual US$2.5 million grant—partly financed by William H. Gates Foundation, Fondation Mérieux, and the governments of Japan, Ireland, Switzerland and the USA. CVI is disbanded after eight "troubled" years and soon replaced by Gavi.

1990 From July to December, 53 deaths are reported to VAERS, with 30 reports being in babies under 6 months old. Nine deaths alone are attributed to HibTITER®, yet there is no investigation into the deaths.

1990 In July, the Vaccine Safety Datalink (VSD)—initially called the Large Linked Database (LLDB)—is activated by the CDC in collaboration with national healthcare organizations to monitor the safety of vaccines and to study their adverse effects.

1990 ✪ BCG vaccine (TICE strain) by Organon Teknika Corp., is licensed in the US for the treatment of "carcinoma-in-situ of the urinary bladder." The TICE strain was developed at the University of Illinois from an original Pasteur Institute strain.

1991 ✪ SmithKline Beecham's hepatitis A vaccine, Havrix®, is introduced in Europe.

1991 The WHO's Global Advisory Group of the Expanded Program on Immunization recommends that all countries introduce hepatitis B vaccine into their national immunization schedules by 1997. In 1992, the World Health Assembly endorses the recommendation.

1991 Every Child By Two (renamed Vaccinate Your Family in 2020) is founded by former First Lady Rosalynn Carter and politician Betty Bumpers. Its aim is to ensure that all American children receive all their vaccines by the age of two. ECBT serves as a catalyst for the Vaccines for Children (VFC) program.

1991 The first attempt is made to inject synthetic messenger RNA into an animal for it to produce proteins. Big problems arise as the injections make the mice sick due to inflammation, with large doses being lethal.

1991 ✪ Conjugated *Haemophilus influenzae* type B vaccine, ProHIBIT® distributed by Swiss Serum and Vaccine Institute Berne (BERNA) for Pasteur Mérieux Sérums et Vaccins, is approved for use in Switzerland. It will be withdrawn in 1999.

1991 ✪ ACEL-IMUNE® DTaP vaccine by Lederle is licensed in December by the FDA for use as fourth and/or fifth doses in the total vaccination series among children aged 15 months–6 years who previously received three or four doses of diphtheria and tetanus toxoids combined with whole-cell pertussis vaccine. In 1996, ACEL-IMUNE® is recommended for the complete five-dose series of DTP vaccination. The vaccine is discontinued in 2001 by Wyeth-Ayerst, who acquires American Home Products/Lederle-Praxis in 1994.

1991	The FDA issues recommendations that vaccine makers avoid using bovine-derived materials from cattle born, raised, or slaughtered in countries where BSE is known to exist.
1991	The Vaccine Damage Payments Act in the UK increases compensation to £30,000.
1991	Hepatitis B vaccine Engerix-B® by SmithKline Beecham, containing both aluminum and thimerosal, is recommended by the ACIP to all newborns within the first 12 hours of life, despite zero placebo control studies, insufficient safety data, and low risk of disease to infants. During the clinical trials, infants vaccinated with hepatitis B vaccine are actively monitored for only 5 days.
1991	The mass vaccination of soldiers deployed during the Gulf War takes place in the US and UK, including vaccines (some experimental and unlicensed) against anthrax, bubonic plague, cholera, typhoid, yellow fever, and hepatitis A and B.
1991	Seven private European vaccine makers produce the majority of vaccines used in Europe and much of the rest of the world. These are Behringwerke (Germany), Immuno (Austria), Medeva-Evans (United Kingdom), Pasteur Mérieux Sérums et Vaccins (France), Sclavo (Italy), SmithKline Beecham (United Kingdom), and Swiss Serum and Vaccine Institute (Switzerland). These seven companies form the European Vaccine Manufacturers—a special group within the European Federation of Pharmaceutical Industries and Associations (EFPIA), an influential lobby in Brussels. In March, the EFPIA organizes the First European Conference on Vaccinology in Annecy, France, on the property of Fondation Mérieux.
1991	Vaccines Europe is created as a specialized vaccine group within the EFPIA, to provide a voice for the vaccine industry in Europe.
1991	✪ Mammalian cell-derived DNA recombinant GenHevac-B® vaccine by Pasteur Mérieux Sérums et Vaccins is approved in Europe, containing 20 mcg aluminum hydroxide and made of proteins produced in Chinese hamster ovary cells.
1991	✪ TBE vaccine Encepur® is licensed in Europe by Behringwerke (later Chiron-Behring).

1991	A case of pertussis is redefined by the WHO as a patient with a cough for at least 21 days and a laboratory confirmation of *Bordetella pertussis*. Like with polio, this redefinition in itself eliminates many reported cases.
1991	In a memo written to the president of Merck's vaccine division, Dr. Maurice Hilleman warns against giving thimerosal-containing vaccines to newborns, highlighting that by 6 months, babies would receive 87 times the amount considered safe. He states that a single 1 mL dose contains 50 mcg of mercury, 1.7x greater than the daily allowance of 30 mcg. His warning is ignored.
1992	The total number of Investigational New Drug (IND) applications submitted to the Center for Biologics Evaluation and Research of the FDA increases dramatically, from 66 INDs in 1980, to just under 558 in 1992.
1992	The first clinical trial of MF59®-adjuvanted influenza vaccine by Biocine is conducted. The product that emerges, Fluad®, is licensed first in Italy in 1997 and by 2014 (under Novartis) is approved in 30 countries.
1992	About 20 million doses of DTP and OPV are distributed in the USA with 19 million doses of MMR (following the recommendation of a second dose in the schedule).
1992	Good manufacturing practices (GMP) for biological products are first published by WHO.
1992	✪ Japanese encephalitis vaccine made with inactivated mouse-brain derived virus JE-Vax®, is licensed in the USA. It uses the Nakayama-NIH strain and is produced by the Research Institute of Osaka University (BIKEN) and distributed by Pasteur Mérieux Sérum et Vaccins for travelers to Asia.
1992	✪ A new conjugate vaccine for *Haemophilus influenzae* type b (Hib) is introduced in Canada by Pasteur Mérieux Sérums et Vaccins. This is integrated with pediatric whole-cell pertussis combined vaccines DPT-Hib and PENTA (DPT-Polio-Hib).
1992	✪ RA 27/3 (human diploid fibroblast) strain rubella vaccine, Ervevax® by SmithKline Beecham, is licensed in the UK. It will be discontinued in 2003.

A Chronology of Vaccination

1992 ✪ A whole-cell killed oral cholera vaccine Dukoral®, or WC-Rbs is licensed in Europe, manufactured by French company, Valneva.

1992 ✪ In the UK, DT vaccine by Medeva is licensed, and will be withdrawn in 1999.

1992 According to a Mercer Management Report for Gavi (published in 2002), the global vaccine market grows from 1992 to 2002 at an annual rate of 10% from US$ 2.9 billion to US$ 6 billion, and is forecasted to continue. This growth is predominantly driven by high-income country demand for pricier vaccines.

1992 Serum Institute of India's measles vaccine becomes the first vaccine manufactured by a developing country to be WHO-prequalified. A year later its DTP vaccine is prequalified and will be widely distributed by UNICEF.

1992 The FDA institutes "Accelerated Approval Regulations" in order to speed up the availability of new drugs to treat HIV/AIDS, though the designation has also been used for cancer drugs. These regulations allow drugs for serious conditions, that fit an unmet medical need, to be approved based on a surrogate endpoint.

1992 ✪ Hepatitis A vaccine and two *Haemophilus influenzae* type b conjugate vaccines (HibTITER® by Lederle-Praxis and ActHIB® by Pasteur Mérieux Sérums et Vaccins), are introduced in the UK. Hib vaccine is added to the routine childhood immunization schedule. HibTITER® and ActHIB® will be withdrawn from the UK in 2001 and in 2002 respectively.

1992 The magazine *Rolling Stone* publishes a controversial article by journalist Tom Curtis called "The Origin of AIDS." He suggests that HIV may have been triggered by the use of an experimental oral polio vaccine in Africa that was contaminated by a simian virus carried by African green monkeys. The WHO strongly rejects this theory.

1992 ✪ Pneumococcal polysaccharide vaccine (PPV23) Pneumovax® by Merck Sharp & Dohme is licensed in the UK.

1992 SmithKline Beecham issues an urgent stop to the use of MMR vaccine Pluserix® in the UK. Both Immravax® by Pasteur Mérieux Sérums et Vaccins and Pluserix® are suspended and replaced with Merck's MMR-II®. Pluserix® is also taken off the Swiss market.

1992 Pasteur Mérieux Sérums et Vaccins and SmithKline Beecham are the two largest international vaccine manufacturers and the largest suppliers to UNICEF. On a much smaller scale is the private company Swiss Serum and Vaccine Institute, who also manufactures vaccines for the Swiss market and UNICEF.

1992 ✪ Tripedia® DTaP vaccine by Connaught is licensed in August by the FDA for use as fourth and/or fifth doses. It is the second acellular pertussis vaccine to be licensed in the USA. The purified acellular pertussis vaccine component is produced by Japanese company BIKEN/Tanabe Corporation, and is combined with the diphtheria and tetanus toxoids manufactured by Connaught Laboratories. Tripedia®'s package insert mentions "SIDS" and "autism" as adverse events reported during postmarketing of the vaccine. Production of Tripedia® will be discontinued in February 2011.

1992 The total worldwide value of human vaccines sold is estimated to be as high as US$ 3 billion, of which only US$ 65 million represents UNICEF purchases for vaccines used in developing nations.

1992 The National Vaccine Advisory Committee (NVAC), in collaboration with the Ad Hoc Working Group for the Development of Standards for Pediatric Immunization Practices—a working group representing public and private agencies with input from state and local health departments, physician and nursing organizations, and public and private providers—develops a set of standards as to what constitutes the most essential and desirable vaccination policies and practices.

1992 The Prescription Drug User Fee Act (PDUFA) is passed in the USA, requiring that vaccine and pharmaceutical manufacturers submit fees to the FDA of US$100,000 or more per application. These funds are used to complement the FDA's budget—previously solely funded by taxpayers—and dedicated to expediting the drug approval process. PDUFA enables the FDA to reduce drug review time from the 30-month average to 15 months. PDUFA is reauthorized every five years and by 2024, fees have increased to US$4,048,695 (for applications requiring clinical data), US$2,024,348 (for applications not requiring clinical data) and US$416,734 for the annual program fee.

1992 ☼ *Haemophilus influenzae* type b vaccine PedvaxHIB® by Merck Sharpe & Dohme is licensed in Switzerland, only to be discontinued four years later.

1992 ☼ Hib vaccine is introduced in Australia for children from 18 months old.

1992 The WHO passes a resolution at the annual assembly to recommend universal hepatitis B vaccination (HBV) to babies. This leads to an increase in the numbers of countries with HBV programs, from 31 to 179 by 2011.

1993 The US National Immunization Program (NIP) is created as a separate project designed to provide leadership and services to local and state public health departments involved in vaccination-related activities. The NIP reports directly to the CDC Office of the Director.

1993 The Cochrane Collaboration is founded as an organization that synthesizes medical research and publishes independent systematic reviews about the benefits and harms of healthcare interventions.

1993 ☼ The combined DTP and HbOC *Haemophilus influenzae* type b vaccine Tetramune® is licensed in the USA by Lederle-Praxis, containing 1:10,000 thimerosal, aluminum and 25 mcg of CRM197 protein. It is a combination of registered vaccines Tri-Immunol® (DTP) and HibTITER® and is recommended for use in infants as a 4-dose regimen at 2, 4, 6, and 15–18 months. It will be licensed in Canada a year later. This marks the first novel combination vaccine to be licensed in the USA since the MMR vaccine in 1971. It will be withdrawn by the end of 2000.

1993 ☼ TBE vaccine Encepur®K by Behringwerke is licensed in Europe, including Switzerland, for children 1–11 years old.

1993 Excluding the cost of vaccine, the charges associated with administering the complete series of pediatric vaccines run from as little as US$25 at a public health clinic to more than US$200 at a private physician's office. The total price of fully vaccinating a child in the US ranges from almost US$147 in the public sector to more than US$448 in the private sector. By 2018, the public sector cost to vaccinate every child increases to US$1,948.

TABLE 4-6 Cost and Price (including Excise Tax) of the Basic Series of Childhood Vaccines in the United States, as of March 31, 1993

Vaccine	Price ($)		No. of Doses	Cost ($)	
	Public sector	Private sector		Public sector	Private sector
DTaP	11.01	16.33	2	22.02	32.66
DTP	5.99	10.04	3	17.97	30.12
Hib-CV	5.37	15.13	4	21.48	60.52
MMR	15.33	25.29	2	30.66	50.58
OPV	2.16	10.43	4	8.64	41.72
Hepatitis B	6.91	10.71	3	20.73	32.12
Total			18	121.50	247.72

https://nap.nationalacademies.org/catalog/2224/the-childrens-vaccine-initiative-achieving-the-vision

A Chronology of Vaccination

1993

The Immunization Action Coalition (IAC) is set up by Deborah Wexler, MD, with a mission to increase vaccination rates across the USA. The IAC and CDC establish a close working relationship with millions provided in grants over the years, specifically for hepatitis B education for healthcare professionals from 1995. Other funders will include the American Pharmacists Association, AstraZeneca, Bavarian Nordic, CSL Seqirus, GSK, Merck, Sanofi US, Pfizer, Dynavax, Moderna, Valneva and VBI Vaccines. American pediatrician and vaccinologist Dr. Paul Offit serves on the advisory board of the organization.

1993 The Australian government approves a genetically modified recombinant myxoma virus as an immunocontraceptive vaccine to induce sterility in the wild rabbit populations.

1993 ✪ In Australia, Hib vaccine is introduced for infants from 2 months old, and hepatitis A vaccine Havrix® is licensed.

1993– 1996 ✪ The WHO (via UNICEF) launches a national tetanus campaign in Nicaragua, then in 1994 in Mexico and in 1995 in the Philippines. It draws attention as it specifically targets pregnant women and females from age 14–44, requiring 2–5 shots in a single year. Human Life International and its Mexican affiliate, the Comite Pro Vida claim to detect hCG in several vials and high levels of anti-hCG antibodies in vaccinated women. The WHO and UNICEF strongly deny the allegations of using an anti-hCG vaccine. In the Philippines, a court injunction succeeds in temporarily halting the use of tetanus toxoid and the vaccine campaign is also suspended in Nicaragua.

1993 In total, forty vaccines and toxoids as well as an additional 10 immune globulins and antitoxins are licensed and available for use in the US.

1993 ✪ Conjugated Hib vaccines ActHIB® by Pasteur Mérieux Vaccins and OmniHIB® by SmithKline Beecham are licensed in the US. They are recommended as 4-dose regimens for babies at 2, 4, 6, and 12–15 months old. The vaccines use the first protein conjugate technology developed by the NIH.

1993 ✪ Genetically modified single acellular pertussis vaccine Acelluvax® by Biocine, is licensed in Italy.

1993 American molecular biologist Phillip A. Sharp, PhD, of the Massachusetts Institute of Technology, and one of the four founders of Biogen, receives the Nobel Prize for his discovery of split genes and RNA splicing.

1993 MMR vaccine is removed from the routine infant vaccination schedule in Japan following reported cases of aseptic meningitis—three deaths are registered and eight children are permanently handicapped. The MMR includes BIKEN's mumps strain that is later discontinued due to safety concerns.

1994 ✪ Subunit polysaccharide typhoid vaccine is first licensed in the US, made with purified Vi capsular polysaccharide from the Ty2 Salmonella Typhi strain. Typhim VI® by Pasteur Mérieux Sérum et Vaccins and Typherix® by SmithKline Beecham are launched.

1994 Mandatory vaccination in Japan is abolished.

1994 ✪ In the UK, Pasteur Mérieux Sérum et Vaccins launches an adult dose Td vaccine (reduced diphtheria toxoid and tetanus) called Diftavax®. It will be discontinued in 2006.

1994 France launches their national hepatitis B vaccine program. Following safety concerns, the campaign is suspended four years later.

1994 The WHO and UNICEF declare the diphtheria epidemic in the Newly Independent States of the former USSR, an international health emergency. As a result of the epidemic, it is estimated that over a million working days are lost—a significant loss in productivity.

1994 ✪ The FDA licenses a plague vaccine by Greer Laboratories, previously manufactured by Cutter. Production is discontinued in 1999.

1994 The Wellcome Trust Centre for Human Genetics is founded at the Nuffield Department of Medicine in the Medical Sciences Division, University of Oxford.

1994 Jackie Fletcher, mother to a 13-month-old son who suffered from epilepsy and subsequent brain damage 10 days after his MMR vaccine, sets up Justice Awareness Basic Support (JABS) as a parent support group for vaccine injured children based in the UK. Following a 16-year campaign for justice, she is awarded compensation of £90,000 in 2010.

1994 The US Vaccines for Children Program (VCP) is created by the Omnibus Budget Reconciliation Act of 1993, providing free (federally-funded) ACIP recommended vaccines to all children who are Medicaid-eligible, uninsured, or underinsured children, including American Indian/Alaska Natives.

1994 The Western Hemisphere is certified as polio-free by the WHO's International Commission for the Certification of Polio Eradication.

1994 The Oxford Vaccine Group (OVG), based in the Deptartment of Pediatrics at the University of Oxford, is founded by Prof. Richard Moxon and led by Prof. Andrew J. Pollard from 2001 (who chairs the JCVI since 2013). OVG conducts pre-clinical and clinical studies of new and improved vaccines for both children and adults.

1994 The first genetically modified food, Calgene's Flavr Savr™ tomato—engineered for delayed ripening—is approved for commercial production in the USA. The FDA declares that genetically engineered foods are "not inherently dangerous" and does not require special regulation as they are considered to have "substantial equivalence." Flavr Savr™ is withdrawn in 1997 due to poor sales.

1994 In November, the UK launches a nationwide measles and rubella vaccine campaign aimed at all schoolchildren.

1994 UNICEF releases the first study (by Mercer Management Consulting) on vaccine economics, establishing "a common language and principles to lay the groundwork for future market-shaping initiatives."

1994 The Global Program for Vaccines and Immunization is created by the WHO, merging two WHO programs—the Expanded Program on Immunization and the former Program for Vaccine Development. It includes the addition of a new unit for Vaccine Supply and Quality. A year later, the WHO program adds a Vaccine Trial Registry, of trials mainly in developing countries.

1994 ✪ The first and only aluminum-free hepatitis A vaccine, Epaxal®, is licensed in Switzerland by Crucell. It is a proprietary virosomal adjuvanted vaccine, formulated using MRC-5 cells and inactivated by formaldehyde—it will be approved in 40 countries by 2010.

1994 ✪ In Germany, SmithKline Beecham launches Oralvac®, a sublingual desensitizing allergy vaccine.

1995 Medeva Holdings applies for a patent on their HBV (hepatitis B) surface antigen particles, prepared by recombinant DNA technology. The patent US6100065A is granted in 2000 and expires in 2017.

1995 ✪ The first inactivated hepatitis A vaccine Havrix® by SmithKline Beecham is licensed in the USA.

A Comparison of the Composition of Adult Doses of Vaqta and Havrix*

	Vaqta	Havrix
Hepatitis A virus antigen†	50 U	1440 ELISA U
Manufacturing process residuals		
MRC-5 cell proteins	<0.1 µg	≤5.0 µg
MRC-5 cell DNA	<4×10⁻⁶ µg	...
Formaldehyde	<0.8 µg	≤100 µg
Other chemical residuals	<10 ppb	...
Preservative	None	0.5% phenoxyethanol, wt/vol
Excipients		
Inorganic salts	0.007% Sodium borate in 0.9% sodium chloride	Phosphate-buffered saline
Aluminum hydroxide	450 µg	500 µg
Other	None	50 µg polysorbate 0.3% amino acids, wt/vol

*Data from product circulars. Ellipses indicate data not available from product circular; ELISA, enzyme-linked immunosorbent assay; and MRC, the (British) Medical Research Council.
†For Vaqta, 1 U is approximately equivalent to 1 ng of viral protein. The units used for the 2 vaccines are not the same.

1995 The US abandons the Sabin live polio vaccine (OPV) and reverts back to using Salk's inactivated vaccine (IPV), as Sabin's live oral vaccine was acknowledged to sometimes cause polio. OPV will continue to be used in vaccination campaigns in LMICs.

A Chronology of Vaccination

1995	✪ Merck's live-attenuated Oka strain varicella vaccine Varivax® (against chickenpox) is licensed in the USA, from age 12 months.
1995	The ACIP, the American Academy of Pediatrics, and the American Association of Family Physicians issue their first ever "harmonized" national routine childhood vaccination schedule, combining the recommendations of all three groups.
1995	The first complete genome sequencing of a free-living organism—the bacterium *Haemophilus influenzae*—is achieved by the Institute for Genomic Research, founded in 1992 by American biologist John Craig Venture. Genomic and proteomic technologies are used to identify the 1,830,137 base pairs.
1995	✪ Thimerosal and aluminum-containing hepatitis B vaccine Hepavax-Gene®, by Berna Biotech Korea Corp, is launched in Korea and subsequently introduced worldwide for active vaccination against HBV infection in neonates, children, and adults. The HBsAg protein antigen is expressed by the methylotropic yeast *Hansenula polymorpha.* A thimerosal-free version will be marketed from 2004.
1995	In March, the US government changes the rules of the vaccine court "in a way that made cases more contentious, protracted and harder for petitioners to win." Following the IOM 1991 *Adverse Effects of Pertussis and Rubella Vaccines* and the 1994 *Adverse Events Associated with Childhood Vaccines* reports, the Vaccine Injury Table is amended to "make it consistent with current medical and scientific knowledge regarding adverse events associated with certain vaccines." This places the burden of proof on petitioners, resulting in prolonged proceedings and reduced compensation.
1995	British teenager, Stephen Churchill, age 19, becomes the first victim of a new version of Creutzfeldt-Jakob Disease (vCJD). There are a total of three vCJD deaths in 1995.

1995	A comprehensive multilateral agreement of the World Trade Organization on intellectual property, the Trade-Related Intellectual Property Rights (TRIPS) Agreement, comes into effect. It is a legally binding trade instrument that mandates patent protection for pharmaceutical products (including vaccines) for up to 20 years, with any violations resulting in trade sanctions.
1995	The Edward Jenner Institute for Vaccine Research is established as a registered charity in Berkshire, UK. It is a public-private partnership between the government, the Medical Research Council, the Department of Health, the UK Institute for Animal Health and pharmaceutical giant Glaxo Wellcome. Its mission is to promote public health through vaccine research.
1995	The CDC Foundation is set up as a nonprofit authorized by Congress to mobilize philanthropic partners and private-sector resources to support the CDC's critical public health protection work.
1995	The IOM's Vaccine Safety Forum is established to "examine critical issues relevant to the safety of vaccines used in the US and to discuss methods for improving the safety of vaccination programs." In November, a workshop is convened on "detecting and responding to adverse events following vaccination." A second workshop takes place in April 1996 to discuss "the various immunologic and genetic factors that might influence individuals' responses to vaccines." The workshops' summary document states that the objective is to "illuminate issues, not to resolve them."
1996	Basic health insurance coverage (LaMal) becomes obligatory for all residents in Switzerland. Monthly premiums initially cost the average citizen less than CHF 140, rising to over CHF 500 by 2023.
1996	✪ Formalin-inactivated hepatitis A vaccine Vaqta® by Merck, is licensed in the USA for children from age 2 years and older.

1996

Pub Med

The gold standard for medical literature is developed: PubMed. It is maintained by the National Center for Biotechnology Information and the US National Library of Medicine, supported by NIH.

1996
Australian immunologist Peter C. Doherty and Swiss immunologist Rolf M. Zinkernagel receive the Nobel Prize in Physiology or Medicine for their discoveries concerning the specificity of the cell-mediated immune defense.

1996
✪ SmithKline Beecham's Infanrix® DTaP and Hib vaccine is licensed in Switzerland.

1996
✪ Hib vaccine Hiberix® by SmithKline Beecham is granted its first marketing authorization in Germany.

1996
✪ ACEL-IMUNE® by Wyeth-Lederle is licensed in the US for the complete five-dose regimen of DTaP vaccination of infants from 2 months old.

1996
The FDA approves DTaP vaccine for the complete childhood vaccine schedule, replacing the whole-cell DTP versions.

1996
✪ Combined hepatitis B and Hib chimeric vaccine Comvax® by Merck, is licensed in the USA and will be withdrawn in 2014.

1996
Diphtheria and tetanus toxoids and acellular pertussis (DTaP) vaccine Tripedia® is licensed in the USA for primary and booster vaccination of infants.

1996
✪ Combined hepatitis A and B vaccine Twinrix® by SmithKline Beecham is licensed in the the EU for adult use, and a year later as a 3-dose regimen for pediatric use.

1996
SmithKline Beecham applies for a patent on its novel vaccine formulation, comprising a hepatitis B component, particularly hepatitis B surface antigen, in combination with aluminum phosphate and 3−O-acylated monophosphoryl lipid A. In 1999, the patent US5972346A is granted and expires in 2016.

1996
The first mammal is cloned—Dolly the sheep lives 6 years before being euthanized due to lung and joint problems.

1996
South African president, Nelson Mandela, launches the *Kick Polio Out of Africa* vaccination campaign, aiming to administer polio vaccine to 50 million children in one year.

1996
✪ A new combination DTaP and Hib vaccine called TriHIBit® by Aventis Pasteur is licensed in the USA for the fourth dose in the DTaP and Hib series.

1996
The International AIDS Vaccine Initiative (IAVI) is established to accelerate the development of AIDS vaccines, supported by the Rockefeller Foundation, Alfred P. Sloan Foundation, Fondation Mérieux, the World Bank, and UNAIDS. IAVI is one of the first public-private partnerships. Seth Berkley, formerly at the Rockefeller Foundation, becomes IAVI's president and CEO in 1997. Seth Berkley becomes CEO of Gavi, the Vaccine Alliance from 2011–2023.

1996
✪ Connaught's five-component acellular pertussis DTaP vaccine is licensed in the USA. Considered the most efficacious pertussis vaccine by key opinion leaders globally, it also causes significantly fewer reactions than its thimerosal-containing whole-cell predecessors.

1996
A landmark decision is made by the British House of Lords on a patent infringement case *Biogen v. Medeva,* filed in 1992. It is the first time that genetic engineering in the context of patent law is considered—the case being about Medeva's third-generation hepatitis B vaccine, that Biogen claimed was an infringement on its patent. Biogen's appeal was dismissed and Medeva won the case, but never marketed its vaccine.

1997
✪ Formaldehyde-inactivated hepatitis A vaccine Avaxim® by Pasteur Mérieux Connaught is licensed in Europe for persons over 12 years old.

1997
✪ Priorix® MMR vaccine is first licensed in Germany by SmithKline Beecham.

1997
✪ SmithKline Beecham launches its thimerosal-free DTaP vaccine for infants, Infanrix®. It is licensed in the US for the first 5 doses in infants from 2 months old and will be introduced in Australia in 1999.

A Chronology of Vaccination

1997  **International Vaccine Institute**

The International Vaccine Institute is established in Seoul, South Korea, by the UN Development Program as an independent, nonprofit organization based on the belief that children's health in developing countries "can be dramatically improved by the use of new and improved vaccines."

1997 ✪ Pasteur Mérieux Connaught's Pentacel® is launched and includes acellular pertussis vaccine as well as diphtheria, tetanus, Hib and polio (IPV), becoming the one-shot universal pediatric vaccine in Canada.

1997 The WHO officially recommends Hib conjugate vaccines to children worldwide as part of the routine vaccination schedule.

1997 ✪ Chiron-Biocine (later acquired by Novartis) launches Fluad® containing MF59® squalene adjuvant and is first licensed in Italy, becoming the first novel adjuvant approved since aluminum in the 1920s.

1997 The FDA, WHO, and European Medicines Agency (EMA) increase the upper limit for the amount of residual DNA permitted in each vaccine dose to 10 ng/dose (nanogram). There is no scientific basis for the increase of permitted genetic materials.

1997 American neurologist and biochemist Stanley B. Prusiner is awarded the Nobel Prize in Physiology or Medicine "for his discovery of Prions—a new biological principle of infection."

1997 The FDAMA legislation requires that clinical trials for pharmaceuticals and biologics be registered on the NIH-run website clinicaltrials.gov that goes online in 2000.

1997 The International Coordinating Group (ICG) on Vaccine Provision is established following major outbreaks of meningitis in Africa. It is a mechanism to manage and coordinate the provision of emergency vaccine supplies and antibiotics to countries during major outbreaks.

1997  **Institute for Vaccine Safety** Johns Hopkins Bloomberg School of Public Health

The Institute for Vaccine Safety is created in the Department of International Health at the Johns Hopkins University School of Public Health—now named the Johns Hopkins Bloomberg School of Public Health. Its mission is "to provide an independent assessment of vaccines and vaccine safety to help guide decision makers and educate physicians, the public and the media about key issues surrounding the safety of vaccines."

1997 In the WHO's article *Strategies for minimizing nosocomial measles transmission*, it is stated that "Since there are virtually no contraindications to measles vaccination, measles vaccine should be administered regardless of the patient's health status."

1997 ✪ Combined inactivated hepatitis A and recombinant hepatitis B vaccine Twinrix® by SmithKline Beecham is licensed in the UK and "widely promoted in the lay press for travelers to insanitary regions of the world."

1997 Secretary of Defense William Cohen orders mandatory anthrax vaccination for all active and reserve troops, leading to some well-publicized refusals. This massive human experiment is halted in 2003 by a federal judge, Emmet G. Sullivan, who rules that the mandate is illegal.

1997 The National Vaccine Information Center (NVIC) brings together more than 500 parents, doctors, scientists, health officials, lawyers, ethicists, journalists, and consumer activists from 34 states and five countries for the First International Public Conference on Vaccination to discuss why and how vaccines cause injury, death, and chronic illness.

1997 The Oviedo Convention—Convention for the Protection of Human Rights and Dignity of the Human Being with regard to the Application of Biology and Medicine—is formulated by the Council of Europe.

1997

RabAvert
RABIES VACCINE

✪ Chiron-Behring's RabAvert® rabies vaccine is licensed in the USA for pre- and post-exposure to rabies, for all age groups. It is a second generation vaccine, produced by cell culture techniques—propagated on chick embryo fibroblasts.

1997 The first transgenic cow, Rosie, is genetically engineered to produce milk enriched with a human protein called alpha-lactalbumin.

1997 A web-based influenze surveillance tool, FluNet, is launched by the WHO to track the movement of flu viruses worldwide.

1997 Under the FDA Modernization Act (FDAMA), a comprehensive review of the use of thimerosal (ethylmercury) in childhood vaccines is carried out. It is discovered that infants could be exposed to cumulative levels of mercury that exceed EPA recommended guidelines—up to 187.5 mcg by six months old and up to 275 mcg by two years old. As a precautionary measure, vaccine manufacturers are urged by health authorities to reduce or eliminate thimerosal in vaccines as soon as possible.

1997 In the USA, under the FDAMA, regulation of direct-to-consumer (DTC) advertising of pharmaceutical and other medical products is relaxed, making it easier for drug companies to broadcast television advertising. As a result, DTC drug marketing increases from US$700 million in advertising costs in 1996 to US$5.4 billion by 2006.

1997 The first outbreak of avian influenza A(H5N1) virus in humans occurs with 18 associated human cases and six deaths in Hong Kong. The outbreak is contained by the territory-wide slaughter of more than 1.5 million chickens by the end of the year.

1997 According to the WHO, the global vaccine market is worth over US$3 billion.

1998 Wyeth voluntarily discontinues manufacturing its triple DTP vaccine Tri-Immunol®, which was first licensed in the US in 1970.

1998 The term metagenomics is first coined by American microbiologist Jo Handelsman, referring to the function-based analysis of mixed environmental DNA species.

1998 While France suspends its hepatitis B vaccine program, Switzerland launches its national hepatitis B vaccination campaign aimed at adolescents aged 11–15 years old. The three vaccines available are Engerix-B® by SmithKline Beecham, Gen H-B-Vax® by MSD, and Heprecomb® by Swiss Serum & Vaccine Institute, containing aluminum and thimerosal.

1998 ✪ Connaught launches Adacel®, a combined vaccine containing an adult "booster" dose of acellular pertussis vaccine along with the adult (reduced amount) formulation of diphtheria and tetanus toxoids.

1998 The Human Rights Act is established in the UK, intended to reinforce rights and freedoms guaranteed under the European Convention on Human Rights.

1998 Genentech's spin-off company VaxGen launches the first HIV-1 human clinical trials for AIDSVAX®, until 2003. VaxGen will receive US$8 million in government funding and then hire the former HIV chief Dr. William Heyward. He admits in 2001 to violating the anti-graft law and pays an out-of-court settlement of US$32,500. VaxGen continues to receive significant government funding, notably in 2004 with US$100 million granted to develop a recombinant anthrax vaccine. Stock value plummets in 2006 after the company loses a US$877.5 million government contract for a recombinant anthrax vaccine.

1998 ✪ DTaP vaccine Certiva® by North American Vaccine (later Baxter) is licensed in the US for primary and boosters shots for children given at 2, 4, 6, 15–20 months, and at age four to six. It is withdrawn in 2000.

1998 French neurologist Prof. Romain Gherardi publishes a seminal paper describing a newly recognized histologic entity called macrophagic myofasciitis (MMF), a condition linked to intramuscular injection of aluminum hydroxide adjuvant-containing vaccines, notably hepatitis B.

A Chronology of Vaccination

1998

Thanks to a US$100 million gift from Bill and Melinda Gates, the Children's Vaccine Program is established at the WHO's Program for Appropriate Technology in Health (PATH). The first vaccines purchased for developing nations are Hib, hepatitis B, rotavirus and pneumococcal vaccines, not typically used in developing countries.

1998

In Jordan, a mass diphtheria-tetanus vaccination campaign in school is launched. Of the nearly 20,000 children vaccinated in September, over 800 children report side effects and 122 children are admitted to hospitals with symptoms of headache, fever, nausea, dizziness, and fainting. While some minor reactions are attributed to the vaccine, the WHO and Jordanian Ministry of Health conclude that the culprit is mass psychogenic illness—vaccine anxiety.

1998

✪ LYMErix™ recombinant OpsA vaccine against Lyme disease by SmithKline Beecham is licensed in the USA for 15–70 year olds. SKB announces it will no longer manufacture or distribute the Lyme disease vaccine in February 2002 citing "poor sales."

1998

In February, the infamous Wakefield 12-children case study, suggesting a possible link between Merck's MMR-II® and bowel disorders in autistic children, is published in *The Lancet*. During a press conference at the Royal Free Hospital in London, British gastroenterologist Dr. Andrew Wakefield states that he cannot in "good conscience" encourage MMR, but recommends the single vaccines instead, until further studies are done. In August, the UK Department of Health withdraws the licenses for single vaccines despite high demand, and Merck ceases production in the USA by 2009.

1998

✪ In the US, a license is reissued to Aventis Pasteur for the BCG vaccine Mycobax®.

1998

✪ Bharat Biotech launches its first genetically engineered hepatitis B vaccine in India, Revac-B+® as the "world's first cesium-chloride (CsCl) free recombinant hepatitis B vaccine." The HBsAg manufactured by recombinant DNA technology is "extracted from the culture media by density gradient centrifugation using cesium salts" that may remain in the vaccine in residual amounts. Bharat's patented HIMAX technology for purification of HBsAg and ultracentrifugation eliminates the use of heart-damaging toxic metal salts such as CsCl.

1998

The European Vaccine Initiative (EVI) is founded in Heidelberg, Germany, as a nonprofit Product Development Partnership "with the goal of supporting and accelerating the development of effective and affordable vaccines for global health." EVI works "closely with academic researchers, the private sector, governments, and other organizations to spearhead vaccine development." Funding is provided by CEPI, the EU, and the WHO among others.

1998

✪ Live oral tetravalent rotavirus vaccine RotaShield® by Wyeth-Lederle Vaccines (a unit of American Home Products) is licensed for infants at 2, 4, and 6 months. The vaccine is suspended a year later following cases of intussusception—severe bowel obstruction. In total, VAERS reports 112 cases and one death—Wyeth-Lederle voluntarily withdraws RotaShield® and recalls all doses.

1998

Bill Gates's computer company Microsoft is charged by the US Department of Justice with violations of the Sherman Anti-Trust Act. The court initially rules that Microsoft should be broken up into two separate entities. Microsoft appeals, manages to avoid being broken up, and settles the lawsuit in 2002–2004 for an undisclosed amount.

1998 Created as part of the US National Library of Medicine, the US National Center for Biotechnology Information website is set up as the largest resource for biomedical information.

1998 Retired US Air Force Colonel and Biodefense Programs Director at the Department of Homeland Security Dr. Robert P. Kadlec, writes an internal strategy report for the Pentagon "promoting the development of pandemic pathogens as stealth weapons that the Pentagon could deploy against its enemies without leaving a fingerprint." Kadlec states that "biological weapons under the cover of an endemic or natural disease occurrence provides an attacker the potential for plausible denial. Biological warfare's potential to create significant economic losses and consequent political instability, coupled with plausible denial, exceeds the possibilities of any other human weapon."

1998–1999 All but 4 states in the USA have vaccine mandates for students entering kindergarten through to high school. Religious exemptions are permitted in all but 2 states, and 22 states allow philosophical exemptions. (Missouri, Indiana, and Nebraska lost their philosophical exemptions in 1992, 1993, and 1994, respectively). In 1999, legislation is introduced in at least 8 states to eliminate state childhood vaccination laws or add philosophical exemptions.

1998–1999 ✪ In the USA, the following influenza vaccines are available: Fluzone® by Connaught Labs (whole-virus or split-virion); Fluvirin® by Evans Medical (purified surface antigen); Fluogen® by Parkedale Pharmaceuticals (split); and Flushield® by Wyeth-Ayerst Laboratories (split).

1999 Former UN official and BBC correspondent Edward Hooper publishes his book *The River: A Journey to the Source of HIV and AIDS,* asserting the monkey kidney cells in the oral polio vaccine were contaminated with simian immunodeficiency virus, causing the modern AIDS epidemic. His claim is strongly refuted by public health authorities and the WHO.

1999 The FDA approves Roche's Tamiflu (oseltamivir phosphate), an oral antiviral for the treatment of type A and B influenza.

1999 ✪ Allergy Therapeutics launches Pollinex® Quattro in Germany, a short-course hay fever allergy vaccination containing the adjuvant monophosphoryl-lipid A (MPL).

1999 ✪ Combined Hib and hepatitis B (HepB) vaccine Procomvax® by Aventis Pasteur MSD is first licensed in Europe.

1999 The WHO publishes a pandemic planning framework emphasizing the "need to enhance influenza surveillance, vaccine production and distribution, antiviral drugs, influenza research and emergency preparedness."

1999 The first "self-spreading" animal vaccine against rabbit hemorrhagic disease and myxomatosis is tested in Spain.

1999 The Global Advisory Committee on Vaccine Safety (GACVS) is established by the WHO to advise on vaccine-related safety issues.

1999 ✪ The UK is the first country to introduce a serogroup C meningococcal (MenC) polysaccharide-protein conjugate vaccine to be used in routine childhood vaccination at 2, 3, and 4 months. It is also included as a two-dose catch-up campaign for children 5–11 months old, and as a single dose for those aged 1–17 years old (extended to 24 years old in 2002). By 2000, three aluminum-adjuvanted MenC vaccines are licensed in the UK against meningococcal meningitis; Menjugate® by Chiron, Meningitec® by Wyeth-Ayerst and NeisVac-C® by Baxter.

1999 In the US, the FDA determines that infants may be exposed to cumulative doses of injected ethylmercury—from the thimerosal in over 30 vaccines on the recommended childhood schedule—that exceed EPA federal safety guidelines. The established guidelines are limited to the ingestion of methylmercury, another form of organic mercury, as none exist for ethylmercury. In July, the American Academy of Pediatrics (AAP) and the US Public Health Service (USPHS) jointly recommend the removal of thimerosal from vaccines as soon as possible. They also recommend the temporary suspension of giving thimerosal-containing hepatitis B vaccine at birth for infants born to low-risk mothers, until a thimerosal-free version is available.

A Chronology of Vaccination

1999–2000

1999 ✪ A "thimerosal-free" hepatitis B vaccine, Recombivax-HB® by Merck & Co., Inc., is approved in the USA, replacing the previous 1:20,000 thimerosal-containing vaccine given to newborns since 1991.

1999 The ACIP recommends that only inactivated polio vaccine (IPV) be used in the USA for vaccination against polio, due to safety concerns around Sabin's oral live poliovirus vaccine (OPV)—which continues to be used in developing countries.

1999 Silesian German and American biologist Günter Blobel receives the Nobel Prize in Physiology or Medicine "for the discovery that proteins have intrinsic signals that govern their transport and localization in the cell."

1999 The Advanced Course in Vaccinology (ADVAC) is created by Fondation Mérieux in collaboration with the University of Geneva, to provide a comprehensive overview of the latest advances in the field of vaccines. The course is led by Swiss vaccinologist Prof. Claire-Anne Siegrist, MD, and Dr. Stanley A. Plotkin, both of whom have significant conflicts of interest due to their close ties to the pharmaceutical industry.

1999 The British Medical Association publishes a report called *Biotechnology, Weapons and Humanity* that considers whether new biological weapons—made possible by the mapping of the human genome—could enable the attack on the specific genetic constitution of a national or ethnic group. A delayed ethnic weapon could be possible that is designed to target future generations.

1999 The BSL-4 Jean Mérieux Laboratory is opened by Fondation Mérieux in Lyon, France, in order to "give the international research community a state-of-the-art platform for the fight against emerging pathogens." It will be handed over to the French National Institute of Health and Medical Research (Inserm) in 2004.

1999 Bill and Melinda Gates set up the Bill & Melinda Gates Institute for Population and Reproductive Health at the Johns Hopkins University School of Public Health, with seed funding of US$ 20 million, and further financing of approximately US$ 970 million until 2022. The institute focuses on research in family planning, reproductive health and population dynamics.

1999 ✪ In the UK, SmithKline Beecham launches combined DTaP-Hib vaccine, Infanrix®. It will be discontinued in 2003. Infanrix® DTaP vaccine is licensed in Canada.

1999 With an initial grant of US$ 750 milion from Bill and Melinda Gates, the Global Fund for Children's Vaccines is created as a "new financing resource for low-income countries to strengthen their vaccine services and to purchase new and under-used vaccines." By 2000, it will receive over US$ 1 billion in funding, and will be officially launched as the "Global Fund" in January 2002.

1999 In India, 9,576 cases of acute flaccid paralysis (AFP) are reported to the national surveillance system. Of these cases, 181 are recognized as caused by vaccine-derived poliovirus (VDPV) following administration of oral polio vaccine.

2000–2009

2000s Intensive research and development into nanotechnology delivery systems for vaccines begins, and accelerates from 2010.

2000 The CDC declares that measles is no longer endemic in the US following elimination campaigns (largely driven by vaccination) that began in 1967.

2000 ✪ Priorix® MMR vaccine is first licensed in Switzerland by SmithKline Beecham.

2000 ✪ Meningococcal C (MenC) conjugate vaccine NeisVac-C® is introduced in the UK by Baxter.

2000 ✪ Combined hepatitis B and chimeric Hib vaccine Comvax® by Merck is introduced in Australia.

2000

BILL & MELINDA GATES *foundation*

In Seattle, USA, the Bill & Melinda Gates Foundation (BMFG) is founded by the merging of the William H. Gates Foundation (established in 1994) and the Gates Learning Foundation. They have US$69 billion in assets as of 2020 and become an important funder of the WHO and vaccination programs, financing projects in health and education, fighting poverty and disease.

2000 The Vaccine Damage Payments Act in the UK increases payment to £100,000.

2000 ✪ Wyeth's pneumococcal conjugate PCV7 Prevnar® vaccine is introduced in the USA, using technology developed at the University of Rochester, USA. Prevnar® is recommended as a 4-dose regimen at 2, 4, 6, and 12–15 months old. Costing US$232 for all four doses, it will be the first vaccine to reach blockbuster sales status.

2000 ✪ Hepatitis B vaccine, Hepacare® by Medeva Pharma Ltd., is licensed in Europe, originally developed with Janssen-Cilag International. It is the first recombinant protein product to receive EU-wide approval, however the product is never marketed and approval is withdrawn in 2002.

2000 A document is published by the nonprofit and educational organization the Project for the New American Century called *Rebuilding America's Defenses: Strategy, Forces and Resources for a New Century*. On page 60, it is written that "advanced forms of biological warfare that can 'target' specific genotypes may transform biological warfare from the realm of terror to a politically useful tool."

2000 ✪ Combined DTaP, inactivated polio, Hib and HepB hexavalent 6-in-1 vaccine, Infanrix Hexa®, by GSK is licensed in Germany.

2000 Australia adopts the policy of vaccinating all newborns with hepatitis B vaccine.

2000 PATH invents the Uniject needle, an easy-to-use, single-dose, autodisabling, pre-filled syringe, notably for use in the tetanus vaccine campaign for developing nations.

2000

GAVI Global Alliance for Vaccines and Immunization   Gavi The Vaccine Alliance

The launch of the Global Alliance for Vaccine and Immunization (GAVI) is officially announced at the World Economic Forum in Switzerland as a coalition of organizations "in response to stagnating global immunization rates and widening disparities in vaccine access." It includes representatives from UNICEF, the World Bank Group, the Bill & Melinda Gates Foundation, the WHO, national governments, the International Federation of Pharmaceutical Manufacturers Association (IFPMA), the Rockefeller Foundation, and Dr. E. Borst-Eilers, PhD, Swiss Minister of Health, Welfare and Sport. GAVI, since 2014 named Gavi, the Vaccine Alliance, receives seed funding of US$750 million from the BMGF.

2000

Brighton *collaboration*

The Brighton Collaboration is officially launched during a meeting in Verona, Italy, with the WHO as observer and funding agency. Its mission is to standardize the definitions and diagnostic criteria of Adverse Events Following Immunization (AEFI).

2000 The Developing Countries Vaccine Manufacturers Network (DCVMN) is founded as an international alliance of manufacturers "to distribute surveillance against identified and emergent contagious infections, with the goal of raising the obtainability and quality of vaccines inexpensive to everyone."

2000

INFOVAC LA PLATEFORME D'INFORMATION SUR LES VACCINATIONS

Under the patronage of the Swiss Federal Office of Public Health and the Swiss Society of Pediatrics, the University of Geneva and Fondation Mérieux launch the official website infovac.ch. It becomes the authoritative reference in Switzerland on vaccines.

2000 The first application of whole genome sequencing in vaccine research: formulation of the reverse vaccinology approach.

A Chronology of Vaccination

2000 Dr. Thomas Verstraeten of the CDC Division of Epidemiology, Vaccine Safety and Development Branch, presents the results of his yet unpublished analysis of data on 400,000 children from the Vaccine Safety Datalink (VSD), extracted between 1991–1997. The conclusion of the analysis is that there is an "increased risk of developmental neurologic impairment after high exposure to thimerosal-containing vaccine in first month of life." This study is the primary focus of the Simpsonwood meeting.

2000 The CDC holds a secret meeting on June 7–8 at the Simpsonwood Conference Center in Norcross, Georgia, to discuss Verstraeten's problematic data on early exposure to thimerosal in vaccines. Top public health authorities—including Dr. John Clements of the WHO's Expanded Program on Immunization—discuss the scientific review of the Verstraeten's VSD data at this meeting, chaired by Dr. Walter Orenstein, director of the CDC National Immunization Program. At the meeting, Verstraeten concludes "the screening analysis suggests a possible association between certain neurologic developmental disorders. Namely tics, attention deficit disorder, speech and language disorders, and exposure to mercury from thimerosal containing vaccines before the age of six months." Verstraeten's damning data will be "massaged" into statistical insignificance by the time it is published in the journal *Pediatrics* in 2004.

2000 Wholesale manufacture of whole-cell pertussis vaccine ceases in the USA. Whole-cell pertussis in DTP vaccine campaigns will continue to be recommended by the WHO. DTP remains in production and given to infants in LMICs in Asia, Africa and South America.

2000 ✪ GSK's DTaP, IPV and Hib pentavalent 5-in-1 vaccine Infanrix® is licensed in Switzerland.

2000 The US HHS National Vaccine Program Office sponsors an international workshop on aluminum in vaccines in Puerto Rico. It is admitted that there is "no data on the potential toxicity of the mixture of mercury and aluminum" and "scant data about risk levels for injected aluminum." It is identified that the following areas on aluminum adjuvants should be more thoroughly studied: toxicology and pharmacokinetics, specifically the processing of aluminum by infants and children; the mechanisms of interaction with the immune system; and the necessity of aluminum adjuvants in booster doses. Despite this, the published workshop summary states "based on 70 years of experience, the use of salts of aluminum as adjuvants in vaccines has proven to be safe and effective."

2001 ✪ CRM197-conjugated meningococcal serogroup C vaccine Meningitec®, by Wyeth-Ayerst, is introduced in Australia. It is replaced in 2013 with GSK's combined Hib and MenC conjugate vaccine Menitorix®.

2001 ✪ Wyeth's pneumococcal conjugate PCV7 vaccine Prevenar® is introduced in the UK.

2001 In May, more than 120 countries sign the Stockholm Convention—a global treaty that restricts the production and use of Persistent Organic Pollutants by banning 12 chemicals. It goes into effect in 2004; however, it is not ratified by the USA, Israel, or Malaysia.

2001 The training tabletop exercise of a covert US smallpox attack "Dark Winter," takes place under the Johns Hopkins Center for Health Security. It is largely written and designed by Tara O'Toole, Thomas Inglesby of Johns Hopkins, along with Randy Larsen and Mark DeMier of the Analytic Services (ANSER) Institute for Homeland Security. O'Toole, Inglesby, and Larsen would be directly involved in the response to the anthrax attacks that took place after September 11. O'Toole will be involved in the Clade X simulation, and Inglesby in the Event 201 simulation in 2019.

2001 ✪ Wyeth's HibTITER® vaccine is withdrawn in Australia, replaced by Merck's Pedvax®.

2001 Between September 18 and October 12, anthrax-laced letters are sent to several US news media offices and senators Tom Daschle and Patrick Leahy (who were holding up the speedy passage of the PATRIOT Act), killing five people and infecting seventeen others. Dr. Bruce Ivins—co-inventor of the anthrax vaccine—microbiologist and vaccinologist employed by the US biodefense labs at Fort Detrick since the 1970s, will be identified as the main suspect. In 2008, he overdoses on Tylenol before being formally charged by the FBI. No direct evidence of Dr. Ivins's involvement has ever been uncovered.

2001 The US September 11 terrorist attacks occur, involving four hijacked planes, the destruction of the Twin Towers in New York, and damage to the Pentagon—the day before Secretary of Defense Donald Rumsfeld announced that US$ 2.3 trillion in Pentagon transactions could not be tracked.

2001 On September 13 *The NEJM* publishes an editorial entitled "Sponsorship, Authorship, and Accountability," written by several top medical journal editors calling out the use of clinical trials primarily for marketing, issues of conflict of interest, and the need to strengthen publication ethics.

2001 The Clinical Immunization Safety Assessment (CISA) Project is established by the CDC as "a national network of vaccine safety experts from CDC, eight medical research centers and other partners" charged with conducting vaccine safety research and consulting on "complex vaccine safety questions."

2001 ✪ Combined inactivated hepatitis A and recombinant hepatitis B vaccine Twinrix® by SmithKline Beecham is licensed in the USA. It is recommended as a 3-dose regimen for people over 18 years old.

2001 The American Red Cross, the CDC, the UN Foundation, UNICEF, and the WHO launch a global partnership called the "Measles & Rubella Initiative." The goal is to increase uptake of measles and rubella vaccines while providing improved monitoring and evaluation of outbreaks.

2001 ✪ GSK's Hib vaccine Hiberix® is purchased by the NHS in the UK, until July 2006.

2001 ✪ Combined DTaP-IPV-HepB pentavalent 5-in-1 vaccine Infanrix® by GSK is licensed in the USA.

2001 The Meningitis Vaccine Project is launched, funded by the BMGF in partnership with the WHO and PATH, who allocate US$ 70 million to develop new Group A meningococcal conjugate vaccines for sub-Saharan Africa.

2001 *The Uniting and Strengthening America by Providing Appropriate Tools Required to Intercept and Obstruct Terrorism Act*, known as the 342-page US PATRIOT Act, is hastily signed on October 26 in response to the September 11 attacks and anthrax-laced letters—drastically increasing surveillance of the government over its citizens.

2001 The Global Pertussis Initiative is financed by Aventis Pasteur (later Sanofi) as a "scientific advocacy group with a large breadth and depth of expertise in the management of pertussis." GPI's mission is to evaluate the ongoing global pertussis challenges and recommend appropriate control strategies, like vaccines.

2002 ✪ Combined DTaP-IPV-HepB vaccine Pediarix® by GSK is licensed in the USA.

2002 ✪ DTaP vaccine Daptacel® by Aventis Pasteur is licensed in the USA.

2002 ✪ GSK's combined aluminum-adjuvanted, inactivated hepatitis A and recombinant hepatitis B vaccine, Ambirix®, is approved by the EMA for children 1–15 years old.

2002 The first formal adult US vaccination schedule is published and updated annually.

2002 On April 19, the University of North Carolina at Chapel Hill files a patent for *Methods for Producing Recombinant Coronavirus*—an infectious replication-defective clone of coronavirus. Three American inventors are listed on this patent US# 7279327, including epidemiologist Ralph Baric, PhD, an expert on coronaviruses.

2002 In September, Switzerland becomes a full member of the United Nations.

A Chronology of Vaccination

2002

swiss**medic**

Swissmedic replaces the Swiss state-governed agency and becomes a public-private medical regulatory agency in Switzerland, responsible for licensing medicines, medical devices, and vaccines. Its financing is sourced primarily from the pharmaceutical industry.

2002 The National Vaccine Injury Compensation Program (NVICP) sets up the Omnibus Autism Proceeding (OAP) to aggregate 5,400 claims submitted by families claiming that vaccines had triggered their children's autism. Just six test cases will be examined. Hannah Poling, the daughter of a Johns Hopkins pediatric neurologist who suffered encephalopathy after receiving five injections for nine diseases in one day, is selected as a test case by petitioners. Five years later, before the OAP is heard, the US government settles her case and seals it, taking her out of the running. It takes two more years for both sides to agree on compensation. The settlement is leaked out to the press, revealing that the Poling family receives US$1.5 million plus US$500,000 per year, acknowledging that the "vaccines aggravated an underlying mitochondrial disorder." Ignoring much of the scientific evidence, the Special Masters rule against the petitioners in all the test cases, and in 2009 they dismiss all 5,400 claims.

2002 The animal efficacy rule, or Animal Rule, is finalized by the FDA and authorized by the US Congress following the September 11 attacks amidst concerns of bioterrorism. The regulation takes effect in July and allows for the approval of medical products, including vaccines, based on "adequate and well controlled" animal studies that suggest the products are "reasonably likely to provide clinical benefit in humans" and shown to have an acceptable safety profile.

2002 The WHO declares Europe "polio-free."

2002

⑤ The Global Fund

The Global Fund is founded—a multi-stakeholder international organization who receives seed money from the BMGF. The Global Fund is to become the world's largest financier of AIDS, tuberculosis, and malaria prevention, treatment, care programs, and vaccines. As of June 2019, the organization has paid out more than US$41.6 billion to support these programs.

2002 President G.W. Bush announces a major smallpox vaccination campaign to protect the country against a potential biological attack. Fewer than 40,000 healthcare workers and first responders get vaccinated.

2002 The first SARS coronavirus pandemic in Asia is declared by the WHO, beginning in China around November. China is accused of hiding information about SARS, contributing to its spread. However, according to US State Department senior Chinese policy advisor Miles Yu, Chinese officials privately suspect that SARS is a US-engineered pathogen. At the end of the pandemic in 2003, 774 deaths are reported worldwide.

2002 The US House of Representatives passes the Homeland Security Bill that creates the Homeland Security Department. Tucked away in that bill is a clause that dismisses all lawsuits against vaccine makers for having thimerosal in childhood vaccines. House Majority Leader Dick Armey tells CBS News he did it "to keep vaccine-makers from going out of business under the weight of mounting lawsuits." According to congressional testimony from Robert Kennedy Jr., "In 2002, the day after [Senate Majority Leader Bill] Frist quietly slipped a rider known as the 'Eli Lilly Protection Act' into a homeland security bill, the company contributed US$10,000 to his campaign and bought 5,000 copies of his book on bioterrorism." Bill Frist, also a board certified surgeon, had already received US$873,000 in contributions from the pharmaceutical industry.

2002 In May, the BMGF purchases over US$200 million worth of shares in nine pharmaceutical companies including Johnson & Johnson, Merck, and Pfizer.

2002 The Clinton Health Access Initiative (CHAI) is launched with the aim to help people living with HIV/AIDS, by bringing down the price of medication. Five years later the CHAI expands to include malaria and by 2011, extends its work to lowering costs and increasing access to vaccines, alongside the BMGF.

2002 American infectious disease expert Dr. Julie Gerberding becomes the first female CDC director. During her tenure, she ushers through Merck's fast-tracked 3-dose Gardasil® HPV vaccine. She also oversees the CDC post-licensure studies exonerating Merck's MMR from any safety concerns. She resigns in 2009, and in December the same year (waiting exactly as long as federal law requires) is named President of Merck's US$5 billion global vaccine business. In 2009, she encourages people to get vaccinated against the new H1N1 virus while she is a consultant for Edelman, a public relations firm with many Big Pharma clients, including Merck. In 2014, Merck announces her appointment as executive VP for strategic communications, global public policy, and population health. In 2022, she announces her retirement from Merck to assume the role as CEO of the NIH Foundation.

2002 China makes the practice of giving hepatitis B vaccine to all neonates—at birth—mandatory, with a schedule of 3 doses of either 10 mcg of recombinant yeast vaccine or 20 mcg of Chinese hamster oocyte-derived vaccine. Ten years later, Gavi estimates that HepB coverage rates of neonates to be over 75% in all but one province of China.

2003 The first genetically modified animal is commercialized in the USA called a GloFish®—a Zebra fish given a fluorescent gene so that it glows under UV lighting.

2003 ✪ Dryvax® smallpox vaccine is made available to first responders in the USA.

2003 ✪ The first intranasal live-attenuated trivalent influenza A and B strain vaccine FluMist® by MedImmune is licensed in the US, indicated for healthy, non-pregnant people from 5–49 years old. The same vaccine branded Fluenz Tetra® in the UK is also approved for use.

2003 A human monkeypox outbreak occurs in the USA, with a total of 72 suspected or confirmed cases and no deaths.

2003– 2005 Thimerosal is removed from vaccines in the UK and no longer an ingredient in any of the childhood or adult vaccines routinely administered in developed countries. Before 2005, thimerosal was present in diphtheria- and tetanus-containing vaccines, as well as Hib and hepatitis B vaccine, and some influenza vaccines.

2003 In the UK, the Medicines Control Agency and Medical Devices Agency merge to form the Medicines and Healthcare products Regulatory Agency (MHRA), who are responsible for ensuring that medicines and medical devices licensed in the UK are effective and acceptably safe. Funded by the UK Department of Health and Social Care for the regulation of medical devices, the costs of regulating medicine are met through pharmaceutical industry fees.

2003 The WHO initiates the Vaccine Safety Net Project (VSNP)—a global network of websites, established by the WHO, that provides "reliable information on vaccine safety." VSNP's mission is to counter websites providing "unbalanced, misleading and alarming vaccine safety information [...] prompting a wave of undue fears." A key player is the GACVS who are mandated to respond promptly, efficiently, and with scientific rigor to vaccine safety issues.

2003 NIAID awards US$81 million to support candidate HIV vaccines as part of its public-private partnership to accelerate HIV vaccine development. Partner companies include AlphaVax Human Vaccines, Epimmune, Novavax, and Progenics Pharmaceuticals.

A Chronology of Vaccination

2003–2005

2003 The US federal court outlaws the military's mandate that army personnel submit to unlicensed, untested, and dangerous anthrax vaccine—implicated in the Gulf War syndrome in hundreds of thousands of veterans. The US District Judge writes "the United States cannot demand that members of the armed forces also serve as guinea pigs for experimental drugs."

2003 Allergy Therapeutics launches Oralvac Plus® in Europe, a sublingual allergy vaccine for the treatment of severe allergies.

2003 The ACIP recommends that infants age 6–23 months old get an annual influenza vaccine, with implementation scheduled for 2004.

2004 In April, "Biodefense for the 21st Century" is signed by President George W. Bush as a presidential directive providing a comprehensive framework for US biodefense. It is the first ever national strategy against biological threats.

2004 ✪ Indian vaccine company BioMed Ltd., is the first company in India to produce its own *Haemophilus influenzae* type b conjugate vaccine, Peda Hib™ and meningococcal polysaccharide vaccine groups A, C, Y, W-135, Quadri Meningo™.

2004 ✪ Decavac® by Aventis Pasteur, a preservative-free tetanus and diphtheria toxoid adsorbed vaccine, is licensed in the USA for adult use.

2004 ✪ DTaP combined with Hib and inactivated polio 5-in-1 vaccine (IPV) is introduced in the UK under the brand name Infanrix®. Infanrix® DTaP-Hib vaccine and 6-in-1 DTaP-Hib-IPV-HepB Infanrix Hexa® are approved in Canada. The 6-in-1 vaccine will be used as part of the routine infant schedule in British Columbia.

2004 The FDA approves Applied Digital Systems' Verichip, an implantable radiofrequency identification (RFID) device for patients—about the size of a grain of rice—allowing doctors to access personal medical records.

2004 ✪ Repevax® by Aventis Pasteur is approved in the UK as a Tdap-IPV booster every 5 to 10 years, to be used from age 3 years old—also recommended to pregnant women. Td-IPV vaccine Revaxis®, by Aventis Pasteur, is also licensed in the UK as a booster for teenagers.

2004 ✪ *Haemophilus influenzae* type b and hepatitis B vaccine Procomvax® by Aventis Pasteur MSD, initially licensed in 1999 in the EU, becomes the first approved vaccine reformulated to contain Merck's proprietary adjuvant amorphous aluminum hydroxysulfate (AAHS). For "commercial reasons" Aventis Pasteur MSD allows the EU marketing authorization to expire in May 2009.

2004 In July, US President G.W. Bush signs the Project BioShield Act. It authorizes US$5.6 billion over 10 years for the government to buy and stockpile pharmaceuticals and vaccines to fight anthrax, smallpox, and other potential agents of bioterror as well as to "help incentivize private industry." The act gives the government new authority to accelerate research and development on promising medicines to defend against bioterror. It also changes the process of authorizing and deploying medical defensive "countermeasures" in a crisis, allowing patients to quickly receive treatments in an emergency. The act brings into existence the "Emergency Use Authorization" legalizing the distribution of fast-tracked vaccines developed during declared epidemics without FDA safety or efficacy testing.

2004 Swiss pediatrician Prof. Claire-Anne Siegrist, MD, is the first president of the newly established Swiss Federal Commission for Vaccination (FCV), who is responsible for recommending vaccine schedules. At the same time she is the Vaccinology Chair at the University of Geneva, a position created and financed by Fondation Mérieux. She keeps her FCV presidency until 2014 despite her numerous financial conflicts of interest with vaccine makers, notably Aventis Pasteur, Wyeth, and GSK.

2004	Prevnar® surpasses US$1 billion in sales worldwide, becoming the first ever blockbuster vaccine product.
2004	The CDC and the American College of Obstetricians and Gynecologists recommend that all pregnant women be vaccinated against influenza—using the inactivated vaccine in any trimester.
2005	✪ Meningococcal groups A, C, Y and W-135 polysaccharide diphtheria toxoid conjugate vaccine (MCV-4), Menactra® by Sanofi Pasteur is licensed in the USA, for people from 11–55 years old.
2005	The Public Readiness and Emergency Preparedness Act (PREP Act) is signed into law by American President George W. Bush to provide a tort liability shield to manufacturers of medical countermeasures—including vaccines, in the event of a public health emergency. It also establishes the Countermeasures Injury Compensation Program (CICP), a federal program under the Secretary of HHS, that provides benefits for serious injuries that occur following administration of a covered countermeasure (includes vaccines, antivirals, drugs, biologics or medical devices). Vaccines covered by CICP include Ebola, pandemic influenza type A, smallpox, anthrax, and future COVID-19 vaccines.
2005	✪ Chickenpox (varicella) is added to the combined live-attenuated MMR-II® vaccine by Merck and sold as ProQuad®, licensed in the USA for children from 12 months old.
2005	In the US, a new federal Medicare rule becomes effective, requiring all long-term care facilities to offer annual influenza vaccines and a one-time pneumococcal vaccine to all residents as a condition of participation in the Medicare insurance program.
2005	American virologist Ralph Baric, PhD, is invited to speak about synthetic coronaviruses at the DARPA-MITRE sponsored event, *Biohacking: Biological Warfare Enabling Technologies* in Washington, DC.
2005	✪ The EMA approves GSK's 5-in-1 vaccine Quintanrix® (DTwP-Hib-HepB) which will be withdrawn in Europe in 2008.

2005	 The smallpox bioterrorist attack tabletop exercise "Atlantic Storm" takes places at the Center for Biosecurity of the University of Pittsburgh Medical Center in collaboration with Johns Hopkins University.
2005	The WHO adopts the International Health Regulations (IHR) for its member states, which goes into effect in 2007. It introduces the notion of a Public Health Emergency of International Concern (PHEIC), declared by the Director-General and defined as "an extraordinary event which is determined to constitute a public health risk to other States through the international spread of disease and to potentially require a coordinated international response." All declarations so far have only been for viral illnesses. A PHEIC can also activate government contracts with vaccine manufacturers.

is still a PHEIC as of 2024-09-18

2005-2024	PHEIC timeline
2009-2010	Swine flu / H1N1
2014-present*	Poliomyelitis
2014-2016	Ebola
2016	Zika
2019-2020	Kivu Ebola
2020-2023	COVID-19 / SARS-CoV-2
2022-2023	Monkeypox
2024	Mpox

2005	✪ The FDA lowers the age limit from 24 months to 12 months for Merck's inactivated hepatitis A vaccine Vaqta® (pediatric dose) and GSK's hepatitis A vaccine Havrix®.
2005	✪ Combined meningococcal group C polysaccharide (MenC) and Hib vaccine Menitorix® by GSK is licensed in the UK, for infants from 2 months to 2 years old.
2005	The WHO recommends the inactivated influenza vaccine for all pregnant women during the influenza season.
2005	✪ Adult formulation Tdap vaccine Boostrix® by GSK is licensed in the USA as a booster shot for 10–18 year olds.

A Chronology of Vaccination

2005 The Bill & Melinda Gates Foundation invests US$ 50 million in the project "Innovative Vector Control Consortium" in collaboration with large agrochemical companies in an effort to combat mosquito-vector diseases, notably malaria, using novel molecules and genetically engineered insects.

2005 The entire genome of the 1918–1919 H1N1 pandemic influenza virus is sequenced. The same year, *The US Government National Strategy for Pandemic Influenza* is published.

2005 The European Centre for Disease Prevention and Control (ECDC), based in Sweden, is established as an EU agency with the aim of strengthening Europe's defenses against infectious diseases.

2005 The Edward Jenner Institute for Vaccine Research is replaced by the Jenner Institute, a partnership between the University of Oxford and the UK Institute for Animal Health. Associated with the Nuffield Department of Medicine at Oxford University Health, funding is complemented by the BMGF. The institute, directed by Irish-British vaccinologist Adrian Hill, develops vaccines and runs clinical trials for diseases including malaria, tuberculosis, Ebola, influenza, and MERS-coronavirus.

2005 The CIOMS / WHO Working Group on Vaccine Pharmacovigilance is established as a joint initiative to address vaccine safety issues (AEFI) and to "increase awareness and dissemination of general guidelines developed by the Brighton Collaboration."

2005 UNESCO's Universal Declaration on Bioethics and Human Rights is approved, including the principle regarding consent, detailed in Article 6: "Any preventive, diagnostic and therapeutic medical intervention is only to be carried out with the prior, free and informed consent of the person concerned, based on adequate information. The consent should, where appropriate, be express and may be withdrawn by the person concerned at any time and for any reason without disadvantage or prejudice."

2005 ✪ Fendrix® rDNA hepatitis B vaccine, by GSK, is licensed in Europe by the European Commission. It is the first vaccine to be approved with the adjuvant system AS04, consisting of 0.5 mg aluminum phosphate and 50 mcg monophosphoryl lipid A (MPL). The hepatitis B surface antigen is adjuvanted, adsorbed and produced in yeast cells *Saccharomyces cerevisiae* by recombinant DNA technology (rDNA).

2005 ✪ GSK's inactivated influenza vaccine, Fluarix® is licensed in the USA for adults.

2005 ✪ Adacel® Tdap vaccine by Sanofi Pasteur is licensed in the US from age 11 to 64.

2005 HHS awards a contract to Sanofi Pasteur worth US$ 97 million to develop a cell culture based influenza vaccine for the USA.

2005 US Congress passes the "Biodefense and Pandemic Vaccine and Drug Development Act" nicknamed "BioShield Two." It aims to speed up the development process for new vaccines and drugs in case of a pandemic. It adds a further layer of protection for the pharmaceutical and biotech industry from legal liability for injuries incurred from their countermeasure products.

2006 ✪ Pneumococcal conjugate PCV7 vaccine Prevenar® by Wyeth is introduced in the UK.

2006 ✪ Seasonal, inactivated, split-virion, trivalent influenza vaccine (TIV) FluLaval® by GSK is licensed and distributed in the US.

2006 The ACIP recommends a second dose of varicella (chickenpox) vaccine for children.

2006 Delta Technology's proprietary recombinant human albumin (rHA) Recombumin®, replaces human serum albumin in Merck's MMR product. It is the only difference between MMR-II® and MMRVaxPro®.

2006 ✪ Merck Sharp & Dohme's MMRVaxPro® is licensed in Switzerland and Europe, replacing MMR-II® which is removed from the Swiss and European markets.

2006 The first use of proteomics to identify surface exposed proteins in the screening of vaccine candidates: reverse vaccinology approach.

2006

⭐ Merck launches quadrivalent (HPV strains 6, 11, 18, 18) Gardasil® with an award-winning "One Less" TV advertising campaign. Gardasil® the first HPV "anticancer" vaccine is fast-tracked by the FDA and licensed in the USA, initially recommended only to females aged 9–26 years old. It contains Merck's proprietary adjuvant, amorphous aluminum hydroxyphosphate sulfate (AAHS).

2006

The ACIP recommends that teenagers get a booster dose of tetanus, diphtheria, and pertussis vaccines, otherwise known as Tdap—with a reduced dose of antitoxins.

2006

All thimerosal containing vaccines are removed from the Swiss market, though it remains in several vaccines available around the world, notably in LMICs.

2006

American biologists Andrew Z. Fire and Craig C. Mello receive the Nobel Prize in Physiology or Medicine "for their discovery of RNA interference—gene silencing by double-stranded RNA."

2006

⭐ Live-attenuated oral pentavalent rotavirus vaccine RotaTeq® by Merck is licensed in the USA for infants 6–32 weeks old.

2006

In the US, the Administration for Strategic Preparedness and Response (ASPR) is founded as a government agency operating under HHS. Its functions include "preparedness planning and response; building federal emergency medical operational capabilities; countermeasures [including vaccines] research, advance development, and procurement; and grants to strengthen the capabilities of hospitals and healthcare systems in public health emergencies and medical disasters."

2006

The International Finance Facility for Immunization (IFFIm) is founded as a collaborative financial tool between the WHO, the World Bank, and Gavi to facilitate funding of vaccine programs, notably in LMICs.

2006

⭐ Pneumococcal 13-valent conjugate vaccine Prevnar-13® by Wyeth, is introduced worldwide for infants and adults, replacing the previous 7-valent version.

2006

The WHO recommends that newborns should receive hepatitis B vaccine (HBV) within 24 hours after birth. In the next two years, newborn HBV vaccination increases from 27% to 69% worldwide.

2006

⭐ From 2006 to 2010 a total of US$ 910 million, including a US$ 110 million grant from UNICEF "to Support Vaccination Programs In The Developing World" is awarded to Dutch biopharma company Crucell (a global biopharmaceutical company ostensibly "focused on research development, production and marketing of vaccines, proteins and antibodies that prevent and/or treat infectious diseases"), for their new fully liquid pentavalent vaccine Quinvaxem™, co-developed with Novartis Vaccines, Diagnostics of Chiron and Berna Biotech in Korea. The much touted vaccine "for protection against five childhood diseases" is prequalified by the WHO in September 2006, enabling Crucell to bypass normal safety procedures. Quinvaxem™, the pentavalent DTwP-HepB-Hib vaccine containing "trace amounts" of thimerosal, is withdrawn in 2008.

2006

⭐ Zostavax® live-attenuated shingles vaccine by Merck is approved in the USA and recommended to people over the age of 60, as a single shot.

2006

Dow AgroSciences receives the first US regulatory approval for its Newcastle disease plant-based vaccine for poultry from the US Department of Agriculture Center for Veterinary Biologics—using genetically engineered tobacco.

A Chronology of Vaccination

2007 ✪ Quadrivalent HPV vaccine Gardasil® by Merck is licensed in Switzerland, targeted at young girls from age eleven.

2007 Queensland University of Technology in Australia begins development of a vaccine against chlamydia.

2007 In the US, HHS grants US$ 132.5 million over five years to Sanofi Pasteur and MedImmune to retrofit existing domestic vaccine production facilities to manufacture pandemic influenza vaccines.

2007 The Vaccine Damage Payments Act in the UK increases payment to £ 120,000.

2007 ✪ The first avian H5N1 influenza vaccine by Sanofi Pasteur is licensed in the USA.

2007 The ACIP recommends the intranasal live-attenuated influenza vaccine, FluMist®, to children from 2–5 years old.

2007 ✪ The FDA approves the use of Sanofi Pasteur's quadrivalent ACWY meningococcal vaccine Menactra® (MCV-4) in children age 2–10 years old. The ACIP recommends this vaccine for all children from 11–18 years old. In 2011, FDA approves the vaccine in children from 9–23 months old.

2007 Merck voluntarily recalls 1.2 million doses of PedvaxHIB® and Comvax® in the US after "quality-control checks found production equipment may not have been properly sterilized." Manufacture of its Hib-containing vaccines is temporarily suspended.

2007 Three years after it voluntarily withdraws its dangerous blockbuster arthritis drug Vioxx® from the market, Merck agrees to pay US$ 4.85 billion to settle nearly 27,000 lawsuits that claim the painkiller caused heart attacks. Merck recorded more than US$ 11 billion in Vioxx sales during its time on the market, from mid-1999 to September 2004, with an estimated 20 million Vioxx recipients. By 2011, Merck will pay nearly US$ 6 billion in litigation settlements.

2007 The Institute for Health Metrics and Evaluation (IHME) is founded at the University of Washington, Seattle. Launched with a core grant of US$ 105 million, mainly funded by the BMGF, the IHME specializes in measuring global health and utilizes computer modeling to try to establish causal relationships. A decade later, the BMGF provides another grant of US$ 279 million.

2007 ✪ The FDA approves a new inactivated influenza vaccine Afluria® by CSL Biotherapies for people age 18 and over.

2007 ✪ In Canada, Sanofi Pasteur's MCV-4 vaccine Menactra® is approved for use in people from 2–55 years old.

2007 The Quantitative Immunization and Vaccines Related Research (QUIVER) advisory committee is set up "to put modeling evidence into both methodologic context and best practices for the Strategic Advisory Group of Experts on Immunization (SAGE) and the WHO's Immunization, Vaccines, and Biologicals (IVB) Department." Today called the Immunization and Vaccines Implementation Research Advisory Committee (IVIR-AC), it is the principal advisory group to the WHO "providing independent appraisal of, and advice on implementation research related to vaccines and immunization, to inform public health decisions."

2008 ✪ Gardasil® (quadrivalent) by Merck Sharp & Dohme is licensed in the UK, and is approved for use in over 80 countries, with an estimated revenue of US$ 1.4–1.6 billion.

2008 ✪ Live oral monovalent rotavirus vaccine Rotarix® by GSK—given in a 2-dose regimen to infants 6–24 weeks old—is licensed in the USA.

2008 ✪ Quinvaxem® 5-in-1 (DTwP-HepB-Hib) vaccine, produced by Crucell for developing countries, is withdrawn from India following serious safety concerns after two years of use and an estimated 100 million doses administered in 50 countries.

2008

The organization Voices for Vaccines (V4V) is founded, upon suggestion of Dr. Stanley A. Plotkin, as a group of laypeople favorable to vaccination and providing a reliable source of information supportive of vaccines. V4V is administratively housed within the Task Force for Child Survival and Development. Dr. Paul Offit, along with Dr. Plotkin, will become scientific advisors to the nonprofit group. In 2018, during the deposition of Dr. Plotkin by Aaron Siri, Esq, it is revealed that V4V receives funding from vaccine manufacturers Merck and Pfizer. Professor of Law, Dorit Rubinstein Reiss, frequent contributor to blogging platforms pushing vaccination, is on the Parent Advisory Board.

2008 The US Government Accountability Office estimates that 1 in 15 young adults between 18–26 years old are "seriously mentally ill."

2008 The FDA approves an extended indication for GSK's Boostrix® Tdap vaccine in people aged 10–64 years old.

2008 The GSK Vaccines Institute for Global Health is established as an institute in Italy, within GSK, to develop effective and affordable vaccines for high-impact infectious diseases in the world's poorest communities.

2008 The ACIP recommends that all children age 6 months to 18 years old get an annual influenza vaccine.

2008 The US military shifts its smallpox vaccine stock from freeze-dried Dryvax® vaccine by Wyeth, to Acambis' (division of Sanofi Pasteur) ACAM2000®, based on the vaccinia virus grown in cell culture. The CDC announces its distribution to civilian laboratory personnel and the military.

2008 ✪ The FDA approves DTaP-IPV Kinrix® by GSK for use in children 4–6 years old.

2008 ✪ Tetanus-diphtheria toxoid vaccine Tenivac® by Sanofi Pasteur, is licensed by the FDA for people over 60 years old. The original approval was for 7–59 years old.

2008 The Children's Hospital Foundation (parent company of The Children's Hospital of Philadelphia® or CHOP) and Royalty Pharma announce that the Foundation is selling its worldwide royalty interest of RotaTeq® sales to Royalty Pharma for US$182 million. RotaTeq® was created as a result of research jointly performed between 1980–2006 by Merck, CHOP, and The Wistar Institute in Philadelphia. Dr. Stanley A. Plotkin and Dr. Paul Offit (who allegedly received approximately US$6,000,000) have inventor patent rights on RotaTeq® vaccine and receive an undisclosed percentage of the royalties.

2008 German virologist Harald zur Hausen is awarded the Nobel Prize in Physiology or Medicine "for his discovery of human papillomaviruses causing cervical cancer." A few days before the official ceremony, the National Anti-Corruption Unit in Sweden, led by prosecutor Nils-Erik Schultz, announces its investigation into suspicions of collusion, implicating the pharmaceutical company AstraZeneca and two members of the Nobel Assembly. AstraZeneca bought MedImmune in 2007 and became the owner of the patents linked to the HPV vaccine. Therefore AstraZeneca has financial interests directly linked to the commercial success of the vaccine. AstraZeneca also sponsors two communication companies linked to the Nobel Foundation: Nobel Media and Nobel Web. Two members of the Nobel Assembly have ties to AstraZeneca; Bertil Fredholm and Bo Angelin. Since 2007, Bo Angelin has a seat on the board of directors at AstraZeneca. Conflicts of interest, not being considered illegal, mean that there is no basis for Schultz to launch a criminal investigation.

2008 The FDA decides that clinical trials outside the USA—used to support applications for US product registration—no longer have to conform to the Declaration of Helsinki.

2008 In the October edition of *Journal of Infectious Diseases*, Anthony Fauci acknowledges that the majority of deaths during the "Spanish flu" pandemic of 1918 were likely due to bacterial pneumonia, not the flu virus.

A Chronology of Vaccination

2008 A new combination 5-in-1 vaccine (DTaP-IPV-Hib) Pentacel® by Sanofi Pasteur is approved by the FDA in the US, for use in children 6 weeks to 4 years old.

2008 The CDC receives FDA approval for a highly sensitive influenza PCR assay that can "detect influenza with high specificity" to enhance diagnosis and treatment options.

2008 The Novartis Vaccines Institute for Global Health is created with the nonprofit mission of developing effective and affordable vaccines for neglected infectious diseases in developing nations. NVGH is developing new vaccines against salmonella (typhoid fever, paratyphoid fever, and non-typhoid salmonella), shigella, and related diseases, and receives funding from the BMGF.

2008 French virologists Françoise Barré-Sinoussi and Luc Montagnier receive the Nobel Prize in Physiology or Medicine "for their discovery of human immunodeficiency virus."

2009 ✪ GSK's Hib vaccine Hiberix® is licensed by the FDA as a booster dose for children aged 15 months through 4 years. It is approved in just 5.1 months under the Accelerated Approval Regulations, in response to a Hib vaccine shortage that lasted from December 2007 to July 2009.

2009 In April, a new H1N1 virus is detected in the US. Eight days later, the WHO declares the H1N1 avian influenza a public health emergency of international concern (PHEIC).

2009 ✪ The EMA approves GSK's pneumococcal vaccine Synflorix® against invasive disease and acute otitis media caused by *Streptococcus pneumoniae* in infants and children from 6 weeks up to 2 years of age. The same 10-valent vaccine is launched in India under the brand name Streptorix®.

2009 HHS allocates US$487 million in funding to Swiss pharmaceutical company Novartis Vaccines and Diagnostics to build a cell-based influenza vaccine production facility.

2009 ✪ HPV vaccine is introduced in Japan.

2009 **USAID | PREDICT**

USAID launches its PREDICT global surveillance program as a project to identify viruses with pandemic potential, in order to rapidly respond to outbreaks. The USAID PREDICT program runs from 2009 to 2020 and receives more than US$200 million in funding. Financing of China's Wuhan Institute of Virology begins via EcoHealth Alliance. Other partners include Wildlife Conservation Society, Metabiota (a San Francisco startup that compiles global data to predict disease outbreaks), the Smithsonian (the world's largest museum and research complex), as well as the University of California Davis's One Health Institute.

2009 In May, the WHO changes the definition of "pandemic" to no longer include "high numbers of deaths and severe disease." Declaring pandemics allows lucrative commercial contracts between government and vaccine manufacturers to be activated, while ensuring producers are not held liable for vaccine injuries. Soon after, in June, the first pandemic (H1N1, swine flu) of the 21st century is declared by Margaret Chan, Director-General of the WHO.

2009 The US Department of Health and Human Services (HHS) creates the Post-Licensure Rapid Immunization Safety Monitoring (PRISM) program, that uses data from national health insurance plans to monitor the safety of the H1N1 influenza vaccines.

2009 HHS directs US$1 billion in funding towards the development of a vaccine against H1N1 influenza type A.

2009 ✪ The second HPV vaccine, bivalent Cervarix® by GSK, is licensed in the USA for prevention of cervical cancer. This vaccine is formulated with GSK's proprietary novel adjuvant ASO4 and is the first new adjuvant to be featured in an approved childhood vaccine since aluminum salts in the 1920s.

2009 The largest healthcare fraud settlement in history is paid in the USA by American pharmaceutical and vaccine manufacturer Pfizer (and its subsidiary Pharmacia & Upjohn Company). Pfizer agrees to pay US$ 2.3 billion for "intent to defraud or mislead" of which US$ 1 billion is paid to resolve allegations under the civil False Claims Act that the company illegally promoted four drugs: Bextra® (withdrawn in 2005), Geodon®, Zyvox®, and Lyrica®. Pfizer enters into a Corporate Integrity Agreement.

2009 The genome editing tool (genetic scissors) called CRISPR-Cas9 is pioneered by American biochemist Jennifer Doudna and French microbiologist Emmanuelle Charpentier. CRISPR is simpler, faster, cheaper, and more accurate than older genome editing methods, making it easier than ever to edit DNA. For this discovery, they are awarded the Nobel Prize in Chemistry in 2020, "for the development of a method for genome editing." This is the first scientific Nobel Prize to ever be won by only two women.

2009 Samples of viral vaccine material supplied by Baxter International's research facility in Austria, are found to be contaminated with live avian flu virus (influenza A virus subtype H5N1). Samples with the less harmful seasonal flu virus H3N2 are found mixed with the deadly H5N1 strain. Before mass administration, the vaccine is tested and kills animals in a lab in the Czech Republic; however, the product is shipped to several other countries including Slovenia and Germany.

2009 In October, Merck announces that it will not resume production of monovalent measles (Attenuvax®), mumps (Mumpsvax®), and rubella (Meruvax-II®) vaccines in the USA. Only Measles Live Pro® by Serum Institute of India—commercialized by Emergent BioSolutions—is available in Switzerland, India, and other countries.

2009 The FDA approves Merck's Gardasil® HPV vaccine for boys to prevent genital warts.

2009 ✪ The European Commission grants the first ever marketing authorization for a pandemic H5N1 influenza vaccine (split virion, inactivated, AS03-adjuvanted) called Adjupanrix® by GSK, that also contains 5 mcg thimerosal as a preservative.

2009 ✪ Both seasonal and pandemic flu vaccines are recommended to pregnant women in the USA and Europe—according to independent analysis, there is a seven-fold increase in the number of preterm births and miscarriages in double-vaccinated women.

2009 ✪ By December, several H1N1 vaccines are brought to market; Focetria® by Novartis, Pandemrix® by GSK. Manufacturers refuse to supply governments unless they are indemnified against any claim for injury. About 60 million people receive the vaccines, including many children. Pandemrix® will be associated with narcolepsy and cataplexy.

2010–2019

2010– 2018 During this 9-year period, US government funding of pandemic preparedness and health security exceeds US$ 95 billion.

2010 ✪ In the UK, influenza vaccine is routinely recommended to all pregnant women at any stage during their pregnancy.

2010 Bill Gates announces the "Decade of Vaccines" at the World Economic Forum in Davos, Switzerland, pledging US$ 10 billion over a decade to finance vaccine campaigns.

2010 Gardasil® HPV vaccine is approved by the FDA to include the expanded indication of protection against anal cancer.

2010 ✪ Bivalent HPV vaccine Cervarix® by GSK is licensed in Switzerland.

2010 ✪ Combined meningococcal conjugate A, C, Y and W-135 vaccine is licensed in Europe.

2010 Under Project BioShield, the first smallpox vaccine for immunocompromised patients is delivered in the USA.

2010 ✪ In the UK, pertussis vaccine (as Tdap Boostrix® by GSK) is recommended to pregnant women from 28 weeks. Several years later this is changed to 20 weeks.

A Chronology of Vaccination

2010–2011

2010 ✪ Intercell's Japanese encephalitis vaccine Ixiaro®, distributed and marketed by Novartis, is licensed in Switzerland. Ixiaro® is a purified, inactivated vaccine made using tissue culture rather than live organisms.

2010 The infamous 1998 "MMR-autism" paper in *The Lancet* is retracted following the General Medical Council's (GMC) judgment of "professional misconduct" by Dr. Andrew Wakefield and coauthor, gastroenterologist Dr. John Walker-Smith, revoking their medical licenses. Dr. Walker-Smith appeals the GMC ruling and wins his case in 2012. Dr. Wakefield's insurance does not cover his legal fees, therefore he never appeals.

2010 President of Autism Science Foundation, Alison Singer (former vice president of business news programming at NBC), elaborates a new model for talking to parents about vaccines: *Making the C.A.S.E. for Vaccines*. This four-step approach features a framework for health professionals to communicate vaccine safety science to combat vaccine hesitancy: "**C**orroborate: Acknowledge the parents' concern and find some point on which you can agree. Set the tone for a respectful, successful talk. **A**bout Me: Describe what you have done to build your knowledge base and expertise. **S**cience: Describe what the science says. **E**xplain/Advise: Give your advice to patient, based on the science."

2010 ✪ The WHO prequalifies MenAfrivac® by Serum Institute of India, a polysaccharide conjugate vaccine against meningoccocal disease group A for people age 1–29 years in African nations. The vaccine contains both thimerosal and aluminum phosphate.

2010 The first therapeutic vaccine (personalized immunotherapy) to treat prostate cancer is approved by the FDA. Provenge® by Dendreon Pharmaceuticals costs US$93,000 for the full three-course treatment.

2010 The US Patient Protection and Affordable Care Act is signed into law, resulting in healthcare coverage plans that must offer patients ACIP-recommended vaccines for free.

2010  The BMGF provides seed funding for the creation of The Vaccine Confidence Project™, led by British anthropologist Heidi Larson at the London School of Hygiene & Tropical Medicine. The aim is to monitor public confidence in vaccines, developed in response to hesitancy and "misinformation" on vaccination programs. The VCP is supported by GSK, Merck & Co. Inc., and Johnson & Johnson, as well as by the European Federation of Pharmaceutical Industries and Associations.

2010 ✪ The EMA grants a conditional marketing authorization to GSK for its split influenza virus, AS03-adjuvanted, thimerosal-preserved, inactivated H1N1 influenza vaccine Arepanrix™. It is also marketed in Canada from 2009, and from 2013 as a H5N1 pandemic vaccine.

2010 Scientists at the J. Craig Venter Institute announce that they have created the first synthetic bacterial genome—without a membrane or cytoplasm—they call Synthia; it becomes the world's first synthetic lifeform.

2010 ✪ Pfizer's (who acquired Wyeth in 2009) new pneumococcal 13-valent conjugate vaccine (PCV13) Prevenar®13 is introduced in the UK and the USA (as Prevnar®13) to provide broader protection against *Streptococcus pneumoniae* infections.

2010 The US federal government sets up "Healthy People 2020" with the goal of attaining high-quality, longer lives free of preventable diseases, disability, injury, and premature death. Vaccination is cited as a key tool in the preservation of health and prevention of infectious diseases. HP2020 "sets out ambitious objectives of 80–90% coverage for most vaccines, as a benchmark for progress." A similar programme, "Health2020," is established in Switzerland.

2010 The ACIP recommends that everyone age 6 months and older get an annual flu shot.

2010 Two Merck scientists, Stephen Krahling and Joan Wlochowski, file their False Claims Act whistleblower lawsuit claiming Merck, the only company licensed by the FDA to sell a mumps vaccine in the US, skewed tests of the vaccine by adding animal (rabbit) antibodies to blood samples. The whistleblowers claim that Merck "defrauded the government for more than a decade in an ongoing scheme to sell the government a mumps vaccine that is mislabeled, misbranded, adulterated and falsely certified as having an efficacy rate that is significantly higher than it actually is." In 2023, the District Court states that, "even assuming the representations were false, none were material to the CDC's purchasing decision," and grants summary judgment in Merck's favor. That same year, the film *Protocol 7* is released, documenting the events up to and including that lawsuit.

2010 Seven girls die in the HPV vaccine trials in India coordinated by PATH in collaboration with Merck (Gardasil®), GSK (Cervarix®), and the BMGF. A report by the Indian Parliament accuses the BMGF and PATH of conducting "a well-planned scheme to commercially exploit" the nation's poverty and powerless-ness and lack of education in rural India in order to push HPV vaccines.

2010 ✪ Menveo® conjugate vaccine by Novartis is licensed in the USA and Europe for 11–55 year olds to prevent invasive meningococcal disease caused by *Neisseria meningitidis* serogroups A, C, Y, and W-135. It is composed of 10 mcg of A and 5 mcg each of C, Y and W-135 oligosaccharides covalently bonded to the CRM197 protein.

2011 The CIA launches a fake hepatitis B vaccine campaign in Pakistan to provide cover for a DNA collecting operation in an attempt to locate Osama bin Laden or his family.

2011 The CDC declares vaccination one of ten public health achievements of the first decade of the 21st century.

2011 The FDA expands the licensure of Prevnar® PCV13 to include people over the age of 50.

2011 Harvard Pilgrim Health Care, Inc., publishes a final VAERS research project report *Electronic Support for Public Health–Vaccine Adverse Event Reporting System*. It reports to the Agency for Healthcare Research and Quality of the US Department of HHS that "fewer than 1% of vaccine adverse events are reported," highlighting the necessity for a more comprehensive reporting system. Despite the program's apparent success and US$ 999,995 budget, there is no follow-up by the public health authorities.

2011 Following the WHO's 2010 endorsed proposal to convene a Committee with an independent chair to review the 2005 IHR, and the experience gained in the global response to the 2009 influenza pandemic, the Harvey V. Fineberg Committee publishes its report clearing WHO of all wrongdoing in relation to pharmaceutical companies, or exaggerating the severity of the 2009 pandemic. Dr. Harvey V. Fineberg is president of the Institute of Medicine (IOM) and responsible for reviewing the government's response to the 1976 swine flu outbreak, whom he also absolved of any wrongdoing at the time.

2011 ✪ The first meningococcal conjugate vaccine, Menactra® by Sanofi, is approved by the FDA for use in infants and toddlers.

2011 Israeli immunologist Dr. Yehuda Shoenfeld and his colleagues publish their seminal article on ASIA, a term they coin to mean "autoimmune/inflammatory syndrome induced by adjuvants." They review "the current data regarding the role of adjuvants in the pathogenesis of immune-mediated diseases."

2011 The ACIP recommends that all boys age 11–12 receive the Gardasil® HPV vaccine.

2011 ✪ In Canada, Sanofi Pasteur's 5-in-1 vaccine (DTaP-IPV—Quadracel®, with lyophilized Hib—ActHIB®) Pediacel® is licensed, for children aged 6 weeks to 6 years. In 2024, Pediacel® will be replaced with Sanofi's DTaP-IPV-Hib vaccine, Pentacel®.

2011 Dutch virologist Ron Fouchier admits to having "mutated the hell out of H5N1" while working on an NIH-funded study, enabling the H5N1 virus to be transmissible by air.

A Chronology of Vaccination

2011–2013

2011 The European Medicines Agency warns against using the H1N1 influenza vaccine Pandemrix® with squalene-based adjuvant AS03 on those under age 20 due to investigations indicating a 6- to 13-fold heightened risk of narcolepsy in vaccinated children in Sweden and Finland. Manufacturer GSK states that "over 31 million doses of Pandemrix have been administered worldwide in 47 countries. A total of 335 cases of narcolepsy in people vaccinated with Pandemrix have been reported to GSK as of July 6, 2011, with 68% of these cases of narcolepsy originating from Finland and Sweden." The EMA marketing authorization for Pandemrix expires in 2015.

2011 The Administration for Strategic Preparedness and Response (ASPR) is allocated an annual budget of US$ 111 million for pandemic influenza preparation.

2011 Following the magnitude 9.0 Tōhoku earthquake and tsunami that causes a failure in the electrical grid and energy backup, the worst nuclear accident since Chernobyl occurs at the Fukushima Daiichi power plant in Ōkuma—polluting the air, soil, and water with radiation.

2011 The WHO reports there are 19 developed countries providing no-fault compensation for those injured by vaccines.

2011 American immunologist Bruce A. Beutler and Luxembourg-born French biologist Jules A. Hoffmann receive the Nobel Prize in Physiology and Medicine "for their discoveries concerning the activation of innate immunity." Canadian medical researcher at Rockefeller University Ralph M. Steinman also receives the Nobel Prize "for his discovery of the dendritic cell and its role in adaptive immunity."

2011 ✪ The EU approves an influenza A H5N1 virus, AS03-adjuvanted, split-virion monovalent vaccine called Pumarix™, manufactured by GSK, for use in people 18 years old and above.

2011 ✪ The first seasonal influenza vaccine manufactured using cell culture technology, Flucelvax® by Novartis (using monkey "Vero" cells), is licensed in the USA by the FDA.

2012 In the USA, the ACIP votes to recommend that all pregnant women receive Tdap vaccine during each pregnancy, ideally between 27 weeks and 36 weeks of gestation, irrespective of prior Tdap vaccination history.

2012 *WORLD IMMUNIZATION WEEK*

The first international World Immunization Week is celebrated from April 21–30, launched by the WHO.

2012 ✪ Two quadrivalent influenza vaccines, an intranasal live-attenuated quadrivalent influenza vaccine (Q/LAIV) FluMist® by MedImmune, and a split inactivated quadrivalent influenza vaccine (I/QIV), Fluarix® by GSK, are first licensed in the US.

2012 At a meeting of company executives, ex-president of GSK Biologicals, Jean Stéphenne, discloses the company's strategy of purchasing all hepatitis B vaccine patents so that competitors would have to pay GSK to market any HepB vaccines. By combining HepB with other non-patented antigens (like tetanus and IPV) in penta- and hexavalent vaccines, this means that these products would also become patent-protected.

2012 ✪ Menhibrix® by GSK—a new combination Hib and meningococcal serogroups C, Y (HibMenCY) vaccine for infants—is licensed by the FDA and recommended by the ACIP.

2012 The first Annual Vaccine Acceptance Meeting takes place at Les Pensières, property of Fondation Mérieux in Annecy, France.

2012 Canadian plant-based biotechnology company Medicago manufactures 10 million doses of a monovalent influenza vaccine candidate within one month, under contract with DARPA.

2012 GSK breaks a new record, paying out US$3 billion for the largest healthcare fraud settlement in US history. GSK admits to its failure to report certain safety data and to illegally marketing its drugs, Wellbutrin® and Paxil® (both antidepressants), Advair® (an asthma drug), Avandia® (a diabetes drug) and Lamictal® (an epilepsy drug) for off-label use. GSK enters into a Corporate Integrity Agreement.

2012 In April, the WHO's Immunization, Vaccines and Biologicals Department publishes the *Guide to the WHO information sheets on observed rates of vaccine reactions.* It is written: "Randomized placebo-controlled trials serve best to determine vaccine reaction rates. In reality, most clinical trials that are performed prelicensure are often not placebo-controlled or do not have the statistical power to determine the vaccine reaction rate of uncommon or rare reactions. Therefore, vaccine reaction rates are usually determined from post-licensure studies."

2012 ✪ A quadrivalent conjugate vaccine against meningococcus serogroups A, C, Y and W-135, Nimenrix® (developed by GSK and later acquired by Pfizer) is approved in Europe as the first quadrivalent vaccine against invasive meningococcal disease to be given as a single dose for children over one year old. Four years later, the meningococcal vaccine is approved for infants 6 weeks of age and older. It is not licensed in the US.

2012 GSK is fined US$93,000 by an Argentine court over its 2007–2008 clinical trials for pneumonia vaccine Synflorix® where 14 babies died. GSK failed to obtain parental consent to conduct trials on 15,000 Argentine babies, and an additional 9,000 babies from Colombia and Panama.

2012 The UN Foundation launches its Shot@Life campaign, aiming to ensure that children worldwide have access to vaccines.

2012 NIAID provides US$21.7M in biodefense pathogen research funds to American scientist Ralph Baric, PhD, at the University of North Carolina, Chapel Hill.

2012 The Global Vaccine Action Plan (GVAP) for 2011–2020 is endorsed by the 194 member states of the WHO World Health Assembly as a framework to "prevent millions of deaths by 2020 through more equitable access to existing vaccines for people in all communities." It calls on member states to reach ≥90% national coverage for all vaccines on the routine immunization schedule by 2020. The Global Vaccine Safety Blueprint is created alongside GVAP with an implementation mechanism called the Global Vaccine Safety Initiative (GVSI) "to set objectives for building the capacity of vaccine safety in low and middle-income countries (LMICs) to better detect, report, and analyse adverse events."

2012 ✪ The first hepatitis E vaccine Hecolin® (HEV 239) containing 25 mcg thimerosal, manufactured by Xiamen Innovax Biotech Co., Ltd., is launched in China for 16–65 years old. The vaccine platform is based on Innovax's unique *Escherichia coli*-based recombinant vaccine.

2013 CureVac in Germany launches its first mRNA-based biological (rabies vaccine) for an infectious disease, in a human phase I clinical trial on 101 people.

2013 ✪ The EMA publishes its "positive opinion" of Sanofi's 6-in-1 vaccine Hexaxim® for use outside Europe, notably Asia. It is added to the WHO's EPI at 6, 10, and 14 weeks old. The same vaccine known as Hexacima® / Hexyon® is approved for Europeans. It is produced in yeast *Hansenula polymorpha* cells by recombinant DNA technology, cultivated on Vero cells, and contains aluminum hydroxide adjuvant, trometamol, and phenylalanine.

2013 The Ministry for Health of Vietnam suspends the use of Quinvaxem® by Crucell after it is linked to the deaths of twelve children.

2013 ✪ The first recombinant surface protein, highly purified, insect cell cultured and egg-free seasonal influenza vaccine FluBlok® Quadrivalent by Protein Sciences Corporation, is introduced in the US and recommended from 18 years old.

A Chronology of Vaccination

2013–2014

2013 ✪ In November, the FDA approves the first H5N1 inactivated avian influenza (bird flu) AS03-adjuvanted split-virion vaccine for adults. The pandemic Influenza A virus adjuvanted monovalent vaccine—also known as Q-Pan H5N1 influenza vaccine and by the brand names Pumarix™ in Europe and Arepanrix™ in Canada—is developed by GSK.

2013 The FDA expands the indication for Novartis's Menveo® vaccine against meningitis to include children from 2–23 months old.

2013 ✪ Novartis Bexsero® multi-component 2-dose meningococcal group B vaccine (later owned by GSK) is licensed in Europe and the USA (in 2015), becoming the first vaccine developed with rDNA using "reverse vaccinology." This 4CMenB vaccine is introduced in the UK National Immunization Program in 2015 at 2, 4, and 12 months.

2013 ✪ The first live, non-replicating, third-generation smallpox vaccine Imvanex® by Bavarian Nordic is licensed in the EU for the prevention of smallpox in the general adult population. "This is the first EU approval of a novel biodefense vaccine developed through a public-private partnership with the US government." It contains an attenuated (weakened) form of modified vaccinia, Ankara-Bavarian Nordic® (MVA-BN) strain. Product development was funded by NIAID (since 2003) and BARDA (since 2007) when a Project BioShield procurement contract was awarded for advanced development and manufacturing. This non-replicating smallpox vaccine (also called JYNNEOS®) has been delivered to the US Strategic National Stockpile since 2010 for emergency use in immune-compromised people. Although the last case of smallpox in the world was recorded in 1977, "the US government considers smallpox a high-priority bioterrorism threat."

2013 Since the WHO's recommendation in 2006, 183 countries have adopted routine pediatric hepatitis B vaccine programs at birth.

2013 In the US, the Pandemic and All-Hazards Preparedness Reauthorization Act is signed into law. It involves an expanded approach to the procurement of countermeasures (including vaccines), and qualified pandemic or epidemic products.

2013 ✪ The first rotavirus and shingles vaccines are licensed in the UK.

2013 HPV vaccine is added to the Japanese national vaccine program in April. In June, following complaints about severe side effects, the health ministry suspends its recommendation that all girls in their early teens receive the vaccine, causing the vaccination rate to drop from 70% for girls born in the mid-1990s to 1% in 2019.

2013 The WHO revises how AEFI are classified and publishes their first guidelines of "Causality assessment of an adverse event following immunization (AEFI), user manual for the revised WHO classification." It introduces new criteria for the assessment of causality of individual adverse events following vaccination. The algorithmic approach, developed by CISA, makes it challenging to detect signals of serious harm—including deaths—after vaccination. Only reactions (with a valid diagnosis) previously acknowledged in epidemiological studies to be caused by the vaccine can be classified as vaccine-product-related-reactions. Reactions observed for the first time during post-marketing surveillance are not considered "consistent with causal association to immunization." All new serious adverse reactions are labeled as coincidental "inconsistent with causal association, indeterminate" or "unclassifiable," and the association with vaccine can be dismissed. The WHO adds that "causality assessment usually will not prove or disprove an association between an event and the immunization. It is meant to assist in determining the level of certainty of such an association. A definite causal association or absence of association often cannot be established for an individual event."

2013   A significant revision of the Swiss Epidemic Law comes into force, including a clause for mandatory vaccination of "at-risk" groups, and tort moral payouts for injury following administration of recommended vaccines capped at CHF 70,000. Clause 6A is added which specifies that the WHO can declare a public health emergency of international concern (PHEIC) that activates a "special situation" which can suspend parliamentary power and constitutional rights.

2013   The recommendation by the JCVI to extend seasonal influenza vaccination in the UK—using AstraZeneca's quadrivalent intranasal Fluenz Tetra® vaccine (FluMist® in the US)—to all children aged 2–16 years, catalyzes the beginning of the UK childhood influenza vaccination program.

2013   The WHO declares that a single-dose of yellow fever vaccine is sufficient in providing long-term protection—a booster every ten years is no longer necessary.

2013   In the EU, medical products under "additional monitoring" due to limited data on long-term use are indicated with a black triangle (▼).

2013   ✪ AstraZeneca gains EU approval for its intranasal quadrivalent influenza vaccine, Fluenz Tetra®.

2014   ✪ Gardasil®9 (nonovalent) by Merck is licensed in the USA.

2014   An Ebola outbreak occurs in Western Africa—the largest Ebola epidemic in history. The WHO declares a PHEIC in August and reports over 28,000 cases and 11,000 deaths until June 2016. The WHO is highly criticized for its slow response.

2012–2021 **Swiss vaccine campaigns via the media (TV, billboards, magazines), schools, and pharmacies**

Influenza for medical personnel

I get vaccinated–for the well-being of my patients and my team.

Pertussis for pregnant women

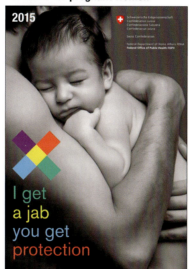

COVID-19 for general population

Source: Federal Office of Public Health (FOPH), Switzerland

Major National Vaccination Campaigns in Switzerland

Smallpox	Early 1800s / 1944-1948	**Measles**	1987 / 2009 / 2015
Polio	1960 / 1970 / 1980 / 1990	**HPV**	2008 / Yearly in schools
Tetanus	Troops during WWII / Civilians in 1963	**Pertussis**	2013 / 2015*
Influenza	Yearly since 1985 / 2009*	**COVID-19**	2021* / 2022**

*including for pregnant women / **bivalent booster from age 16, including pregnant women from 14 weeks

A Chronology of Vaccination

2014	In May, the WHO's secretary-general declares polio a PHEIC.
2014	Andrea Ventura's group at the Memorial Sloan Kettering Cancer Center in New York, uses CRISPR-Cas9 to engineer a respiratory virus that can cause lung cancer in mice.
2014	The World Mosquito Program first begins releasing genetically modified mosquitoes in Brazil to combat dengue fever, essentially "vaccinating mosquitoes against giving humans disease."
2014	As part of a worldwide recall following reports of contamination with iron oxide and/ or other metal particulates in some vials, meningococcal vaccine Meningitec® is taken off the market in Australia.
2014	✪ Meningococcal group B vaccine Trumenba® by Pfizer is approved in the USA on an accelerated approval regulatory pathway due to several outbreaks on college campuses in 2013–2014.
2014	The White House Office of Science and Technology Policy and HHS institute a gain-of-function research moratorium and pause funding on any dual-use research into specific pandemic-potential pathogens (influenza, MERS, and SARS). The NIH issues eighteen cease-and-desist orders to GoF research. This moratorium will be lifted in early 2017.
2014	In October in Switzerland, the online reporting portal ElViS (Electronic Vigilance System), is launched, enabling healthcare professionals to report adverse events of drugs and vaccines online to one of the regional Swiss pharmacovigilance centers.
2014	Brain-damaged victims of the 2009 UK swine flu vaccine receive £60 million in compensation from the government under the Vaccine Damage Payments Scheme. It is subsequently revealed that Pandemrix® could cause narcolepsy and cataplexy in about one in 16,000 people. Across Europe, more than 800 children are known to have been affected.
2014	✪ The FDA approves a quadrivalent formulation of the inactivated influenza vaccine Fluzone® by Sanofi Pasteur.

2014	The Agency for Healthcare Research and Quality for its Effective Health Care Program publishes the 740-page report *Safety of Vaccines Used for Routine Immunization in the United States,* following up from the IOM 2012 report. One of the technical experts is Kathryn M. Edwards, and Frank DeStefano is on the panel of peer reviewers.
2014	The FDA approves the first ever intravenous medication for influenza, peramivir (Rapivab), to treat adults with influenza.
2015	DARPA launches the "Bot Challenge, an open competition for researchers to study the influence of social media bots (automated software programmed to pattern human behavior) on US vaccination conversations."
2015	✪ Meningococcal group B and A, C, W, Y vaccines are approved in the UK.
2015	✪ In the UK, two 5-in-1 vaccines (DTaP-IPV-Hib) are approved, Pediacel® by Sanofi Pasteur, and Infanrix® by GSK, for children aged 2 months to less than 10 years.
2015	The WHO's SAGE Working Group on Vaccine Hesitancy (including Heidi Larson) publishes an article in the August issue of *Vaccine*, entitled "Strategies for Addressing Vaccine Hesitancy – A Systematic Review." It concludes that "given the complexity of vaccine hesitancy and the limited evidence available on how it can be addressed, identified strategies should be carefully tailored according to the target population, their reasons for hesitancy, and the specific context."
2015	The FDA expands the licensure of HPV vaccine Gardasil®9 to include males from 16–26 years old, to protect against cancers of the head, neck, penis, and anus.
2015	✪ The FDA approves a new injectable seasonal trivalent influenza vaccine, Fluad® by Seqirus, for people aged 65+. Like most flu vaccines, the antigens are grown using egg proteins, but it is the first oil-in-water adjuvanted vaccine approved in the US for anyone other than the military (used since the 1950s). Adjuvant MF59 contains squalene, a protein common in the human body.

USA: Pertussis (Tdap) vaccine campaign for pregnant women

Getting your whooping cough vaccine in your 3rd trimester...

helps protect your baby from the start.

Outbreaks of whooping cough are happening across the United States. This disease can cause your baby to have coughing fits, gasp for air, and turn blue from lack of oxygen. It can even be deadly. When you get the whooping cough vaccine (also called Tdap) during your third trimester, you'll pass antibodies to your baby. This will help keep him protected during his first few months of life, when he is most vulnerable to serious disease and complications.

Talk to your doctor or midwife about the whooping cough vaccine.

CDC
U.S. Department of Health and Human Services
Centers for Disease Control and Prevention

Born with protection against whooping cough.
www.cdc.gov/whoopingcough

AMERICAN ACADEMY OF FAMILY PHYSICIANS   American Academy of Pediatrics   AMERICAN COLLEGE of NURSE-MIDWIVES   The American College of Obstetricians and Gynecologists

February 2015

2015	✪ Boostrix® Tdap is recommended for every pregnancy in the US, Australia, and in Switzerland, from 27 weeks gestation—intended as ideal protection against pertussis for the newborn. In Switzerland, several years later, the recommendation is changed to vaccinating from 13 weeks.
2015	✪ Emergent BioSolution's anthrax vaccine, BioThrax®, becomes the first vaccine to receive approval for a new indication (post-exposure prophylaxis) based on the FDA's Animal Rule, which was implemented in 2002 to facilitate rapid licensure of medical countermeasures for "lethal or permanently disabling" conditions.
2015	In parts of Brazil, a wave of microcephaly in newborns is attributed to the mosquito-borne Zika virus. The wave occurs after the initiation of a maternal Tdap program (using up to three doses of GSK's Boostrix®—licensed in Brazil as Refortrix®) with a reported uptake during pregnancy of 44.9%.
2015	✪ A new combined DTaP-IPV vaccine Quadracel®, by Sanofi Pasteur, is licensed by the FDA for use in children 4–6 years old.

2015	✪ The FDA approves the vaccine Bexsero®, by GSK, to prevent disease caused by *Neisseria meningitidis* serogroup B.
2015	✪ GSK's Mosquirix® is a recombinant and adjuvanted vaccine against malaria for infants 6 weeks to 17 months. This RTS,S / AS01E vaccine is approved for use by the EMA in markets outside the EU, notably South America, Africa, and Asia. It is the first time the adujvant AS01E is used in a licensed vaccine. This malaria vaccine also includes hepatitis B surface antigen, HBsAg.
2015	The American Medical Association adopts a policy that supports terminating non-medical vaccine exemptions for all schoolchildren and healthcare professionals.
2015	A measles outbreak of around 145 cases in Disneyland causes a media frenzy and raises fears of the unvaccinated, who are blamed for the outbreak—no deaths or serious complications are reported.
2015	Using the measles outbreak as a catalyst, California bill SB277 is passed, promoted by Senator Richard Pan, making it the third state to eliminate all non-medical (philosophical and religious) vaccine exemptions for entry into school.
2015	The Bill & Melinda Gates Foundation makes a US$52 million equity investment in German clinical-stage biopharmaceutical company CureVac to support continued development of its mRNA technology platform, and the construction of an industrial-scale GMP production facility.
2015	An international team, including two scientists from the Wuhan Institute of Virology, genetically modify a mouse-adapted SARS virus with a spike protein from a bat coronavirus to infect a human cell line (HeLa), resulting in a "chimeric" virus that successfully replicates in human cells.
2015	Irish biochemist William C. Campbell and Japanese biochemist Satoshi Ōmura receive the Nobel Prize in Physiology or Medicine "for their discoveries concerning a novel therapy [ivermectin] against infections caused by roundworm parasites."

A Chronology of Vaccination

2015	Lymphatic vessels are discovered in the central nervous system by a team led by Drs. Antoine Louveau and Jonathan Kipnis of the University of Virginia School of Medicine in the USA. This leads to an improved understanding of the direct connection between the brain and the immune system.
2016	✪ Gardasil®9 (nonovalent) by Merck Sharp & Dohme is licensed in Switzerland.
2016	Healthcare providers in the USA are required by law to get an annual influenza shot to keep their jobs, despite its low effectiveness.
2016	"No Jab No Pay" legislation comes into effect in Australia, allowing only medical vaccine exemptions, while conscientious objections are no longer tolerated. This law withholds three benefits from lower-income parents of children < 20 years old who are not fully vaccinated: Child Care Benefit, the Child Care Rebate, and a part of the Family Tax Benefit.
2016	The FDA expands the approval of Pfizer's pneumococcal conjugate vaccine Prevnar-13® to include adults 18–49 years old.
2016	Robert F. Kennedy Jr. announces the launch of The World Mercury Project as a public health advocacy organization dedicated to ending exposure to neurotoxic mercury in fish, medical products, dental amalgams, and vaccines. In 2018, the organization is renamed Children's Health Defense.
2016	✪ Sanofi Pasteur's hexavalent DTaP-IPV-Hib-HepB vaccine Vaxelis® is first licensed by the EMA in Europe. It is recommended from 6 weeks old and contains Merck's proprietary adjuvant amorphous aluminum hydroxyphosphate sulfate (AAHS).
2016	Moderna Inc., begins its first clinical trial of a lipid nanoparticle-encapsulated mRNA-based vaccine against influenza—using H10N8 and H7N9 strains.
2016	The Cochrane Collaboration accepts a grant from the BMGF worth US$1.15 million.
2016	✪ In Australia, shingles vaccine (herpes zoster), Zostavax®, is licensed for people age 70 years and older.

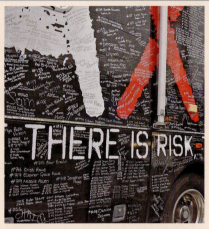

2016–2017	Polly Tommey, English mother of a vaccine-injured child and producer of *Vaxxed: From Cover-Up to Catastrophe*, travels the US by bus documenting video testimony of vaccine injuries and collecting over 7,000 signatures.
2016	In March, the documentary *Vaxxed: From Cover-Up to Catastrophe*, directed by Dr. Andrew Wakefield and produced by Del Bigtree, is pulled from the New York Tribeca Film Festival. This highly publicized event creates a massive buzz, leading to sold-out theatres nationwide. The documentary investigates how the CDC concealed and destroyed data from its 2004 study that showed a link between the timing of the MMR vaccine and autism in African-American boys. The information was disclosed by a CDC whistleblower, senior research scientist William Thompson.
2016	The FDA approves expanding the age range of GSK's inactivated influenza vaccine FluLaval® Quadrivalent to include vaccinating children from 6–35 months old for the prevention of illness by influenza A (H1N1/H3N2) and type B viruses.
2016	The Informed Consent Network, ICAN, is founded by Del Bigtree, former TV producer for the show *Doctors*. Wikipedia claims it "spreads misinformation about the risks of vaccines" while misrepresenting ICAN's many landmark legal and educational activities on vaccine safety.

2016

In addition to Australia's new "No Jab, No Pay" law, Victoria introduces "No Jab, No Play" legislation. This law requires vaccination for enrollment in "early childhood education and care services"—conscientious objection is no longer a permitted exemption. This policy is extended to Western and South Australia by 2020. Amendments to the Queensland Public Health Act 2005 allow the exclusion of an unvaccinated child from early childhood education and care services.

2016

The 17 Sustainable Development Goals (SDGs) of the 2030 Agenda for Sustainable Development—adopted by world leaders at the historic 2015 UN Summit—officially come into force. One of the goals is to "ensure healthy lives and promote well-being for all at all ages."

2016 The Insect Allies program is announced by DARPA as a research project that "aims to protect the US agricultural food supply by delivering protective genes to plants via insects," using an approach of "horizontal environmental genetic alteration agents."

2016 Ralph Baric, PhD, publishes a research article in *Proceedings of the National Academy of Sciences* entitled "SARS-like WIV-CoV Poised for Human Emergence." WIV-CoV is a bat coronavirus collected by the Wuhan Institute of Virology in China.

2016 ✪ Oral cholera vaccine Vaxchora®, produced by Emergent BioSolutions Inc., against the *Vibrio cholerae* serogroup 01 is approved by the FDA in the US.

2016

The Vaccine Knowledge Project, managed by the Oxford Vaccine Group, led by Prof. Andrew J. Pollard from 2001, becomes a part of the WHO's Vaccine Safety Net. The VKP claims to be a source of independent, evidence-based information about vaccines and infectious diseases.

2016 The ID2020 Alliance is founded with seed money from Microsoft, Accenture, PriceWaterhouseCoopers, the Rockefeller Foundation, Cisco, and Gavi. The alliance claims to advocate in favor of "ethical, privacy-protecting approaches to digital ID," with the mission of providing digital IDs for people worldwide by 2030. These digital IDs would be tied to biometric data, demographic information, medical records, education data, financial data, and more. In 2020, the alliance launches the Good Health Pass Collaborative (GHPC), becoming the world's leading digital ID organization.

2016 In an effort to curb mosquito-borne infection, Florida and other US states are sprayed with Naled, an organophosphate insecticide registered since 1959 by the US-EPA. It is estimated that over 16 million acres of land are sprayed every year.

2016 After launching its "A Fair Shot" campaign in 2015, Doctors Without Borders collects 416,000 signatures from people in over 170 countries, and successfully pressures GSK and Pfizer to reduce their pneumonia vaccine price from US$68 to US$5 per dose for children in developing countries.

2016–2017 ✪ Dengvaxia® vaccine by Sanofi Pasteur is given to over 733,000 children and over 50,000 adult volunteers regardless of serology status. Antibody dependent enhancement (ADE) was observed during clinical trials, increasing the risk of dengue disease severity in those previously uninfected. Despite this, the Philippine Department of Health launches a widespread vaccine campaign for schoolchildren—which ends up injuring many children, fourteen of whom die before the vaccine is withdrawn.

A Chronology of Vaccination

2017 The American Academy of Pediatrics (AAP) issues a policy stating that all newborns in the US should routinely receive recombinant hepatitis B vaccine within 24 hours of birth.

2017 At the beginning of January, in his last weeks in office, US President Barack Obama lifts the moratorium on gain-of-function (GoF) research. By December, the NIH resumes funding of GoF research—under the P3CO policy framework—because it is deemed "important in helping us identify, understand, and develop strategies and effective countermeasures [including vaccines] against rapidly evolving pathogens that pose a threat to public health."

2017 At his Georgetown University keynote presentation on January 10, Dr. Anthony Fauci opens and closes his speech with one prediction: "There is no question that there will be a challenge to the coming [Trump] administration in the arena of infectious diseases. There will be a surprise outbreak."

2017 In January, Crépeaux et al., publish an important research article in *Toxicology* called "Non-Linear Dose-Response of Aluminium Hydroxide Adjuvant Particles: Selective Low Dose Neurotoxicity." An unexpected neurotoxicological pattern limited to a low dose of Alhydrogel®—vaccine adjuvant aluminum oxyhydroxide (Al)—is observed. They conclude that "Alhydrogel® injected at low dose in mouse muscle may selectively induce long-term Al cerebral accumulation and neurotoxic effects," potentially overturning "the dose makes the poison" rule of classical chemical toxicology.

2017 The MHRA receives a £980,000 grant by the BMGF to fund work "on improving safety monitoring for new medicines in low and middle-income countries." Following a Freedom of Information request in 2022, the MHRA states that it has received approximately £3 million from the BMGF spanning several years.

2017 ✪ The first typhoid conjugate vaccine (TCV) Typbar-TCV® by Indian company Bharat-Biotech achieves WHO prequalification.

2017

The Merck Foundation is established as a German nonprofit, the philanthrophic arm of Merck KGaA that "aims to improve the health and well-being of people and advance their lives through science and technology."

2017 Meningococcal polysaccharide quadrivalent vaccine, Menomune® A, C, Y and W-135 by Connaught—now owned by Sanofi—is discontinued in the US, replaced by meningococcal conjugate vaccines Menveo® by Sanofi and Menactra® by GSK.

2017 In April, the Indian government blocks the BMGF from further financing the Public Health Foundation of India and other NGOs in an effort to obstruct the BMGF from influencing India's national vaccination program.

2017 Sanofi and AstraZeneca announce an agreement to develop and commercialize Beyfortus®, a recombinant monoclonal antibody for babies to protect them against RSV. This is the first monoclonal antibody therapy that will be marketed as a "vaccine" and added to the childhood vaccination schedule in the USA in 2023.

2017 CureVac's phase I clinical trial of its RNActive® prophylactic rabies vaccine is published in *The Lancet.* The study is the first in-human proof-of-concept clinical trial of a prophylactic mRNA-based vaccine. Subsequently, the mRNA drug substance encoding a rabies virus glycoprotein is given the International Nonproprietary Name (INN) Nadorameran by the WHO as the first drug substance of this new class. A conclusion from CureVac's mRNA rabies vaccine trial is that "intradermal or intramuscular needle-syringe injection was ineffective."

2017 During a TEDx event, Tal Zaks, Moderna's chief medical officer, gives a presentation on mRNA vaccine technology. Moderna can hack "the software of life," claims Zaks, by introducing or changing "a line of code," using bioinformatics to prevent and treat disease.

2017

CEPI

The Coalition for Epidemic Preparedness Innovations (CEPI) is founded by the government of Norway, the BMGF, the Wellcome Trust, the World Economic Forum (WEF), and India's department of biotechnology, with an initial investment of US$ 460 million. Its objective is "to outsmart epidemics by developing safe and effective vaccines against known infectious disease threats that could be deployed rapidly to contain outbreaks, before they become global health emergencies."

2017 Eli Lilly & Company and CureVac announce a global immuno-oncology collaboration focused on the development and commercialization of up to five potential cancer vaccine products based on CureVac's proprietary RNActive® technology. The companies will use messenger RNA (mRNA) technology to target tumor neoantigens for a more robust anti-cancer immune response.

2017 A class action lawsuit is filed in Colombia against the Colombian government and Merck Sharp & Dohme by a group representing 700 individuals who suffered neurological and immune-related adverse events following a Gardasil® vaccine campaign in 2014. They seek approximately US$ 30.5 million in compensation for damages. As of May 2022, plaintiffs are still awaiting judicial results in the 28th Civil Court of the Bogota Circuit. An ongoing class action lawsuit over the HPV vaccine in Japan involves 63 plaintiffs.

2017 Danish medical anthropologist Peter Aaby publishes a paper in *eBioMedicine* called "The Introduction of Diphtheria-Tetanus-Pertussis and Oral Polio Vaccine Among Young Infants in an Urban African Community: A Natural Experiment." This important paper examines the impact of vaccines on child mortality, and finds that DTP vaccinations are associated with increased infant mortality. His findings are dismissed as inconclusive by the WHO, but are confirmed numerous times in subsequent papers.

2017 ✪ In Australia, the 2-dose oral rotavirus vaccine Rotarix® by GSK is introduced from 2 months old, replacing Merck's RotaTeq®.

2017 Novartis's drug Kymriah® is the first gene therapy available in the USA. This T-cell tailor-made treatment genetically modifies a cancer patient's own cells to fight disease. It is approved by the FDA for children and young adults for an aggressive type of blood cancer—B-cell acute lymphoblastic leukemia—and costs US$ 475,000. Novartis claims not to charge patients if they do not respond to therapy within a month.

2017 On May 31, lawyer Robert F. Kennedy Jr., at the request of the White House, has a two-hour meeting with Francis Collins, director, NIH; Anthony Fauci, director, National Institute of Allergy and Infectious Diseases (NIAID); and other government officials at the National Institutes of Health (NIH) in Bethesda to discuss vaccine safety. Kennedy presents some of the significant deficiencies in safety testing and surveillance (both pre- and post-licensure) and the well-documented conflicts of interest at the CDC and FDA. Instead of addressing the valid safety concerns raised, Francis Collins concludes their interactions with "in our view and that of the vast majority of objective experts in medicine and public health, there is overwhelming scientific evidence that supports the safety and exceptional value of vaccinations."

2017 In October, Johns Hopkins Center for Health Security publishes a 76-page report *SPARS Pandemic 2025–2028: A Futuristic Scenario for Public Health Risk Communications.* This is a hypothetical scenario of the emergence of a novel coronavirus—named the St. Paul Acute Respiratory Syndrome Coronavirus (SPARS-CoV, or SPARS)—tested using pancoronavirus RT-PCR. The scenario is "designed to illustrate the public health risk communication challenges that could potentially emerge during a naturally occurring infectious disease outbreak requiring development and distribution of novel and/or investigational drugs, vaccines [Corovax], therapeutics, or other medical countermeasures."

A Chronology of Vaccination

2017–2018

2017 ✪ FDA licenses Shingrix®, the new herpes-zoster vaccine from GlaxoSmithKline, for use in adults age 50 and older. It uses a surface protein of the varicella zoster virus produced by culturing genetically engineered Chinese hamster ovary cells. It will replace the less effective Zostavax® vaccine, which will be discontinued in the US in December 2020. In 2023, GSK reports global Shingrix® vaccine sales of £3.4 billion.

2017 The most expensive year to date for payouts under the US Vaccine Injury Compensation Program. Seven hundred and six awards are paid out, costing the taxpayer over US$280 million per year, an average of US$5.4 million per week or US$772,600 per day to pay for vaccine-injured victims.

2018 Ompattro® by Alnylam is the first FDA-approved RNA lipidnanoparticle (LNP) gene-based therapeutic.

2018 Vaccinologist Dr. Stanley A. Plotkin volunteers to testify as an expert witness in the legal case of *Lori Matheson v. Michael Schmitt*, regarding the vaccination of their daughter, Faith. His January video deposition in Pennsylvania by attorney Aaron Siri Esq. of New York lasts nearly 9 hours. Read further details on pages 152–153.

2018 The Wuhan Institute of Virology's BSL-4 laboratory begins operations.

2018 Tdap vaccine is officially added to the vaccination schedule in Australia (as dTpa) for every pregnancy after 28 weeks, using either Boostrix® or Adacel®.

2018 Australia funds an MMR vaccine catch-up campaign for adults born during or since 1966, and aged ≥20 years without evidence of valid MMR vaccination or serological immunity.

2018 French President Emmanuel Macron expands the childhood vaccine mandates in France from three vaccines (diphtheria, tetanus, polio) to eleven (pertussis, *haemophilus influenzae* type B, measles, mumps, rubella, hepatitis B, meningitis, and pneumococcus).

2018

In May, the WHO and World Bank Group launch the Global Preparedness Monitoring Board (GPMB) as a new "mechanism to strengthen global health security" by providing a comprehensive appraisal of global preparedness for health emergencies. The board is co-chaired by Dr. Gro Harlem Brundtland, former prime minister of Norway and former WHO director-general, and will include Dr. Chris Elias (BMGF/PATH) and Sir Jeremy Farrar, Wellcome Trust director.

2018 Chinese scientist Jiankui He in Shenzhen uses CRISPR-Cas9 to create the first genome-edited human babies, in an attempt to make them resistant to HIV infection. His violation of Chinese laws results in him losing his job and being jailed for 3 years.

2018 Following legal action initiated by the Informed Consent Action Network (ICAN) in the USA, it is discovered that no formal safety report—required under the National Childhood Vaccine Injury Act for safer childhood vaccines—has ever been prepared or filed since 1989.

2018 According to Merck's annual report, global sales for the following vaccines are: Gardasil®—US$3,151,000,000 ProQuad®—US$2,249,000,000 Pneumovax®23—US$907,000,000 RotaTeq®—US$728,000,000 and Zostavax®—US$217,000,000. Merck's total sales for all human vaccines in just one year is US$7.25 billion.

2018 ✪ The FDA licenses the 6-in-1 vaccine (DTaP-Hib-IPV-HepB) Vaxelis® by MCM Vaccine (a partnership between Sanofi Pasteur and Merck), for use in children age 6 weeks through 4 years as a 3-dose series for infants at 2, 4, and 6 months old. The ACIP votes to add Vaxelis® to the Vaccines for Children Program in 2019. Vaxelis® becomes commercially available in the US in 2021.

2018

A day-long novel parainfluenza virus pandemic modeling tabletop exercise "Clade X" is hosted by Johns Hopkins Center for Health Security.

2018 ✪ Dengvaxia® by Sanofi Pasteur is approved in Europe for use in people who have already been infected with the dengue virus and who live in areas where it is endemic. The USA will approve the vaccine a year later. According to Dr. Paul Offit, this dengue vaccine "has been shown to enhance hemorrhagic-shock syndrome upon exposure to wild-type virus in seronegative, vaccinated children."

2018 ✪ Dynavax's Heplisav-B® hepatitis B vaccine becomes the first licensed vaccine in the USA to contain synthetic DNA. It includes a novel synthetic cytosine phosphoguanine (CpG) enriched oligodeoxynucleotide (ODN) phosphorothiate 1018 adjuvant (CpG 1018). It was approved in 2017 by the FDA, ignoring the increased prevalence of heart attacks in the clinical trials (with no true saline placebo). The ACIP/CDC unanimously vote for the recommendation (2-dose regimen) despite having no data on administration with other vaccines, notably other adjuvanted vaccines—they are all assumed safe. The safety signal of higher rates of heart attacks in recipients is noted. The wholesale acquisition price per dose is US$ 115.

2018 In August, ICAN requests from the FDA the clinical trials relied upon to license Merck's MMR-II vaccine in 1978. The FDA responds seven months later with 215 pages confirming there were no placebo-controlled clinical trials prior to licensure of MMR. It is revealed that fewer than 1,000 children were studied, with follow-up of only 42 days.

2018 According to the HPV vaccine-related VAERS declarations, as of mid-August there are a total of 59,226 reports; 8,682 serious; 16,380 emergency room visits; 432 deaths; and 142 cases of cervical cancer. From 2006 to 2018, approximately 33 million people, mainly children, have been vaccinated with Gardasil® in the USA.

2018 Sabin Vaccine Institute and the Aspen Institute receive funding from the BMGF and the Wellcome Trust to form a partnership, the Sabin-Aspen Vaccine Science & Policy Group. In June 2020, this group publishes a report entitled *Meeting the Challenge of Vaccine Hesitancy* that proposes "actionable steps that leaders across healthcare, research, philanthropy and technology can take to build confidence in vaccines and vaccinations."

2018 The Pirbright Institute, formerly the Institute for Animal Health, a government and privately funded research center in England, is granted a patent #US10130701B2 for its genetically engineered (GE) coronavirus that causes avian bronchitis. The patent stipulates that the GE coronavirus may be used in a vaccine "to treat a human, animal, or avian subject."

2018 The WHO revises how AEFI are classified. New evidence on areas such as "substandard and falsified vaccines" (vaccines that deliberately/fraudulently misrepresent their identity, composition, or source—reported for yellow fever, meningitis, rabies and pentavalent vaccines), "immunization anxiety" and "immunization triggered stress responses" are incorporated into the "new guidance documents. See Appendix H.

2018 The seminal research paper "Aluminium in Brain Tissue in Autism" is published in the *Journal of Trace Elements in Medicine and Biology* by British scientists Matthew Mold, Dorcas Umar, Andrew King, and aluminum expert, Dr. Christopher Exley. Although they analyze brain tissue from only 10 subjects with autism spectrum disorder, they report clear evidence of "necrotic cell death and concomitant inflammation" as well as "extraordinarily high" aluminum content of brain tissue in ASD.

2018 **BILL & MELINDA GATES MEDICAL RESEARCH INSTITUTE**
The Gates Medical Research Institute (GMRI) is founded by the BMGF as a subsidiary, a "nonprofit pharmaceutical enterprise" to develop new in-house drugs and vaccines. With US$ 500 million from the BMGF, GMRI works under contract with Merck and GSK.

Stanley Alan Plotkin, MD

Expert Witness Deposed by Aaron Siri, Esq.

In January 2018, New York attorney-at-law Aaron Siri conducted a nine-hour deposition with volunteer expert witness on vaccinology, Stanley A. Plotkin, in the case of *Lori Matheson v. Michael Schmitt.*

After the extensive deposition, Dr. Plotkin withdrew his expert testimony from the legal case, keeping it out of official records. However, the videotaped deposition and written transcript were leaked online.

Born in 1932, medically trained Dr. Plotkin began his career in 1957 as an Epidemic Intelligence Service (EIS) Officer for the US Public Health Service and has worked extensively at the Children's Hospital of Philadelphia and the University of Pennsylvania. He has also been a paid consultant to Sanofi Pasteur, GlaxoSmithKline, Merck, Pfizer, Inovio Pharmaceuticals, Variations Bio, Takeda Pharmaceutical Company, Dynavax Technologies, Serum Institute of India, CureVac, Valneva SE, Hookipa Pharma, and NTxBio. He has served on the board of Dynavax, VBI Vaccines, Inovio, Geovax labs, CureVac, GlycoVaxyn, Adjuvance Technologies, and BioNet Asia among others. He was chairman of the Scientific Advisory Board for biotech company Mymetics, was medical and scientific director for Pasteur Mérieux Connaught (now Sanofi), co-founded CEPI in 2017, and through his company Vaxconsult has advised all major vaccine manufacturers, the ACIP, the CDC, the Bill & Melinda Gates Foundation, and the NIH.

As well as being the editor of the reference bible on vaccines, *Plotkin's Vaccines*—first published in 1988 and currently in its eighth edition—he was involved in the development of vaccines against rubella, polio, rotavirus, varicella, and rabies, receiving millions of dollars in royalties. As such, it is fair to state that his professional career is based upon his ability to confidently assert and assure the community that vaccines are safe and effective while advocating for their universal use in babies and children worldwide.

While in this Michigan case, *Matheson v. Schmitt,* the judge ruled in favor of the vaccinating father, Dr. Plotkin's testimony, given at the behest of Voices for Vaccines, highlights many outstanding serious safety concerns with vaccines. Siri skillfully covers the issues of inadequate data, the absence of saline placebos or appropriate controls in prelicensure clinical trials, and missing investigations in general health outcomes in vaccinated and unvaccinated children.

Over the course of the nine-hour deposition, memorable episodes include Dr. Plotkin's insistence that hepatitis B vaccine Engerix®—given at birth in the USA since 1991, when the vaccine still contained both aluminum and thimerosal—was monitored for safety in babies longer than four days post-administration until confronted with the evidence noted in the manufacturer's package insert. He maintains the MMR vaccine was tested against a saline placebo in clinical trials before it was licensed in 1971 and again is presented with the evidence that it never was. Throughout the deposition, the 1,691-page seventh edition of *Plotkin's Vaccines* sits in front of him; as the references are only listed online, however, Dr. Plotkin is surprisingly unable to cite any studies or scientific data to back up his own statements proclaiming the safety of vaccines.

Dr. Plotkin concedes that The Task Force for Global Health, current guardians of the Vaccine Safety Datalink (not accessible to independent researchers) and the Brighton Collaboration, is funded by the major vaccine manufacturers and that he himself suggested the creation of "its administrative product," Voices for Vaccines. Dr. Plotkin affirms, "I was one of those who suggested that an organization of laypeople, as opposed to scientists, would be a good idea to oppose all of the nonsense that one sees on the web from anti-vaccination organizations." Having found "laypeople who were interested in promoting vaccines," both he and Dr. Paul Offit took seats on the Scientific Advisory

Board of Voices for Vaccines. When confronted with the Institute of Medicine's 2011 report stating that evidence is inadequate to accept or reject a causal link between DTP/DTaP and autism, Dr. Plotkin maintains that one cannot "prove a negative." However, he insists that vaccines have been extensively studied and that telling parents that vaccines do not cause autism is justified, even though scientific data to back up that claim does not exist. Dr. Plotkin agrees there have never been government- or industry-funded studies looking at total health outcomes of children who follow the CDC's vaccination schedule compared to those who are completely unvaccinated. Despite this, Dr. Plotkin maintains there is "abundant evidence that vaccines do contribute to the health of children."

When asked about aluminum adjuvants, Dr. Plotkin admits he is "not aware that there is evidence that aluminum disrupts the developmental processes in susceptible children." Several scientific papers on findings of aluminum migrating to the brain are cited by the legal team, notably the work of Canadian neuroscientist Prof. Christopher Shaw, as well as Israeli doctor and immunologist Yehuda Shoenfeld. Both scientists have highlighted the numerous issues with aluminum toxicity, including neurological and autoimmune disorders likely induced by aluminum adjuvants. Despite his lack of knowledge, Dr. Plotkin largely dismisses these concerns.

In his subsequent April 2019 article, "How to Prepare for Expert Testimony on the Safety of Vaccination" published in *Pediatrics,* Dr. Plotkin states "That lawyer grilled me for ten consecutive hours on subjects ranging from putative vaccine reactions, the potential dangers of adjuvants in vaccines (including aluminum), the causes of autism, possible contaminants in vaccines, and other antivaccination tropes. The opposing attorney was clearly skilled and knowledgeable about vaccination practices." Dr. Plotkin, however, disparages the science brought forward by Siri and his legal team claiming that "in his questioning, the lawyer constantly referred to articles of doubtful scientific quality."

In response to his experience and to help future expert witnesses better prepare to defend vaccine safety, Dr. Plotkin, together with Dr. Paul Offit and Heather Bodenstab, PhD, created an online library at the Children's Hospital of Philadelphia to contain references on vaccine safety from "reliable medical literature" to enable extensive rebuttals and to better arm vaccinology expert witnesses on "how to handle questions that are not grounded in accepted science."

Though Dr. Plotkin's deposition is very long, it is well worth watching, or reading, to appreciate the depth of topics addressed. It was the first occasion—thankfully not the last (Dr. Kathryn M. Edwards is deposed in 2020)—that such an esteemed expert vaccinologist had been subjected to intensive questioning by an educated attorney well-versed in issues of vaccine safety.

An outline of the nine-hour video along with certain timestamped extracts is available on archive.org—please see the link provided below. All links mentioned below were accessed and available on June 13, 2024.

Further Reading

Consult the selected timestamped extracts from the testimony
https://archive.org/details/2015-831539-DM

Read the full 400-page deposition transcript online
https://www.lumenfidei.ie/documents/dr-stanley-plotkin-testimony.pdf

Watch the full nine-hour video, still on YouTube
https://www.youtube.com/watch?v=vhNGu3jFyIU&t=4860s

"How to Prepare for Expert Testimony on the Safety of Vaccination" published in *Pediatrics*
https://pubmed.ncbi.nlm.nih.gov/30842258/

Children's Hospital of Philadelphia – Vaccine Safety References
https://www.chop.edu/centers-programs/vaccine-education-center/vaccine-safety-references

"I explained that I was of the opinion that there were exceptions in which vaccinations could cause autism. More specifically, I explained that in a subset of children with an underlying mitochondrial dysfunction, vaccine induced fever and immune stimulation that exceeded metabolic energy reserves could, and in at least one of my patients, did cause regressive encephalopathy with features of autism spectrum disorder."

—Dr. Andrew Zimmerman

American board-certified, pediatric neurologist and former director of medical research, Center for Autism and Related Disorders, Kennedy Krieger Institute, and Johns Hopkins University School of Medicine.

Medical expert witness in the Vaccine Injury Compensation Program's Omnibus Autism Proceeding (OAP) in 2007. When Department of Justice attorneys learned of his opinion, they fired him as an expert witness, suppressed his opinion, and misrepresented his scientific views in the closing argument in the OAP test case of *Hazlehurst v. HHS.*

Affadvit – September 7, 2018

https://thewilddoc.com/wp-content/uploads/2019/02/Dr.-Andrew-Zimmerman%E2%80%99s-Full-Affidavit-on-alleged-link-between-vaccines-and-autism-that-U.S.-government-covered-up.pdf

"I was involved in misleading millions of people about the possible negative side effects of vaccines. We lied about the scientific findings.

I regret that my coauthors and I omitted statistically significant information in our 2004 article published in the journal Pediatrics. The omitted data suggested that African American males who received the MMR vaccine before age 36 months were at increased risk for autism. Decisions were made regarding which findings to report after the data were collected, and I believe that the final study protocol was not followed."

—Dr. William W. Thompson, PhD

Senior Scientist at the CDC who declared that in 2004, the CDC fraudently manipulated and destroyed safety data showing that African American boys were at an increased risk of autism following vaccination. For more information watch the documentary *Vaxxed: From Cover-Up to Catastrophe.*

A Chronology of Vaccination

2018–2019

2018 Vaccines are produced in 11 European countries across 27 production sites, including 12 research sites leading the discovery of next-generation vaccines.

2018 In Samoa, two infants die following MMR vaccines that are improperly mixed with an expired anesthetic instead of water. Two senior nurses are charged in 2019 with manslaughter and imprisoned. During the investigation the MMR vaccine campaign is suspended, decreasing the coverage rates that are then blamed for the ensuing 2019 measles epidemic. The unusually virulent measles outbreak occurs in three neighboring Pacific islands (Samoa, Fiji, Tonga) around the same time as the delivery of 257,500 doses of measles vaccine by UNICEF.

2018 The Market Information for Access initiative (MI4A) is launched "to contribute to the achievement of Strategic Development Goal 3.8 (Universal Health Coverage target) by enhancing access to safe, effective, quality and affordable vaccines for all." This extensive nation-based dataset collects information, reported via the WHO/UNICEF, on global vaccine markets.

2019 The WHO declares that "vaccine hesitancy" is one of the top 10 public health threats, ahead of cancer and malaria. No scientific data is presented to support this claim. Later in the year during the Vaccine Safety Summit, WHO officials admit to lacking long-term safety data on vaccines and adequate surveillance systems to follow up on adverse events.

2019 The UK Department for Work and Pensions (DWP) states that from 1979 to May 2019, £74,690,000 has been paid out in compensation from the Vaccine Damage Payment fund, with 941 successful claims.

2019 In March, NIAID awards a US$3.6 million grant to University of Chapel Hill researcher Ralph Baric, PhD, for gain-of-function research.

2019 ✪ A newly developed quadrivalent cell culture (grown on dog kidney cells) influenza A/B vaccine Flucelvax® is licensed in Europe and the USA, made by Seqirus.

2019 In March, US biotech company Moderna updates an October 2015 patent approval request for its respiratory virus (including influenza and coronavirus) RNA vaccine platform (US#10702600B1). To date, Moderna has never marketed a single product, and by the end of the year reports an accumulated deficit of US$1.5 billion.

2019 In May, the Wellcome Trust, in collaboration with Dr. Stanley A. Plotkin, holds a private meeting in London to assess some issues of vaccine safety and to "explore opportunities to perform studies that would enlarge knowledge of vaccine safety." Around 30 experts participate in the 2-day event, including author and anthropologist Heidi Larson, and vaccinologists Dr. Kathryn M. Edwards, Dr. Neal Halsey, Prof. Andrew Pollard, Dr. Paul Offit, Dr. Peter Hotez, Dr. Ulrich Heininger, and Dr. Paul Lambert, as well as director of the CDC Immunization Safety Office, Dr. Frank DeStefano. They highlight that "vaccination remains a highly positive procedure to maintain the health of populations" and the necessity of maintaining "public confidence in the vaccine enterprise." They identify that "there remain substantial challenges to studying the safety of the schedule." One of the ideas proposed is the creation of a "DNA biobank for people with vaccine reactions to assess genomic risk factors. Moreover, a registry of genetic samples from patients with severe adverse reactions to vaccines would permit attempts to correlate adverse reactions and genetic predispositions." As vaccines are not studied for their capacity to cause gene mutations, this is seen by some as a way to blame victims' "bad genes" for vaccine reactions rather than really address vaccine safety.

2019

In May, HHS conducts "Crimson Contagion" as a functional exercise of a novel avian influenza virus (H7N9) outbreak originating in China.

2019 In June, New York State repeals religious exemptions to vaccination as the result of a "highly irregular vote." The bill passes and is signed by Governor Andrew Cuomo "mere hours after the Committee vote, without a single public hearing in either the Assembly or Senate." Lawmakers in at least ten US states attempt to take away or limit "personal belief" exemptions.

2019 In July, the CDC suspends operations at Fort Detrick laboratories due to serious safety concerns and violations of protocol. Operations are resumed in March 2020.

2019 ✪ The third-generation smallpox vaccine and first live, non-replicating monkeypox vaccine JYNNEOS® (Imvanex® in the EU, Imvamune® in Canada) by Bavarian Nordic is approved by the FDA in the USA.

2019 ICAN submits a FOIA request to the CDC asking for all the "studies relied upon by the CDC to claim that the DTaP vaccine does not cause autism." ICAN also submits the same request for hepatitis B (Engerix®/Recombivax-HB®), Hib, Prevnar® (PCV13), and IPV vaccines, "as well as requesting the CDC provide studies to support [that] cumulative exposure to these vaccines during the first six months of life do not cause autism." The CDC does not submit any studies, and ICAN ends up having to file a lawsuit to get access to the information. Finally, the Department of Justice supplies a list of 20 studies in response to ICAN's FOIA request. None of the studies cited involve the vaccines given in the first six months of life.

2019 In September, the GPMB publishes its first annual report titled, *A World At Risk,* providing "a snapshot of the world's ability to prevent and contain a serious global health threat." It also predicts the immediate threat of a lethal respiratory pandemic that could kill millions of people and devastate the global economy due to a "naturally emergent or accidentally or deliberately released" pathogen.

2019 In September, following a raid by the Chinese Communist Party-led government, the Wuhan Insitute of Virology removes all previously available viral sequences from its online database.

2019 In September, BMGF acquires 3 million shares of future COVID-19 vaccine manufacturer BioNTech, based in Germany.

2019 In cooperation with the WHO, the European Commission organizes a Global Vaccination Summit in Brussels on September 12. The event gathers around 400 participants from all over the world, "including political leaders, high-level representatives from the UN and other international organisations, health ministries, leading academics, scientists and health professionals, the private sector and NGOs." The overall mission is to give high-level visibility and the political endorsement of vaccination "the most successful public health measure of modern times," that prevents "an estimated 2.5 million deaths worldwide each year." The summit showcases the global commitment to vaccination, "stepping up action to increase vaccine confidence" and engaging "political leaders and leaders from scientific, medical, industry, philanthropic and civil society in global action against the spread of vaccine misinformation."

2019 Neuroscientist Christopher Shaw, PhD, and lawyer Mary Holland, JD, submit a FOIA request to the CDC, NIH, and the Agency for Toxic Substances and Disease Registry (ATSDR), seeking copies "of any human or animal studies involving the subcutaneous or intramuscular injection of aluminum adjuvant relied upon by the CDC to establish the safety of injecting infants and children with aluminum hydroxide, aluminum phosphate or amorphous aluminum hydroxyphosphate sulfate." Initially the CDC denies the request stating that it is "too broad." After appeal, it is conceded that none of the agencies cited are able to provide any documents pertaining to this request.

2019 Chairman of the US House Intelligence Committee, Adam Schiff, sends letters to the CEOs of Facebook, Google, and Amazon, pressuring them to censor "vaccine misinformation" on their respective platforms.

2019 ✪ Sanofi Pasteur launches its DTaP-IPV-Hib-HepB 6-in-1 vaccine Vaxelis® in Switzerland, which is recommended from 6 weeks old.

December 2–3, 2019

The World Health Organization
Global Vaccine Safety Summit

Quotes from the Global Advisory Committee on Vaccine Safety

The Global Vaccine Safety Summit in Geneva, Switzerland, brings together the highest authorities and leading experts in vaccine safety from the WHO, FDA, CDC, the Vaccine Confidence Project, PATH, the Brighton Collaboration, and the London School of Hygiene & Tropical Medicine.

*"There's a lot of safety science that is needed. Without the good science, we can't have the good communication. **We need much more investment in safety science.** We have a very wobbly health professional frontline that is starting to question vaccines and the safety of vaccines."*

—Prof. Heidi Larson, PhD

Anthropologist, director of Vaccine Confidence Project, and
author of *Stuck: How Vaccine Rumors Start—and Why They Don't Go Away*

*"The primary concern, though, is systemic adverse events rather than local adverse events. **The major health concern which we are seeing are accusations of long-term, long-term effects."***

—Dr. Martin Howell Friede, PhD

Coordinator, WHO Initiative for Vaccine Research

The Global Advisory Committee on Vaccine Safety (GACVS), an independent expert clinical and scientific advisory body, provides WHO with scientifically rigorous advice on vaccine safety issues of potential global importance. GACVS holds their 41st meeting in Geneva, Switzerland, on December 4–5, 2019. The Committee examines data on the safety of vaccines against rotavirus, Ebola virus, and human papillomavirus. It also reviews two generic issues: updating of the global vaccine safety strategy and case reviews of communications on vaccine safety.

"It seems to me they [adjuvants] multiply the reactogenicity in many instances, and therefore it seems to me that it is not unexpected if they multiply the incidence of adverse reactions that are associated with the antigen, but may not have been detected through lack of statistical power in the original studies."

—Prof. Stephen Evans, BA, MSc, CStat, FRCP

Professor of pharmacoepidemiology, London School of Hygiene & Tropical Medicine

"I think we cannot overemphasize the fact that we really don't have very good safety monitoring systems in many countries."

—Dr. Soumya Swaminathan, MD

WHO chief scientist in 2019, pediatrician

Youtube video source of meeting discussion
https://www.youtube.com/watch?v=V4ysYn5TYxc

A Chronology of Vaccination

2019–2020

2019

In October, "Event 201" a 3.5 hour coronavirus pandemic simulation, or tabletop exercise, takes place at the Pierre Hotel in New York hosted by Johns Hopkins Center for Health Security, in partnership with the World Economic Forum and the BMGF.

2019 Less than a week after the pandemic simulation Event 201, Dr. Anthony Fauci and BARDA director Dr. Rick Bright host a meeting of top vaccinologists and virologists at California think-tank, the Milken Institute. They discuss vaccine hesitancy, the need for a universal flu vaccine, and strategies for streamlining vaccine development and accelerating approval processes.

2019 In October, at the same time as Event 201, the 7th International Military World Games takes place in Wuhan, China. Many athletes are reported as being sick with symptoms similar to COVID-19, and may have inadvertently spread the virus upon their return home.

2019 ✪ The FDA approves an influenza A (H5N1) virus, AS03-adjuvanted, split-virion monovalent vaccine, manufactured by ID Biomedical Corporation of Quebec (a subsidiary of GSK), for use in persons 6 months and older. The adult formulation (approved since 2013) includes 5 mcg thimerosal as a preservative and residual formaldehyde (≤12.5 mcg), with the pediatric version having half the amount of ingredients. The AS03 adjuvant is composed of 10.69 mg squalene, 11.86 mg DL-α-tocopherol and 4.86 mg polysorbate 80.

2019 Legislation in Switzerland is created to allow Swissmedic to grant time-limited authorizations (TLA) for medicinal products as long as certain conditions are met. They must 1. be intended to diagnose, prevent or treat a disease that may result in severe disability, life-threatening suffering, or short-term death 2. have no equivalent substitute available 3. bring significant benefit.
TLAs will be granted for COVID-19 vaccines.

2019 The results of the 2017 European Vaccine Initiative's "ThinkYoung" research, designed to investigate the awareness level of vaccination among young Europeans, reveal that only 65% of the young people interviewed consider that the benefits of vaccination outweigh the risks. As a result, the EVI "ThinkYoung" project builds the Young Coalition for Prevention and Vaccination (YC4PV), composed of young health professionals to advocate "for a greater engagement of civil society when it comes to sustainable behavior change."

2019 On November 1, full 5G technology is first rolled out in 50 major Chinese cities. Home to thousands of base stations and China's first smart highway, the highly-polluted city of Wuhan (inhabited by 11 million people) is one of the first pilot cities of the 5G network.

2019 Based at the University of Auckland, New Zealand, the multinational Global Vaccine Data Network™ (GVDN®) is set up with seed funding from the BMGF. This investigator-led research network's mission is to coordinate the global effort to assess vaccine safety and efficacy through 31 sites in over 26 countries on 6 continents.

2019 On November 17, *South China Morning Post* reports the first "official" case of COVID-19 in a 55-year-old man from the Hubei province. Official statements by the Chinese government to the WHO report that the first ever confirmed case of COVID-19 is diagnosed on December 8.

2019

The WEF and the UN sign a strategic partnership framework to expedite the 2030 Agenda for Sustainable Development. Two-hundred and eighty-nine organizations and 27 individuals from all regions of the world sign a letter which includes the warning that "this agreement formalizes a disturbing corporate capture of the UN. It moves the world dangerously towards a privatized and undemocratic global governance."

2019 In December, the Massachusetts Institute of Technology (MIT) announces a new technology that "could enable the rapid and anonymous detection of patient vaccination history to ensure that every child is vaccinated." Delivered by a microneedle patch, a new dye that consists of nano-crystals called quantum dots emit near-infrared light which can be detected by a specially equipped smartphone. The project is funded by the BMGF, the Koch Institute, and the National Cancer Institute.

2019–2020 According to US health officials, a mosquito-borne virus, eastern equine encephalitis (EEE), outbreak in Massachusetts leaves 7 people dead among 17 infected.

2019 The Jenner Institute and the Kennedy Institute of Rheumatology at the University of Oxford, with colleagues at the University of Berne, and the Latvian Biomedical Research & Study Center, develop and test a virus-like particle vaccine (in mice) that could be used to treat chronic pain caused by osteoarthritis.

2019 The WHO's newly formed advisory committee for international governance on human genome editing declares that modifications to human germlines are "irresponsible." The committee proposes a central registry of human editing research to facilitate insight and oversight of the risks.

2019 ✪ The FDA announces the approval of Ervebo® (rVSV-ZEBOV), the first FDA-approved vaccine for the prevention of Ebola virus in people from 18 years old, marketed by Merck and developed by NewLink Genetics. The EMA in Europe grants a conditional authorization for this genetically engineered, replication-competent, live-attenuated vaccine. The WHO prequalifies an Ebola vaccine for the first time, which is the fastest vaccine prequalification process ever conducted. Prequalification means that the vaccine meets WHO standards for quality, safety, and efficacy and can be distributed by UNICEF.

2019 In this year alone, the US Vaccine Injury Compensation fund pays out awards of US$196,217,707.64 to 653 vaccinated victims and their families.

2019 ✪ The European Commission grants a marketing authorization of Janssen-Cilag's first ever vaccine. The Ebola vaccine two-dose regime is composed of Zabdeno® (Ad26.ZEBOV) and Mvabea® (MVA-BN-Filo). It is recommended for the prevention of Ebola caused by the *Zaire ebolavirus* in individuals aged one year and above.

2019 ✪ On December 1, the People's Republic of China implements mandatory vaccination for adults under the Vaccine Administration Law adopted at the 11th Meeting of the Standing Committee of the Thirteenth National People's Congress.

2019 On December 30, China reports a severe pneumonia outbreak in Wuhan caused by a "mysterious new coronavirus." The next day Chinese authorities alert the WHO to the 27 cases of pneumonia-like disease of unknown etiology in Wuhan.

2020–2024

2020

On January 7, the first complete genome of a novel coronavirus is identified in samples of bronchoalveolar lavage fluid from a patient in Wuhan, China, using a combination of Sanger, Illumina, and nanopore sequencing. Three distinct strains are identified and designated as 2019-nCoV—renamed SARS CoV-2 on February 11.

2020 On January 23, Chinese authorities institute a draconian citywide lockdown of Wuhan, the epicenter of the coronavirus disease outbreak.

2020 On January 30, the WHO declares the novel coronavirus disease to be a Public Health Emergency of International Concern (PHEIC).

2020 On January 31, HHS declares a US public health emergency under the Public Health Service Act in response to COVID-19.

2020 ✪ The FDA approves adjuvanted Fluad® quadrivalent influenza vaccine by Seqirus, for adults 65+ years.

A Chronology of Vaccination

2020

2020	✪ In February, the FDA approves Audenz® by Seqirus, designed to protect against H5N1 in the event of an influenza A pandemic. It is the first-ever adjuvanted (MF59), cell-based pandemic vaccine that is approved for children aged 6 months or older.
2020	The WHO reports 959 human cases of circulating vaccine-derived poliovirus type 2 (cVDPV2) and 411 cVDPV2-positive environmental samples in 27 countries, of which 21 are in the African region.
2020–2023	WHO works with YouTube to enhance its "COVID-19 Misinformation Policy" and to provide guidelines for content providers to "ensure no medical misinformation related to the virus proliferates on their platform." Policy updates such as these result in the removal of 850,000 YouTube videos related to "harmful or misleading" COVID-19 misinformation from February 2020 to January 2021. From 2021, general vaccine "misinformation" is targeted, and from August 2023, high priority will be placed on removing content related to misinformation on cancer treatment.
2020	In February, USAID announces a 3-year financial commitment of US$ 1.16 billion for Gavi to support its vaccination efforts.
2020	Development of new vaccine platforms (mRNA, VLPs, replicons) is accelerated thanks to international organizations and institutions dedicated to fast-track approval processes through regulatory agencies.
2020	On March 11, the WHO reclasses the novel SARS coronavirus disease, named COVID-19 in February, as a global pandemic. Over 118,000 cases in 114 countries have been reported. This declaration comes one day after Bill Gates proclaims the COVID-19 pandemic and announces the infusion of US$ 50 million into a WHO-partnered venture called "COVID-19 Therapeutics Accelerator."
2020	Fast-tracked clinical trials are launched for COVID-19 vaccines, all sponsored by the Bill & Melinda Gates Foundation in collaboration with governments and industry, shortening the development of vaccines by a decade.
2020	In March, the Coronavirus Aid, Relief, and Economic Security (CARES) Act, a US$ 2.2 trillion "stimulus bill" is signed into law by President Donald Trump. Direct financial relief of US$ 130 billion is provided to medical and hospital industries. By September 2021, US$ 14.8 billion is used to support COVID-19 vaccine development and distribution.
2020	Under the covert guidance of Dr. Fauci, five scientists author a letter "The proximal origin of SARS-CoV-2" published in *Nature*, claiming the virus could not have come from a laboratory.
2020	From March, the 2019-approved Measles Protection Act becomes effective in Germany. It mandates measles vaccination for all kids (from 12 months old) and adolescents in community facilities. They are obliged to receive two doses of MMR or MMRV, as single measles vaccine is not available. Refusal may result in fines up to € 2,500. It is the first obligatory vaccine in Germany since smallpox.
2020	In March, the Brighton Collaboration in partnership with CEPI creates a priority list of potential adverse events of special interest (AESI) relevant to COVID-19 vaccines, endorsed by the WHO. These potential side effects are not communicated to the public.
2020	American virologists Harvey J. Alter and Charles M. Rice, with British scientist Michael Houghton, are awarded the Nobel Prize in Physiology or Medicine "for their discovery of hepatitis C virus." By this date, Inovio Pharmaceuticals has launched preclinical trials for a synthetic DNA hepatitis C vaccine.
2020	In May, Operation Warp Speed is announced by President Donald Trump as a project that aims to deliver 300 million doses of a "safe and effective vaccine" for COVID-19 by January 2021. Under Operation Warp Speed, HHS leads the mission to execute a "well-defined portfolio of COVID-19 Medical Countermeasures candidates to maximize probability of having safe and effective diagnostics, therapeutics and vaccines as fast as possible for mass distribution."
2020	Influenza vaccine is recommended to babies from 6 months old in Switzerland.

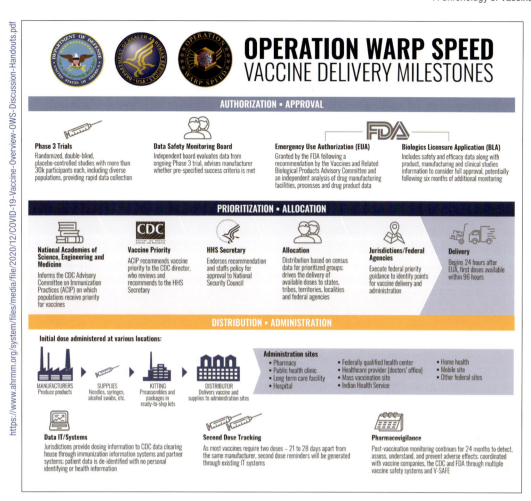

2020 BARDA contributes US$955 million to Moderna's late-stage clinical development of COVID-19 vaccine mRNA-1273 (Spikevax®). Working closely with Operation Warp Speed and the NIH, the funding covers the costs of a 30,000 participant phase III trial in the US. By November, BARDA's cumulative investment in research, development, manufacturing, and procurement of COVID-related vaccines, therapeutics, and diagnostics is US$14 billion.

2020 In May, the US EPA gives the go-ahead to release hundreds of millions of genetically modified mosquitoes in the Florida Keys, in an effort to curb mosquito-borne infections like malaria, dengue, and Zika. The mosquitos are created by Oxitec, a UK-based, US-owned biotechnology company that "develops genetically modified insects in order to improve public health and food security through insect control."

2020 In May, the Sabin-Aspen Vaccine Science & Policy Group publishes a 173-page report entitled *Meeting the Challenge of Vaccination Hesitancy.* "Although vaccination remains a well-accepted norm worldwide, a combination of factors—including misinformation spread on social media; decreased trust in institutions including government, science and industry; and weaknesses within health systems—has emerged to diminish confidence among some populations."

2020 **WHO Foundation**

In May, the WHO Foundation is created to support and complement the efforts of the WHO to mobilize resources and funding. It is directed by former Swiss Health Minister (1991–2009) Prof. Dr. Thomas Zeltner.

A Chronology of Vaccination

2020–2021

2020 The WHO's World Health Assembly adopts the "global strategy towards eliminating cervical cancer." This requires the introduction of HPV vaccine in all countries with a set target of 90% coverage, requiring large investments in LMICs to reach the 2030 targets.

2020 COVID-19 Vaccines Global Access (COVAX) is a worldwide initiative aimed at equitable access to COVID-19 vaccines directed by Gavi, CEPI, and the WHO, with significant involvement of the BMGF (whose staff are placed on various boards and working groups). It is one of the four pillars of the Access to COVID-19 Tools Accelerator, an initiative begun in April by the WHO, the European Commission, and the government of France as a response to the COVID-19 pandemic. COVAX coordinates international resources to enable low-to-middle-income countries equitable access to COVID-19 tests, therapies, and vaccines. UNICEF is the key delivery partner, leveraging its experience as the largest single vaccine buyer in the world and working on the procurement of COVID-19 vaccine doses as well as logistics, country readiness, and in-country delivery. By October 19, 184 countries join COVAX. It is reported that in June 2021, COVAX send "twice as many vaccines to the UK as it did to the entire continent of Africa." By early 2023, the manufacturers involved make US$ 13.8 billion on the vaccines distributed through the initiative, and *The New York Times* reports that COVAX "paid out US$ 1.4 billion to pharmaceutical companies for vaccine orders never delivered."

2020 Swissmedic, the Swiss surveillance and regulatory agency, admits that only around 10% of vaccine injuries are ever declared.

2020 ✪ In August, Sputnik V® by the Gamaleya Research Institute of Epidemiology and Microbiology in Russia, is the world's first registered COVID-19 vaccine. Based on a human adenovirus vector platform, it will be approved for use in 71 countries.

2020 According to leaked EMA documents on the Pfizer-BioNTech COVID-19 vaccine, the limit of residual DNA/RNA permitted in each dose is ≤ 330 ng. It is estimated that 300 ng is equivalent to the total amount of genetic material in over 44,000 cells, well over the WHO established limit of 10 ng/dose.

2020 In August, US attorney Aaron Siri, Esq., deposes the "Godmother" of vaccinology, Dr. Kathryn M. Edwards—a US$ 500 per hour expert witness for the defendant—in the case *Hazlehurst v. Hayes*. Hazlehurst's claim with the NVICP had taken 9 years—from 2002 to 2011 and was dismissed, so Yates' father and lawyer Rolf Hazlehurst brings a civil liability suit against the vaccinating doctor, Dr. Carlton Hayes. After over an hour highlighting Dr. Edwards's numerous conflicts of interest with vaccine manufacturers—which she is initially rather reluctant to divulge—Siri questions her on the lack of safety studies. She assures Siri that the NIH has a "pot of money set aside for studies on vaccines safety." When Siri asks her, "Do you know of any specific study involving injection of aluminum, into humans or animals, that showed aluminum adjuvants are safe?" She replies, "It is really such a ridiculous question, I'm not going to answer it." Dr. Edwards insists several times that "autism is a prenatal event." The trial is the "first time in 35 years for a jury to hear evidence in a court of law regarding whether a vaccine injury can cause neurological injury, including autism." The jury decides in favor of the Defendant, citing that there was no case of medical negligence or reckless misconduct.

2020 In September, the Serum Institute of India collaborates with Gavi and the BMGF to accelerate the production and distribution of up to an additional 100 million doses of COVID-19 vaccines for LMICs, as part of the Gavi COVAX project.

2020 In October, pharmaceutical manufacturer Gilead Sciences agrees to supply the European Union with US$1.2 billion worth of remdesivir (Veklury®) to treat COVID-19. Four years prior, the European Commission granted this drug "orphan designation" (EU/3/16/1615) for the treatment of Ebola.

2020 ✪ In November, type 2 novel oral polio vaccine (nOPV2) is "the first vaccine to be authorized for use under the WHO Emergency Use Listing (EUL) mechanism. The use of nOPV2 is restricted to outbreak response for type 2 circulating vaccine-derived polioviruses (cVDPV2)" in Africa and Asia.

2020 ✪ In December, the first ever mRNA experimental gene-based products against COVID-19 (by Pfizer-BioNTech and Moderna, both developed in under a year) are licensed by the FDA under an Emergency Use Authorization (EUA), with only two months of safety and efficacy data.

2020 On December 2, the MHRA becomes the first regulatory agency to approve an mRNA vaccine. It awards conditional and temporary authorization to Pfizer-BioNTech's COVID-19 vaccine BNT162b2, later branded Comirnaty®. This approval enables the start of the UK's COVID-19 vaccination program.

2020 American pediatrician Dr. Paul Thomas publishes a study "Relative Incidence of Office Visits and Cumulative Rates of Billed Diagnoses Along the Axis of Vaccination" in the *International Journal of Environmental Research and Public Health.* The study shows that unvaccinated patients have significantly less cumulative office visits for asthma, allergic rhinitis, eczema, dermatitis, urticaria, respiratory infections, other infections, otitis media, behavioral issues and ADHD. Dr. Thomas gathered the data because his medical board demanded he prove his alternative schedule was "as safe" as the CDC's in order to keep his license. The data proved it was much safer, yet the Oregon medical board suspended his license 5 days after publication.

2020 Zostavax® by Merck is discontinued in the USA. By the end of 2023, it is withdrawn in the UK and Australia. This shingles vaccine remains available in the EU and Switzerland.

2021

✪ For the first time, experimental mRNA-based products are marketed as "vaccines" and widely administered to populations worldwide—notably young children and even pregnant women (from May), who were not included in clinical trials.

2021 ✪ The UK is the first country in the world (shortly followed by Australia) to administer the Oxford University/AstraZeneca ChAdOx1 viral vector COVID-19 vaccine, Covishield®/Vaxzevria®. This vaccine is never approved in the USA or in Switzerland, but both countries purchase doses for delivery to Asian and African nations under the COVAX-CEPI UN program. As of January 2022, more than 2.5 billion doses will have been distributed to more than 170 countries worldwide.

2021 In February, the FDA issues an EUA for Johnson & Johnson's (Janssen) single-dose COVID-19 vaccine for people aged 18 years and over. A 10-day pause is recommended in April, and by July, information sheets warn of thrombosis with thrombocytopenia syndrome (TTS) and Guillain-Barré syndrome among vaccine recipients. In May 2023, the manufacturer requests the voluntary removal of the EUA, and the vaccine is withdrawn in the US and Europe.

2021 ✪ Sanofi Pasteur's 6-in-1 vaccine Vaxelis® is licensed in the UK, and a year later will be approved in Australia.

2021 Merck confirms an agreement with UNICEF to establish the world's first global Ebola vaccine stockpile with Ervebo®.

A Chronology of Vaccination

2021

2021 In February, less than three months after the COVID-19 vaccination campaign begins, Pfizer produces an internal and confidential report *5.3.6. Cumulative Analysis of Post-Authorization Adverse Event Reports* for its BNT162b2 (marketed as Comirnaty®) COVID-19 vaccine. It reports an alarming number of adverse event reports, of which 1,223 are fatal outcomes and 9,400 are "unknown." The 9-page appendix of this report includes over 1,236 different diseases potentially linked to the vaccine. This document and thousands of others are obtained under a FOIA request filed in August 2021 by the US nonprofit Public Health and Medical Professionals for Transparency—a group of 200+ doctors, scientists, professors, and public health professionals. The FDA initially proposes releasing 500 pages per month, which would mean taking 55 years to release 450,000+ pages worth of documents—despite the fact that the FDA approved the vaccine for emergency use in under three months.

2021 In March, the Nuclear Threat Initiative at the Munich Security Conference sponsors a virtual "high-consequence biological threat" simulation exercise of a monkeypox pandemic that emerges in May of 2022.

2021 The Vaccine Safety and Confidence-Building Working Group (VacSafe WG) at Columbia University's Center for Pandemic Research at the Institute for Social and Economic Research and Policy, is initiated in March and concluded in April 2023. The objective of the VacSafe WG is to "catalyze and support projects that generate, scale and analyze actively-collected vaccine safety surveillance and pharmacovigilance data in Africa," to increase public trust and confidence in vaccines.

2021 At the end of June, the FDA revises the EUA fact sheets for Pfizer-BioNTech's and Moderna's mRNA COVID-19 vaccine, suggesting an increased risk of myocarditis and pericarditis following vaccination.

BNT162b2
5.3.6 Cumulative Analysis of Post-authorization Adverse Event Reports

Table 1 below presents the main characteristics of the overall cases.

Table 1. General Overview: Selected Characteristics of All Cases Received During the Reporting Interval

Characteristics		Relevant cases (N=42086)
Gender:	Female	29914
	Male	9182
	No Data	2990
Age range (years):	≤ 17	175[a]
0.01 -107 years	18-30	4953
Mean = 50.9 years	31-50	13886
n = 34952	51-64	7884
	65-74	3098
	≥ 75	5214
	Unknown	6876
Case outcome:	Recovered/Recovering	19582
	Recovered with sequelae	520
	Not recovered at the time of report	11361
	Fatal	1223
	Unknown	9400

a. in 46 cases reported age was <16-year-old and in 34 cases <12-year-old.

As shown in Figure 1, the System Organ Classes (SOCs) that contained the greatest number (≥2%) of events, in the overall dataset, were General disorders and administration site conditions (51,335 AEs), Nervous system disorders (25,957), Musculoskeletal and connective tissue disorders (17,283), Gastrointestinal disorders (14,096), Skin and subcutaneous tissue disorders (8,476), Respiratory, thoracic and mediastinal disorders (8,848), Infections and infestations (4,610), Injury, poisoning and procedural complications (5,590), and Investigations (3,693).

https://phmpt.org/wp-content/uploads/2021/11/5.3.6-postmarketing-experience.pdf

2021 The GVDN™ is awarded a US$5.6M federal grant from the CDC/HHS to implement, host, and manage a project called *Assessing safety of COVID-19 vaccines across large and diverse populations using the 17-country Global Vaccine Data Network Consortium,* also referred to as the Global COVID Vaccine Safety (GCoVS) project. The aim is to coordinate data and "rapidly evaluate rare vaccine safety concerns."

2021 ✪ Johnson & Johnson (Janssen) launches its genetically modified adenovirus viral vector COVID-19 vaccine in the US and Europe, being the only available single shot.

2021 As of July 2021, more doses of the inactivated viral COVID-19 vaccine, CoronaVac®, made by Chinese company Sinovac, are distributed than any other, with 943 million doses delivered worldwide.

2021 ✪ In August, the FDA approves Ticovac® by Pfizer, a TBE vaccine to prevent tick-borne encephalitis in individuals one year old and older.

2021 Forty-four US states and the District of Columbia grant vaccine exemptions based on religious objections. Six states that do not recognize such an exemption are California, Connecticut, Maine, Mississippi, New York, and West Virginia. Meanwhile 15 states allow personal, conscientious, philosophical, or other exemptions to vaccines.

2021 ✪ The FDA approves VBI Vaccines' new 3-antigen recombinant hepatitis B vaccine PreHevbrio®—derived from Chinese hamster ovary cells (CHO) instead of yeast and including 500 mcg aluminum hydroxide—for people age 18 and above.

2021 In September, the CDC changes its definition of a vaccine from "a product that stimulates a person's immune system to produce immunity to a specific disease; protecting that person from the disease," to "a preparation that is used to stimulate the body's immune response against diseases."

2021 The first COVID-19 vaccine Comirnaty® by Pfizer-BioNTech is approved by the FDA for individuals 16 and older.

2021 ✪ The WHO recommends RTS,S/AS01 (RTS,S) protein subunit malaria vaccine Mosquirix® by GSK for children in sub-Saharan Africa and in regions with moderate to high *P. falciparum* malaria transmission. The vaccine is tested in a pilot study in three African nations: Ghana, Kenya, and Malawi. Over 800,000 children participate in the pilot study.

2021 Estimates by the WHO-UNICEF show the largest sustained decline in childhood vaccination in almost thirty years.

2021 The COVID-19 vaccination certificates are implemented worldwide and vaccine mandates are enacted in numerous countries (despite being still in phase III clinical trials and being unable to prevent transmission or infection), forcing millions of citizens to get vaccinated or face penalties such as unemployment, exclusion from schools and society, with some countries even implementing fines for non-compliance.

2021 In October, the USAID launches the project *Discovery & Exploration of Emerging Pathogens – Viral Zoonoses* (DEEP VZN), to identify animal viruses that may harm humans—replacing the previous PREDICT project. The US$125 million virus research program is canceled in July 2023.

2021 Both the Swiss and American military make COVID-19 vaccination compulsory.

2021 ✪ Merck's 15-valent pneumococcal conjugate vaccine Vaxneuvance® is licensed in the US and the EU, for babies from 6 weeks old as a 4-dose regimen. Containing 125 mcg of aluminum (Al^{3+}), this vaccine is licensed in Switzerland in 2023, interestingly only recommended to people over 65 years of age and with the mention "safety and efficacy of Vaxneuvance in children and adolescents less than 18 years of age have not been established."

2021 The Linux Foundation launches a trust network called the Global COVID Certificate Network (GCCN) to provide countries with a global trust registry for interoperable COVID-19 certificates.

A Chronology of Vaccination

2021–2022

2021 ✪ Pfizer's 20-valent pneumococcal conjugate vaccine Prevnar®20 against twenty *Streptococcus pneumoniae* serotypes, is approved in the US for adults. In 2023, it will be approved for pediatric use from 6 weeks to 17 years of age.

2021 The UK announces the establishment of a UK Animal Manufacturing and Innovation Center in Surrey, intending to accelerate vaccine development for livestock and to control the spread of viral diseases, including coronavirus. The UK government will contribute US$24.79 million and the BMGF will contribute US$19.43 million to establish the center.

2021 Australia, New Zealand, Canada, and the USA require vaccination against COVID-19 for all incoming travelers (non-citizens/non-residents) and mandate the injection to certain populations. Despite this policy, COVID-19 cases continue to increase.

2021 Millions of injuries and thousands of deaths are reported following receipt of COVID-19 injections, including neurological and cardiovascular conditions, as well as blood, immune, and menstrual cycle disorders.

2021 Under the US COVID-19 Vaccine Provider Incentive Program, bonus payments up to US$250 per newly vaccinated member are provided if 75% of members receive at least one dose by the end of the year.

2021 The Vaccine Credential Initiative, a US consortium of major tech and healthcare companies including Microsoft, Apple, Oracle, Salesforce, MITRE, and the Mayo Clinic, is set up to "harmonize the standards and support development of implementation guides needed to issue, share and validate vaccination records bound to an individual identity."

2021 ✪ Vaccine manufacturer in India, Bharat Biotech International, is the first to develop Indian COVID-19 vaccines, whole inactivated virus-based vaccine, COVAXIN®, and novel adenovirus vectored nasal drop vaccine, iNCOVACC®.

2021 The Rockefeller Foundation awards a three-year US$7.5 million grant to an initiative called The Mercury Project, run by the Social Science Research Council (SSRC). The project aims to set up a research consortium to drive acceptance and uptake of COVID-19 vaccination efforts and provide insights to counter health dis- and misinformation.

2021 In November, Bill Gates admits that the COVID-19 vaccines cannot prevent viral transmission, thereby negating the scientific validity behind COVID certificates, which are still widely implemented worldwide.

2021 European Commission's (EC) Health Emergency Preparedness and Response Authority (HERA) is set up. Its core mission is to prevent, detect, and rapidly respond to health emergencies through working closely with other EC and national health agencies, industry and international partners to improve Europe's readiness for health emergencies.

2021 ✪ The patent-free protein subunit COVID-19 vaccine, Corbevax®—developed by Baylor College of Medicine and Texas Children's Hospital Center for Vaccine Development—is approved in India for adults and children from age 12. The vaccine, containing adjuvants aluminum hydroxide (750 mcg) and CpG 1018 (750 mcg), is manufactured by Indian company Biological E, who will receive a WHO-EUL for Corbevax® in January 2024.

2021 According to the WHO, approximately 16 billion doses of 47 distinct vaccines (not just COVID-19 vaccines) are distributed worldwide, worth an estimated US$141 billion.

2021 In December, the WHO launches its proposed Convention or Agreement (CA+) for an International Treaty on Pandemic Prevention, Preparedness and Response—"Pandemic Treaty"—for discussion by member state delegates. The WHO also proposes the revision of the International Health Regulations that may grant the WHO executive powers over member states' public health policies during epidemics.

2022 From January, foods containing GMOs must be labeled "bioengineered" or "derived from bioengineering," under standards set by the USDA's National Bioengineered Food Disclosure Standard.

2022 ✪ Shingrix® shingles-zoster vaccine is officially recommended in Switzerland for "at-risk" individuals from 18 years old.

2022 In partnership with Gavi and the national health service, Ghana becomes the first country to use contactless biometrics for its national vaccination campaign.

2022 The US joins a list of 30 countries that meet the WHO criteria for circulating vaccine-derived poliovirus (cVDPV).

2022 The FDA amends the Emergency Use Authorizations of Moderna and Pfizer-BioNTech COVID-19 vaccines to allow bivalent formulations for use as a single booster dose at least two months following primary or booster vaccination.

2022 ✪ Pfizer's 20-valent pneumococcal conjugate vaccine Apexxnar®, is approved in the EU and the UK for adults.

2022 As of March 1, 585 HPV vaccine injury claims have been filed with the US Vaccine Injury Compensation Program (VICP). There have been 17 claims of death and 155 total cases compensated, including the case of Christina Tarsell, who died in 2018.

2022 ✪ In March, Japanese encephalitis vaccine Imojev® by Sanofi is approved in Victoria, Australia, in response to a regional outbreak. It is given to children from 9 months old.

2022 ✪ In June, a second MMR vaccine, Priorix® by GSK, is approved by the FDA for pediatric and adult use in the US market.

2022 A federal judge in Pennsylvania throws out a four-year MDL lawsuit against Merck that alleges its shingles vaccine Zostavax® caused shingles in 1,189 plaintiffs. The plaintiffs appeal the case dismissal in early 2023. An MDL is a mass tort, multidistrict litigation that is similar to a class action lawsuit, allowing federal courts to streamline thousands of cases into one claim.

2022 In June, *The Lancet* publishes an article online—funded by the BMGF, Gavi, and the WHO, among others—called "Global Impact of the First Year of COVID-19 Vaccination: A Mathematical Modeling Study." This Imperial College London study claims an "estimated 19.8 million deaths from COVID-19 [were] averted as a result of vaccination, based on excess mortality estimates of the impact of the pandemic." AEFIs are not considered, and this study will be cited at the University of Geneva as fact—proof that COVID-19 vaccine saved millions of lives.

2022 In June, Valneva receives an EU approval for its inactivated CpG 1018-adjuvanted COVID-19 vaccine, VLA2001. In October 2023, the authorization will be withdrawn (at Valneva's request) for commercial reasons.

2022 In July, WHO director-general, Dr. Tedros Adhanom Ghebreyesus, defies the expertise of his own scientific panel and declares monkeypox a PHEIC.

2022 ✪ In July, the approval of Bavarian Nordic's smallpox vaccine for adults Imvanex® (EU), JYNNEOS® (US), is extended to include protection against monkeypox. In Australia, the vaccine is granted an "emergency use provision" for people over 18 years old.

2022 In Switzerland, monkeypox vaccination is performed in the absence of authorization, so-called "no-label use." No-label use means that the vaccine is not authorized by Swissmedic, and vaccination with JYNNEOS® is carried out without country-specific prescribing patient information.

2022 The Department of Defense fails its fifth audit and is unable to account for over half of its assets, in excess of US$3.1 trillion. The DOD fails to implement a comprehensive approach to combat department-wide fraud. The national US debt stands at US$33 trillion.

2022 ✪ A quadrivalent conjugate vaccine against meningococcus serogroups A, C, Y and W-135, Menquadfi® by Sanofi is licensed in the USA for children from 2 years old, and in Switzerland from 12 months old. Two years later, this vaccine will be licensed in Australia.

A Chronology of Vaccination

2022 ✪ Novavax launches its COVID-19 vaccine, Nuvaxovid®, containing novel adjuvant Matrix-M® for use in adults from age 18.

2022 In the UK, the NHS Business Services Authority, the body that handles the Vaccine Damage Payment Scheme (VDPS), confirms that as of May 20, it has received 1,681 claims in relation to COVID-19 vaccines. The VDPS has a 60% disability eligibility criteria, a historical acceptance rate of 1.7% with a maximum cap of £120,000.

2022 ✪ In June, the FDA authorizes (EUA) both Pfizer-BioNTech and Moderna mRNA COVID-19 vaccines for children from 6 months old. The CDC endorses the ACIP's recommendation to vaccinate all children against COVID-19, even if they have already had the disease.

2022

In June, the World Bank approves the creation of the Financial Intermediary Fund for Pandemic Prevention, Preparedness and Response (PPPR). Officially launched as the Pandemic Fund at the G20 meeting in Indonesia, Rockefeller Foundation, the Wellcome Trust, the European Commission, the BMGF, and many governments pledge donations amounting to US$1.4 billion for the WHO to implement "disease surveillance, laboratory systems, health workforce, emergency communication and management, and community engagement." The World Bank and the WHO calculate the cost of PPPR at US$31.1 billion per year for "vacccines, therapeutics, and fortified and centralized surveillance."

2022 Bill Gates estimates that over the next decade, all governments combined need to spend US$15-20 billion per year to develop the necessary vaccines, infection-blocking drugs, treatments, and diagnostics to prevent pandemics.

2022 The WHO's IHR updates its list of potential pathogenic agents (all viruses) to include COVID-19, hemorrhagic fever Crimea-Congo, Ebola, Marburg, Lassa fever, MERS-CoV, SARS, Nipah henipaviral fever, Rift Valley fever, Zika, and "Disease X"—the latter being defined as representing the knowledge of a serious international epidemic caused by a pathogen that is yet unknown.

2022 Serum Institute of India manufactures more than 40,000 doses of Oxford's Ebola vaccine in just 60 days and ships them to Uganda, where the WHO has declared an outbreak.

2022 The Vaccine Monitoring Platform (VMP) is founded as a collaboration between the European Medicines Agency (EMA) and the European Centre for Disease Prevention and Control (ECDC), aimed at generating "real-world evidence on the safety and effectiveness of vaccines in the EU and EEA."

2022 ✪ In Canada, the world's first plant-based recombinant COVID-19 vaccine COVIFENZ®—containing Coronavirus-Like Particle (CoVLP) technology composed of recombinant spike glycoprotein expressed as virus-like particles co-administered with GSK's squalene-based pandemic adjuvant AS03—is approved for use in adults from 18–64 years old as a 2-dose regimen. Health Canada orders 76 million doses from the Quebec-based biopharmaceutical firm Medicago, only for the vaccine to be scrapped and the company closed down by parent company Mitsubishi Tanabe Pharma Corporation in February 2023.

2022 The WHO announces that the Wellcome Trust director, Dr. Jeremy Farrar, will become WHO's Chief Scientist from May 2023, replacing Dr. Soumya Swaminathan.

2022 In August, the FDA grants an EUA for Bavarian Nordic's mpox vaccine JYNNEOS® for "individuals less than 18 years of age determined to be at high risk for monkeypox infection." However, the FDA admits that the safety and effectiveness of JYNNEOS® have not been assessed in this age group.

2022 In October, the FDA approves the previously licensed Tdap vaccine, Boostrix® by GSK, as the first vaccine specifically intended to be used during the third trimester of pregnancy to prevent pertussis in infants younger than 2 months.

2022 In October, US President Joe Biden requests US$88 billion in funding for gain-of-function studies and for other biosecurity projects.

2022 In October, the Johns Hopkins Center for Health Security, in partnership with the WHO and BMGF, conducts a pandemic tabletop exercise in Brussels, Belgium, named "Catastrophic Contagion." The pathogen is called Severe Epidemic Enterovirus Respiratory Syndrome (SEERS) and originates in Brazil in 2025.

2022 In October, the ACIP unanimously votes to add the mRNA COVID-19 vaccine to the 2023 child, adolescent, and adult vaccination schedules—including them in the CDC's Vaccines for Children program.

2022 Pfizer becomes the first company in the history of the pharmaceutical industry to top US$100 billion in sales, powered by the US$38 billion in revenue from COVID-19 mRNA vaccine Comirnaty® and US$19 billion from COVID-19 antiviral treatment Paxlovid®.

2023 ✪ Pfizer's pneumococcal conjugate PCV15 Prevenar® vaccine is introduced in Switzerland, replacing the previous 13-valent vaccine.

2023 ✪ Two-dose adsorbed and adjuvanted anthrax vaccine for post-exposure protection, Cyfendus® by Emergent BioSolutions, is licensed by the FDA for use in adults 18–65 years old.

2023 The FDA approves Prevnar®20 by Pfizer for infants and children from 6 weeks old.

2023 The FDA approves Merck's Ebola vaccine Ervebo® for people aged 12 months and over.

2023 ✪ Serum Institute of India commercializes its first indigenously developed quadrivalent HPV vaccine Cervavac®. Two to three doses are recommended to females 9–26 years old, costing 200–400 rupees (€5) per dose.

2023 In February, the government of India and Gavi establish a 3-year collaboration to vaccinate Indian children. "Gavi will allocate US$250 million to identify and vaccinate children who have not yet received any routine vaccines, help existing health systems, and support the introduction of HPV vaccine and typhoid conjugate vaccine into India's national routine immunization schedule."

2023 ✪ The FDA approves the first ever respiratory syncytial virus (RSV) single-shot vaccine, Abrysvo™ by Pfizer Inc., for people over age 60 and pregnant women—to be given between 32–36 weeks gestation. FDA grants "Priority Review," "Fast-Track," and "Breakthrough Therapy" designations for this experimental vaccine, still undergoing phase III clinical trials. Abrysvo™ is a non-adjuvanted stabilized bivalent prefusion F (RSVpreF) subunit vaccine containing RSV preF A (60 mcg) and preF B (60 mcg) recombinant proteins expressed in genetically engineered Chinese hamster ovary cell lines and may also contain residual amounts of host cell proteins ($\leq 0.1\%$ w/w) and DNA (< 0.4 ng/mg of total protein) from the manufacturing process.

2023 By March, it is estimated that 5.55 billion people worldwide have received a COVID-19 vaccine—equal to about 72% of the planet.

2023 In March, the FDA approves bivalent Pfizer-BioNTech COVID-19 vaccine as a booster dose for "certain children" from 6 months to 4 years old.

2023 In April, Project NextGen is announced, led by BARDA and NIAID, and receives an initial investment of US$5 billion. The project aims to "accelerate and streamline the rapid development of the next generation of vaccines and treatments" through public-private partnerships.

2023 On May 4, the WHO's director-general declares that COVID-19 is no longer a public health emergency of international concern (PHEIC) after 1,191 days.

A Chronology of Vaccination

2023 Under the EU4Health program (2021–2027 budget of €5.3 billion), on behalf of the Health Emergency Preparedness and Response Authority (HERA), the European Health and Digital Executive Agency (HaDEA) signs a framework contract with four contractors (in Belgium, Ireland, the Netherlands and Spain—notably Pfizer) setting up the EU FAB network for "sufficient and agile manufacturing capacities for different vaccine types (mRNA-based, vector-based, and protein-based). These capacities will be kept operational and can be activated quickly, securing a total of 325 million doses per year in case of a public health emergency."

2023 A Mississippi lawsuit, filed by Aaron Siri Esq., representing seven parents and funded by Informed Consent Action Network (ICAN), argues that forbidding religious exemptions for childhood vaccinations is a violation of the Constitution. As a result, Mississippi reinstates as of July the religious exemptions that were removed in 1979.

2023 ✪ In July, Cyfendus® by Emergent BioSolutions (anthrax vaccine adsorbed, CPG7909-adjuvanted), previously known as AV7909, is approved by the FDA for post-exposure prophylaxis to inhalational anthrax. This next-generation anthrax vaccine, based on BioThrax®, is developed in collaboration with BARDA as a medical countermeasure for people from 18–65 years old.

2023 From August, YouTube gives "high priority" to removing content related to "misinformation" on non-conventional cancer treatments.

2023 Israeli biotechnological company, ViAqua Therapeutics, will become the first to produce an orally administered RNA-particle vaccine platform for shrimp.

2023 The ACIP recommends use of updated Pfizer-BioNTech and Moderna COVID-19 vaccines (authorized under emergency use) for people aged 6 months and older. The previous bi-valent booster shots are no longer available.

2023 ✪ The WHO recommends a second malaria vaccine R21/Matrix-M™ manufactured by Serum Institute of India, developed through a collaborative effort involving the Jenner Institute at the University of Oxford, the Kenya Medical Research Institute, the London School of Hygiene & Tropical Medicine, and Novavax—who supplies its patented Matrix-M™ adjuvant. The vaccine is added to the list of WHO Prequalified Vaccines, required for UNICEF to procure and Gavi to purchase the vaccine for at-risk children in developing nations.

2022–2023 Beyfortus®, containing the active substance nirsevimab (a new biological entity)—a human immunoglobulin monoclonal antibody produced in Chinese hamster ovary (CHO) cells by recombinant DNA technology—is licensed in the US, the UK, Canada, Europe, Switzerland, and Australia for neonates and infants up to 12–24 months old, during their first RSV season. Though not a vaccine by the traditional definition, this product is considered a "passive immunization" to prevent serious lower respiratory tract disease. It will be added to the childhood vaccine schedule and covered by Vaccines for Children.

2023 On August 25, the Digital Services Act in the EU takes effect for very large online platforms and search engines. The DSA aims to protect digital space against the "spread of illegal content, and the protection of users' fundamental rights." While the ostensible goal is to ensure "a safe, predictable and trustworthy online environment," it enforces censorship and allows government to define acceptable speech.

2023 Casgevy™, a cell-based gene therapy by Vertex Pharmaceuticals, is the first ever FDA-approved treatment utilizing CRISPR-Cas9. This form of genome editing technology modifies the blood stem cells of patients 12 years and older who suffer from severe sickle cell disease. A single course of treatment costs US$2.2 million.

2023 The Nobel Prize in Physiology or Medicine is awarded jointly to Hungarian biochemist and former communist spy Katalin Karikó and American immunologist Drew Weissman for "their discoveries concerning nucleoside base modifications [replacing natural uridine with synthetic Methylpseudouridine] that enabled the development of effective mRNA vaccines against COVID-19." The vice chair of the committee that chose the winners, Olle Kämpe, states that "giving a Nobel Prize for this COVID-19 vaccine can make hesitant people take the vaccine and be sure that it's very efficient and it's safe."

2023 The COVID-19 Origin Act is passed to "mandate the declassification of COVID-19 origin-related intelligence and information." President Joe Biden signs the bill with a statement "to continue to review all classified information and its links to the Wuhan Institute of Virology."

2023 The Nobel Prize in Chemistry is awarded to Moungi G. Bawendi, Louis E. Brus, and Aleksey Yekimov in recognition of their discovery and development of quantum dots. QDs are "engineered core/shell semiconductor nanoparticles with precisely designed and tailor-made particle architectures and surface chemistries, that are so tiny [nanosized] that their sizes determine their properties."

2023 ✪ In November, the FDA approves the first ever chikungunya vaccine IXCHIQ® (VLA1553), produced by French company Valneva with funding from CEPI. This live-attenuated vaccine is indicated as a single-dose regimen for people age 18 and older. The synthetic animal-free recombinant human albumin Recombumin® is included as an excipient. It is the first vaccine for an outbreak disease (transmitted by mosquitos) approved under the FDA's accelerated approval pathway.

2023 During a National Vaccine Advisory Committee (NVAC) meeting, the topic of "countering climate change with vaccines" is discussed with members of the Population Council, CDC, and NIH.

2023 ✪ Both the FDA and EMA approve the first ever adjuvanted respiratory syncytial virus (RSV) vaccine, Arexvy® by GSK, for people from age 60. Adjuvants used include ASO1E composed of 3−O-desacyl-4'-monophosphoryl lipid A (MPL) from *Salmonella minnesota* and QS-21. In the first year alone, GSK boasts £1.2 billion in sales.

2023 Novavax refuses to refund US$700 million to Gavi in advance payments for COVID-19 vaccines it never delivers. A year later, an agreement is made with Novavax paying an initial sum of US$75 million to Gavi, and then making deferred payments of US$80 million annually through December 31, 2028.

2023 By the end of August, there are 103 vaccine candidates in the pipeline, of which 99 are prophylactic vaccines (76 against viral diseases and 25 against bacterial disease), and four are therapeutic vaccines—targeting infectious agents.

2023 ✪ The FDA approves the new recombinant protein polysaccharide pentavalent (A, B, C, W, Y) meningitis vaccine Penbraya® by Pfizer for use in people from 10−25 years old. It includes adjuvant aluminum phosphate (0.25 mg aluminum), polysorbate 80 (0.018 mg), and trometamol (0.097 mg).

2023 In December, Indonesian vaccine producer Bio Farma receives WHO prequalification status and full licensure in Indonesia for its nOPV2 polio vaccine. By this date, over 850 million doses have been given to infants and children in 35 countries.

2023 ✪ Japan approves the first self-amplifying mRNA COVID-19 vaccine ARCT-154 for people over 18 years old, developed by CSL and Arcturus Therapeutics. Self-amplifying mRNA, also known as replicons, instruct the body to make more copies of itself within the cell to produce more antigen-coding mRNA and more target proteins, leading to a prolonged response.

2023 EU countries are reported to have destroyed €4 billion worth of COVID-19 vaccines, representing 215−312 million doses.

A Chronology of Vaccination

2023–2024

2023 Under Project NextGen, agreements are made with CastleVax, Codagenix, and Gritstone Bio to develop self-amplifying mRNA (sa-mRNA) vaccines. Gritstone Bio receives funding from the BMGF and CEPI, and is awarded a US$ 433 million contract with HHS. Despite this substantial funding, Gritstone Bio will file for bankruptcy in October 2024.

2023 By the end of the year, according to the WHO, about 84% of infants worldwide (108 million) have received 3 doses of diphtheria-tetanus-pertussis vaccine. It is worth noting that many infants have received the whole-cell pertussis vaccine containing both aluminum and "trace amounts" of thimerosal.

2024 In January, the European Vaccination Beyond COVID (EUVABECO) project—funded by the European Commission's EU4Health program—is launched "to create the conditions for a new type of vaccination practice throughout the European Union."

2024 ✪ In April, Sanofi announces the launch of inactivated rabies vaccine Verorab® in the UK, for pre- and post-exposure prophylaxis of rabies in all age groups.

2024 The US Department of Health and Human Services, through BARDA/ASPR, awards Moderna with US$ 176 million to develop an mRNA-based pandemic flu vaccine that could be used as protection from H5N1 avian influenza. HHS declares a bird flu emergency in July.

2024 In May, a major "landmark" study based on mathematical modeling is published by *The Lancet* claiming that global vaccination efforts have saved an estimated 154 million lives (101 million infants)—or the equivalent of 6 lives every minute of every year—over the past 50 years. The study, funded by the WHO and the BMGF, is used to show that vaccination is the "single greatest contribution of any health intervention to ensuring babies not only see their first birthdays but continue leading healthy lives into adulthood."

2024 ✪ In May, RSV vaccine mRESVIA® (mRNA-1345) by Moderna—with lipid nanoparticles—is approved by the FDA in the USA for all adults over 60 years old. The ACIP recommends the vaccine from age 75. It is the only respiratory syncytial virus vaccine available in single-dose pre-filled syringes and the second Moderna product to be licensed. The FDA approval is "based on positive data from the Phase 3 clinical trial ConquerRSV, a global study conducted in approximately 37,000 adults ages 60 years or older in 22 countries"—with 3.7 months of median follow-up. By August, mRESVIA® is authorized in Europe for people 60 years and older.

2024 In the US, the ACIP recommends that all adults and children 6 months or older receive a dose of the updated COVID-19 vaccine that targets the Omicron variant XBB.15, replacing the bivalent BA.4/5 mRNA booster.

2024 ✪ In June, the FDA approves Merck's pneumococcal 21-valent conjugate vaccine, Capvaxive®, recommended to people 18 years old and above.

2024 In June, on the last day of the WHA in Geneva, Switzerland, the WHO's Working Group on International Health Regulations approves the amendments proposed. This "approval" does not follow established procedures. Member states have 10 months to voice their objections.

2024 In June, Kansas attorney general, Kris Kobach, files a lawsuit against Pfizer, claiming that the company deceptively marketed its mRNA COVID-19 vaccine product, concealing safety risks and violating the Kansas Consumer Protection Act. Kansas joins Texas, Utah, Mississippi, and Louisiana, all states that are suing Pfizer for concealing myocarditis, pericarditis, failed pregnancies, and deaths after its COVID-19 vaccine.

2024 ✪ In the summer, the first malaria vaccination campaign takes place in Côte d'Ivoire using the new R21/Matrix-M™ malaria vaccine with Novavax's proprietary Matrix-M™ adjuvant. With the support of Gavi, 15 countries in Africa are expected to introduce malaria vaccines during the year, with the goal of vaccinating 6.6 million children by 2025. This malaria vaccine is approved in Ghana, Nigeria, Burkina Faso and the Central African Republic.

2024 In July, Drs. Daniel Salmon (Director of the Johns Hopkins Institute for Vaccine Safety), Stanley A. Plotkin, Walter Orenstein (US National Immunization Program director 1993–2004), and Robert Chen (Brighton Collaboration Scientific Director) publish an article in *The New England Journal of Medicine* stating the "widespread vaccine hesitancy observed during the COVID-19 pandemic suggests that the public is no longer satisfied with the traditional safety goal of simply detecting and quantifying the associated risks **after** a vaccine has been authorized for use." They propose amending the VICP tax code so that the VICP can fund not only the compensation awarded to injured parties, but also to finance important vaccine safety research to rebuild public confidence in vaccination.

2024 A small outbreak of "rare but deadly" mosquito-borne eastern equine encephalomyelitis is reported in the US. Aerial spraying of insecticide is instigated, and people are told to stay inside after 6:00 p.m. "when mosquitoes are most active."

2024 ✪ The WHO prequalifies a new vaccine for dengue by Japanese company Takeda, called TAK-003 or Qdenga®. It is a 2-dose live-attenuated vaccine containing weakened versions of the four serotypes of dengue virus and is recommended for children 6–16 years old in settings with high disease burden. In July, Qdenga® is approved by Swissmedic in Switzerland for tropical travelers from age four.

2024 On August 14, the WHO declares mpox, formerly known as monkeypox, cousin of smallpox, a PHEIC. There are over 14,000 cases reported and 524 deaths, mostly in the Democratic Republic of the Congo (DRC).

2024 On August 22, the FDA approves and grants emergency use authorizations for two updated mRNA COVID vaccines by Pfizer and Moderna from 6 months old, designed to target the KP.2 strain. The FDA maintains that "vaccination continues to be the cornerstone of COVID-19 prevention" and that the "benefits of these vaccines continue to outweigh their risks."

2024 In September, EUVABECO introduces its print and digital European Vaccination Card (EVC), piloted in Latvia, Greece, Belgium, Germany, and Portugal.

2024 A "humanitarian pause" in Gaza is agreed upon during the Israel-Hamas war, to launch an UNICEF-led experimental nOPV2 polio vaccine campaign targeting over 550,000 Palestinian children. The emergency use authorized vaccine is made by state-owned Indonesian manufacturer Bio Farma and was first designed to target "type 2 circulating vaccine-derived polioviruses (cVDPV2)."

2024 In mid-September, the WHO announces Bavarian Nordic's JYNNEOS® mpox vaccine as the first vaccine against mpox to be prequalified for people over 18 years old, as a 2-dose injection. The WHO adds that this vaccine "may be used 'off-label' in infants, children and adolescents, and in pregnant and immunocompromised people," recommended in outbreak settings "where the benefits of vaccination outweigh the potential risks."

2024 In an interview in September, Bill Gates equates vaccine skepticism with inciting violence and endorses the use of artificial intelligence to censor speech and online content that might cause people to not get vaccinated: "We should have free speech. But if you're inciting violence, if you're causing people not to take vaccines, where are those boundaries?"

A Chronology of Vaccination

2024–2050

2024 The UN 2-day summit "Pact for the Future" takes places in September in New York. Described as a "landmark declaration" the pact pledges "action towards an improved world for tomorrow's generations." The lengthy 66-page document includes a pledge "to move faster towards achieving the UN's Sustainable Development Goals (SDGs) and the Paris Agreement commitments on climate change. It speaks of addressing the root causes of conflicts and accelerating commitments on human rights, including women's rights."

2024 ✪ For the first time ever, the FDA licenses a vaccine for self-administration at home, available for the 2025–2026 flu season. FluMist®, AstraZeneca's live-attenuated intranasal quadrivalent influenza vaccine, is approved for eligible adults up to age 49, who can order the vaccine online and give it to themselves and their children from 2 years old. Reports in the medical literature describe cases where FluMist® "caused shedding and transmission of vaccine strain influenza virus." Note that AstraZeneca states that "immune mechanisms conferring protection against influenza following receipt of FluMist Quadrivalent vaccine are not fully understood." The same month, Fluenz Tetra® (the European name for FluMist®) is removed from the market in Switzerland, due to low demand.

2024 By mid-September, South African tech billionaire Elon Musk's SpaceX project has launched over 7,000 low-orbit Starlink satellites into space since 2018. The company is building a vast megaconstellation in low Earth orbit to provide global internet coverage and wireless connectivity to enable the era of "the Internet of Things/Bodies."

Forecast...

2025 The global vaccine market is anticipated to generate revenues worth US$ 100 billion, a ten-fold increase in sales in 20 years.

2030 Several hundred vaccines are available worldwide, many of which are gene-based products using mRNA or DNA viral vector technology platforms. There are also more combination vaccines and several vaccines classified as "new molecular entities."

2030 ✪ Vaccines are given in utero, at birth and repeatedly throughout a lifetime for dozens of diseases. Many vaccines are enforced on certain populations in order to participate in the modern cashless social credit society that is progressively installed from 2020.

2032 The global vaccine market size is projected to rise from US$ 90 billion (in 2024), to reach between US$ 159–186 billion. This growth is largely fueled and accelerated by the 2020 pandemic and the subsequent investments in vaccine technology, development, and pipeline expansion.

2050 Antimicrobial resistance (AMR) is projected to cause 10 million deaths annually.

2050 Enhanced gene medication and therapy is routinely administered to treat diseases and to correct genetic disorders.

2050 Wireless interfaces, the internet of nano-bio things, and SMART (Self-Monitoring Analysis and Reporting Technology) devices dominate society and the health industry through the use of AI, algorithms for diagnostics, and even robot staff to cover specific functions.

Information has been sourced from many, many books, company websites, press articles, pharmaceutical trade as well as medical and academic journals, Wikipedia, and the awesome Wayback Machine.

Vaccines in the Pipeline

- Alzheimer's disease
- Asthma
- Cancer
- Chlamydia
- Cocaine addiction
- COVID-influenza combination vaccine
- Cytomegalovirus
- Dental caries
- Dengue fever
- DNA vector platform
- Diabetes *type 1*
- Eastern equine encephalitis virus
- *E. coli* invasive disease
- Epstein–Barr virus
- Food allergies
- Gonorrhoea
- Hantavirus
- Heart disease
- Hepatitis C
- Herpes simplex viruses 1 and 2
- HIV
- Human metapneumovirus (HMPV)
- *Leishmania*
- Lyme disease (Borreliosis)
- Marburg virus
- Myocarditis
- New molecular entities
- Nipah virus
- Norovirus
- Obesity (anti-ghrelin)
- Psoriasis
- Rhumatoid arthritis
- Rift Valley fever virus
- Self-adjuvanted polymeric nanoparticle delivery system
- Self-amplifying mRNA (sa-mRNA) platform
- Shigellosis
- *Staphylococcus aureus*
- *Streptococcus* infection Group A/B
- Tobacco (nicotine) addiction
- Universal influenza
- Venezuelan equine encephalitis
- West Nile virus
- *Yersinia pestis* (plague)
- Zika virus

References and source materials are available on
www.TheUltimateVaccineTimeline.com

USA Vaccination History
According to Paul A. Offit, MD

1798–2023

These dates represent when vaccines first became available.
Reviewed by Paul A. Offit, MD, on January 25, 2024

Routinely recommended vaccines

1914	Pertussis vaccine
1926	Diphtheria vaccine
1938	Tetanus vaccine
1945	Inactivated influenza vaccine - trivalent (shot; not routinely recommended)
1948	Diphtheria, tetanus, and pertussis vaccines combined to form DTP
1955	Inactivated polio vaccine (shot)
1962	Live polio vaccine (oral)
1963	Measles vaccine
1967	Mumps vaccine
1969	Rubella vaccine
1971	Measles, mumps, and rubella vaccines combined to form MMR
1981	Hepatitis B vaccine (serum-derived)
1985	Hib vaccine
1986	Hepatitis B vaccine (recombinant)
1992	DTaP vaccine
1995	Varicella vaccine
1995–1996	Hepatitis A vaccine
1998	Rotavirus vaccine (Rotashield®) (removed from market 1999)
2000	Pneumococcal conjugate vaccine
2003	Intranasal influenza vaccine — trivalent (not recommended for use between 2016-2018)
2005	Meningococcal (ACWY) conjugate vaccine for adolescents
2005	Tdap vaccine for adolescents
2006	HPV vaccine for adolescent girls (protected against 2 or 4 types of HPV)
2006	Rotavirus vaccine (RotaTeq®)
2006	Shingles vaccine (Zostavax® for adults 60 years & older)

https://www.chop.edu/centers-programs/vaccine-education-center/vaccine-history/vaccine-availability-timeline

2008	Rotavirus vaccine (Rotarix®)
2009	HPV vaccine for adolescent boys (version that protected against 4 types of HPV)
2013	Inactivated and intranasal influenza vaccine — quadrivalent *The intranasal version was not recommended for use between 2016-2018*
2014	Meningococcal B vaccine licensed
2014	HPV vaccine that protects against 9 types of HPV (Gardasil®9)
2017	Second shingles vaccine (Shingrix® for adults 50 years & older)
2020	COVID-19 vaccine
2021	Dengue vaccine (routine recommendation for 9- to 16-year-olds)
2023	Respiratory syncytial virus (RSV) (Arexvy®/Abrysvo® for adults 60+ and pregnant people)
2023	Respiratory syncytial virus (RSV) (nirsevimab, monoclonal antibody for infants)
2023	Meningococcal pentavalent vaccine (A, B, C, W and Y) (Penbraya®)

Vaccines not routinely recommended

1798	Smallpox vaccine (discontinued use for general population in US 1972)
1885	First rabies vaccines
1896	Typhoid fever vaccine
1927	Tuberculosis vaccine (BCG)
1930s	Japanese encephalitis (JE) vaccine
1935	Yellow fever vaccine
1954	Anthrax vaccine
1965	Currently used anthrax vaccine
1986	Currently used JE vaccine
1997	Currently used rabies vaccines
1999–2002	Lyme disease vaccine (no longer available)
2016–2018	Intranasal influenza vaccine
2016	Cholera vaccine
2019	Ebola vaccine
2019	Dengue vaccine
2019	Mpox/Smallpox
2021	Tick-borne encephalitis (TBE)
2023	Chikungunya vaccine

UK Vaccination History
According to the National Authorities

1796–2019

1796	Development of first smallpox vaccine
1851	Obligation to vaccinate against smallpox enforced
1885	Pasteur creates the first live attenuated viral vaccine (rabies)
1893	Diphtheria antitoxin becomes available
1909	Calmette and Guerin create BCG, first live attenuated bacterial vaccine (tuberculosis) for humans
1941	Diphtheria vaccine
1947	Diphtheria tetanus combined shot with H. pertussis (DTwP)
1950	BCG
1956	Inactivated polio
1957	Pertussis
1961	Tetanus
1962	Live oral polio
1968	Measles
1970	Rubella
1988	MMR
1992	Hib conjugate
1994	Adolescent tetanus and diphtheria
1999	Meningococcal C conjugate
2000	Seasonal influenza over 65s
2001	Preschool acellular pertussis
2004	Inactivated polio (DTaP/IPV/Hib or DTaP/IPV or Td/IPV)
2004	Pneumococcal polysaccharide (PPV)
2006	Combined Hib/MenC
2006	Pneumococcal conjugate PCV7
2008	Human papillomavirus (HPV) for girls
2009	Pandemic Influenza (Swine flu H1N1 vaccine)
2010	Pneumococcal conjugate PCV13
2010	Maternal Influenza
2012	Maternal Pertussis (during pregnancy from 27 weeks)
2013	Children's Influenza
2013	Rotavirus
2013	Shingles
2015	Meningococcal B
2015	Meningococcal ACWY
2017	Hexavalent (DTaP/IPV/Hib/HepB)
2019	Human papillomavirus (HPV) Universal Programme

1951 UK Vaccine Schedule

2 months	DTwP
3 months	DTwP
4 months	DTwP
9 months	Smallpox *discontinued in 1971*

https://www.gov.uk/government/publications/vaccination-timeline/vaccination-timeline-from-1796-to-present
Consulted 2024-08-28 / Not yet updated with 2020-2024 vaccines

World Vaccination History
According to Wikipedia

1796–2023

1796	Edward Jenner develops and documents first vaccine for smallpox
1880	First vaccine for cholera by Louis Pasteur
1885	First vaccine for rabies by Louis Pasteur and Émile Roux
1890	First vaccine for tetanus (serum antitoxin) by Emil von Behring
1896	First vaccine for typhoid fever by Almroth Edward Wright, Richard Pfeiffer and Wilhelm Kolle
1897	First vaccine for bubonic plague by Waldemar Haffkine
1921	First vaccine for tuberculosis by Albert Calmette
1923	First vaccine for diphtheria by Gaston Ramon, Emil von Behring and Kitasato Shibasaburō
1924	First vaccine for scarlet fever by George F. Dick and Gladys Dick
1924	First inactive vaccine for tetanus (tetanus toxoid, TT) by Gaston Ramon, C. Zoeller and P. Descombey
1926	First vaccine for pertussis (whooping cough) by Leila Denmark
1932	First vaccine for yellow fever by Max Theiler and Jean Laigret
1937	First vaccine for typhus by Rudolf Weigl, Ludwik Fleck and Hans Zinsser
1937	First vaccine for influenza by Anatoli Smorodintsev
1941	First vaccine for tick-borne encephalitis
1952	First vaccine for polio (Salk vaccine)
1954	First vaccine for Japanese encephalitis
1954	First vaccine for anthrax
1957	First vaccine for adenovirus-4 and 7
1962	First oral polio vaccine (Sabin vaccine)
1963	First vaccine for measles
1967	First vaccine for mumps
1970	First vaccine for rubella
1977	First vaccine for pneumonia (*Streptococcus pneumoniae*)
1978	First vaccine for meningitis (*Neisseria meningitidis*)
1980	Smallpox declared eradicated worldwide due to vaccination efforts
1981	First vaccine for hepatitis B (first vaccine to target a cause of cancer)
1984	First vaccine for chickenpox
1985	First vaccine for Haemophilus influenzae type b (Hib)
1989	First vaccine for Q fever
1990	First vaccine for Hantavirus hemorrhagic fever with renal syndrome
1991	First vaccine for hepatitis A
1998	First vaccine for Lyme disease
1998	First vaccine for rotavirus
2000	First pneumococcal conjugate vaccine approved in the US (PCV7 or Prevnar)
2003	First nasal influenza vaccine approved in US (FluMist)
2003	First vaccine for Argentine hemorrhagic fever
2006	First vaccine for human papillomavirus (which is a cause of cervical cancer)
2006	First herpes zoster vaccine for shingles
2012	First vaccine for hepatitis E
2012	First quadrivalent (4-strain) influenza vaccine
2013	First vaccine for enterovirus 71, one cause of hand, foot, and mouth disease
2015	First vaccine for malaria
2015	First vaccine for dengue fever
2019	First vaccine for Ebola approved
2020	First [gene-based] vaccine for COVID-19
2023	First respiratory syncytial virus vaccine
2023	First vaccine for Chikungunya

https://en.wikipedia.org/wiki/Timeline_of_human_vaccines
Consulted 2024-08-28

Trade Name	Antigen(s)	Years	MANUFACTURER
Acel-Imune	DTaP	1991-2001	Lederle
Attenuvax	Measles (live)	**1965–2009**	Merck
Attenuvax-Smallpox	Measles-Smallpox	1967	Merck
b-CAPSA-1	Hib (polysaccharide)	1985-89	Praxis
Biavax	Rubella-Mumps (live)	**1970–?**	Merck
BioRab	Rabies	1988-2007	MDPH / BioPort
Cendevax	Rubella (live)	1969-79	RIT-SKF
Certiva	DTaP	1998-2000	Baxter
Decavac	Td	1953-2012	Aventis Pasteur
Dip-Pert-Tet	DTP		Cutter Laboratories
Diptussis	Diphtheria-Pertussis	1949-55	Cutter Laboratories
Dryvax	Vaccinia	1944-2008	Wyeth
Ecolarix	Measles-Rubella (live)	**1994–?**	SmithKline Beecham
Flu Shield	Influenza	**1961–2003**	Wyeth
Fluogen	Influenza	**1945–?**	Parke, Davis & Co.
generic	Tetanus-Toxoid (adsorbed)	1937-2014	Pasteur
Heptavax-B	Hepatitis B (plasma derived)	1981-90	Merck
HIB-Immune	Hib (polysaccharide)	1985-89	Lederle
HibTITER	Hib (conjugate)	**1988–2007**	Praxis
HIB-Vax	Hib (polysaccharide)	1985-89	Connaught
JE-VAX	Japanese Encephalitis	1992-2011	BIKEN
Liovax	Smallpox		Chiron / Sclavo
Lirubel	Measles-Rubella (live)	1974-78	Dow
Lirugen	Measles (live)	1965-76	Pitman-Moore / Sanofi
Lymerix	Lyme Disease	1998-2002	GSK
M-Vac	Measles	1963-79	Lederle
M-M-Vax	Measles-Mumps (live)	**1973–?**	Merck
Meningovax	Meningococcal		Merck
Meruvax II	Rubella (live)	1969-79	Merck
Mevilin-L	Measles (live)	**1994–?**	Novartis / Glaxo

MDPH: Michigan Department of Public Health

This Appendix is featured in the CDC's "Pink Book" 2015 *Epidemiology and Prevention of Vaccine-Preventable Diseases* **13ᵗʰ edition. It is no longer included in the latest 14ᵗʰ edition, dated 2021.**

MANUFACTURER	Trade Name	Antigen(s)	Years
Wyeth	MOPV	Polio (live, oral, monovalent, types I, II, & III)	
Merck / Dow	Mumpsvax	Mumps (live)	**1967–2009**
Pasteur Mérieux / SKB	OmniHIB — OmniHIB is exactly the same as ActHIB, only it was re-labeled by SKB/GSK for sale in the US.	Hib (conjugate)	**1993–?**
Lederle	Orimune	Polio (live, oral)	**1963–2000**
Eli Lilly	Perdipigen	Diphtheria/Pertussis	1949-55
Pfizer	Pfizer-Vax Measles-K	Measles (inactivated)	1963-68
Pfizer	Pfizer-Vax Measles-L	Measles (live)	1965-70
Wyeth / Lederle	Pnu-Imune	Pneumococcal (polysaccharide 14- or 23-valent)	1977-83
Merck	Poliovax	Polio (inactivated)	1988-91
AHP-Wyeth	Prevnar	Pneumococcal (conjugate 7-valent)	2000-2011
Connaught	ProHIBIT	Hib (conjugate)	1987-2000
Merck	Purivax	Polio (inactivated)	1956-65
Parke, Davis & Co.	Quadrigen	DTP-Polio	1959-68
Parke, Davis & Co.	Rabies Iradogen	Rabies	1908-57
Wyeth-Lederle	RotaShield	Rotavirus (live oral)	1998-99
Philips Roxane / Parke, Davis & Co.	Rubelogen	Rubella (live)	1969-72
Merck	Rubeovax	Measles (live)	1963-71
Parke, Davis & Co.	Serobacterin	Pertussis	1945-54
Eli Lilly	Solgen	DTP	1962-77
Eli Lilly	Tetra-Solgen	DTP-Polio	1959-68
Lederle Praxis	Tetramune	DTP-Hib	**1993–?**
Merck	Tetravax	DTP-Polio	1959-65
Sharp & Dohme	Topagen	Pertussis (intranasal)	
Lederle	Tri-Immunol	DTP	**1970–1998**
Parke, Davis & Co.	Tridipigen	DTP	
Connaught	TriHIBit	DTaP/Hib	1996-2011
?	Trinfagen No. 1	DT-Polio	Early 1960s
Merck	Trinivac	DTP	1952-64
Aventis Pasteur	Tripedia	DTaP	1992-2011
Wyeth	Wyvac	Rabies	**1980–1985**

The US Institute of Medicine (IOM), now called the National Academy of Medicine, published its first major vaccine-safety report in 1977 and a second in 1988, both focused on polio vaccines. The IOM, founded in 1970, is tasked with reviewing the scientific literature on vaccine injury to determine whether or not the science supports or rejects causation. **In an alarming number of cases the safety science for vaccines is simply lacking.** This is largely because physiological studies into potential biological mechanisms are not funded by CDC and/or NIH, and therefore do not exist.

The Immunization Safety Review Committee (ISRC) was established by the IOM—funded by the CDC—to evaluate the evidence on possible causal associations between vaccinations and certain adverse outcomes, and to then present conclusions and recommendations. The ISRC first convened in the fall of 2000, and was tasked with providing "independent, non-biased advice to vaccine policy-makers, as well as practitioners and the public." The committee's mandate also included assessing the broader societal significance of vaccination safety issues. Members of the ISRC admitted to sharing the view that vaccination is generally beneficial and claimed to have no conflicts of interest.

Timeline Summary of Several Published IOM Reports

Year of IOM Report	Vaccines Reviewed	# of Conditions Studied	Literature **Supports** Causation	Literature **Rejects** Causation	Literature **Inadequate to Accept or Reject** Causation
1991	DTwP + Rubella	20	5	4	**11**
1994a	DT/Td, MM, OPV-IPV, HepB + Hib	54	12	4	**38**
1994b	DTP	Chronic nervous system dysfunction		*The evidence is insufficient to indicate a causal relation between DTP and permanent neurologic damage.*	
2001a	MMR	Autism	*No link between MMR and autism, though the committee does not exclude the possibility that MMR could contribute to autism in a subset of children.*		
2001b	Thimerosal Containing Vaccines (TCV)	NDD	*Evidence is inadequate to accept or reject a causal relationship between thimerosal from childhood vaccines and the neurodevelopmental disorders (NDD) of autism, ADHD, and speech or language delay.*		
2002	Multiple vaccinations DTwP, DTP, MMR (not the complete schedule)	Immune dysfunction (3): asthma; type 1 diabetes; non-specific infections	0	2	**1**
2002	Hepatitis B	Demyelinating neurological disorders (7)	0	1	**6**
2004	Thimerosal Containing Vaccines + MMR	Autism	*No link between TCV/MMR and autism*		
2004	Influenza vaccinations	4	1	–	**4**
2012	DTaP, MMR, Varicella, Influenza, HepB, HepA, HPV + Meningococcal	158	18	5	**135**
2013	Vaccination schedule	**Safety studies on the complete vaccine schedule DO NOT EXIST**			

DT/Td: Diphtheria, Tetanus / DTwP/DTP: Diphtheria, Tetanus, whole-cell Pertussis / DTaP: Diphtheria, Tetanus, acellular Pertussis MMR: Measles Mumps Rubella / OPV: Oral Polio Vaccine / IPV: Inactivated Polio Vaccine / Hib: *Haemophilus influenzae* type b

https://www.nvic.org/nvic-archives/institutemedicine.aspx

Year	Vaccines Reviewed	# of Conditions Studied	Literature Supports Causation	Literature Rejects Causation	Literature Inadequate to Accept or Reject Causation
1991	**Adverse Effects of Pertussis and Rubella Vaccines** DTwP + Rubella	20	5	4	**11**

Supports Causation
DTwP: Anaphylaxis; acute encephalopathy (encephalitis, encephalomyelitis); protracted inconsolable crying and screaming; shock and "Unusual Shock-Like State" with hypotonicity, hyporesponsiveness and short-lived convulsions, usually febrile.
Rubella (MMR): Acute + chronic arthritis

Inadequate to Accept or Reject Causation
DTwP: Aseptic meningitis; permanent and chronic neurological damage; Guillain-Barré syndrome; learning disabilities + attention deficit disorder; hemolytic anemia; juvenile diabetes; erythema multiforme; peripheral mononeuropathy; thrombocytopenia.
Rubella (MMR): Radiculoneuritis + other neuropathies; thrombocytopenic purpura.

https://www.nap.edu/catalog/1815/adverse-effects-of-pertussis-and-rubella-vaccines

Year	Vaccines Reviewed	# of Conditions Studied	Literature Supports Causation	Literature Rejects Causation	Literature Inadequate to Accept or Reject Causation
1994a	**Adverse Events Associated with Childhood Vaccines** Evidence Bearing on Causality DT/Td, MM, OPV-IPV, HepB + Hib	54	15	4	**38**

Supports Causation
DTP/Td: Anaphylaxis; Guillain-Barré syndrome (GBS); death by GBS; brachial neuritis.
MMR/Measles: Death from severe thrombocytopenia and anaphylaxis; anaphylaxis; thrombocytopenia
OPV: Paralytic + non-paralytic polio; GBS; death from GBS and paralytic polio
HepB: Anaphylaxis; death by anaphylaxis. **Hib (unconjugated):** Early Susceptibility to Hib

Inadequate to Accept or Reject Causation
OPV/IPV + Hib: Transverse myelitis; Sudden Infant Death Syndrome SIDS / **IPV + Hib:** GBS / **OPV:** Death
HepB: GBS; other demyelinating diseases; optic neuritis; transverse myelitis; multiple sclerosis; arthritis; death; SIDS. **Hib:** Death; GBS; thrombocytopenia.

https://www.nap.edu/catalog/2138/adverse-events-associated-with-childhood-vaccines-evidence-bearing-on-causality

Vaccine Safety Science IOM Reports

Year	Vaccines Reviewed	Conditions Studied	Literature **Inadequate to Accept or Reject** Causation
1994b	**DPT Vaccine and Chronic Nervous System Dysfunction** A New Analysis DPT (Diphtheria-Pertussis-Tetanus)	Chronic nervous system dysfunction	*The evidence is insufficient to indicate a causal relation between DPT and permanent neurologic damage.*

Conclusion

The committee concluded (a) that the evidence indicated a causal relation between DPT and febrile seizures, (b) that the evidence did not indicate a causal relation between DPT and afebrile seizures, and (c) that there was insufficient evidence to indicate a causal relation between DPT and epilepsy. The inability to determine causality between DPT and permanent neurologic damage centered on the incompleteness of the preliminary findings reported from the 10-year follow-up study of the National Childhood Encephalopathy Study (Madge et al., 1993; Miller et al., 1993).

Madge N, Diamond J, Miller D, Ross E, McManus C, Wadsworth J, Yule W. "The National Childhood Encephalopathy Study: A 10-year follow-up. A report of the medical, social, behavioural and educational outcomes after serious, acute, neurological illness in early childhood." *Developmental Medicine and Child Neurology* 1993; Supplement No. 68;35(7):1–118.

Miller DL, Madge N, Diamond J, Wadsworth J, Ross E. "Pertussis immunisation and serious acute neurological illnesses in children." *British Medical Journal* 1993; 307:1171–1176.

Year	Vaccines Reviewed	Conditions Studied	Literature **Rejects** Causation
2001a	**Immunization Safety Review** MMR and Autism MMR (Measles-Mumps-Rubella)	Autism	*No link between MMR and autism*

Conclusion

The committee concludes that the evidence favors rejection of a causal relationship at the population level between MMR vaccine and autistic spectrum disorder. However, **the committee notes that its conclusion does not exclude the possibility that MMR vaccine could contribute to ASD in a small number of children**, because the epidemiological evidence lacks the precision to assess rare occurrences of a response to MMR vaccine leading to ASD and the proposed biological models linking MMR vaccine to ASD, although far from established, are nevertheless not disproved.

Moreover, because in some cases autistic symptoms emerge after a period of apparently normal development (i.e., regression), usually in the second year of life, the possibility is left open that MMR vaccination may provoke the onset of the disorder.

DT/Td: Diphtheria, Tetanus / DTwP/DTP: Diphtheria, Tetanus, whole-cell Pertussis / DTaP: Diphtheria, Tetanus, acellular Pertussis
MMR: Measles Mumps Rubella / OPV: Oral Polio Vaccine / IPV: Inactivated Polio Vaccine / Hib: *Haemophilus influenzae* type b

https://www.nap.edu/catalog/9814/dpt-vaccine-and-chronic-nervous-system-dysfunction-a-new-analysis

https://www.nap.edu/catalog/10101/immunization-safety-review-measles-mumps-rubella-vaccine-and-autism

Year	Vaccines Reviewed	Conditions Studied	Literature **Inadequate to Accept or Reject** Causation
2001b	**Immunization Safety Review** Vaccines and NDD Thimerosal Containing Vaccines (TCV) and Neurodevelopmental Disorders	Neuro-developmental disorders (NDD)	*The committee concludes that the evidence is inadequate to accept or reject a causal relationship between thimerosal exposures from childhood vaccines and the neurodevelopmental disorders of autism, ADHD, and speech or language delay.*

Conclusion

The committee concludes that although the hypothesis that exposure to thimerosal-containing vaccines could be associated with neurodevelopmental disorders is not established and rests on indirect and incomplete information, primarily from analogies with methylmercury and levels of maximum mercury exposure from vaccines given in children, the hypothesis is biologically plausible.

The committee recommends the use of the thimerosal-free DTaP, Hib and hepatitis B vaccines in the United States. The committee recommends research in appropriate animal models on the neurodevelopmental effects of ethylmercury.

Year	Vaccines Reviewed	Conditions Studied	Literature **Supports** Causation	Literature **Rejects** Causation	Literature **Inadequate to Accept or Reject** Causation
2002	**Immunization Safety Review** Multiple Immunizations and Immune Dysfunction DTwP, DTP, OPV or MMR *not the complete schedule*	Immune dysfunction (3): asthma; type 1 diabetes; non-specific infections	0	2	**1**

Conclusion

The committee concludes that the epidemiological and clinical evidence favors rejection of a causal relationship between multiple immunizations and an increased risk of heterologous (non-specific) infections and type 1 diabetes. The committee concludes that there is strong evidence for the existence of biological mechanisms by which multiple immunizations under the US infant immunization schedule could possibly influence an individual's risk for heterologous infections. The committee concludes that the **epidemiological and clinical evidence is inadequate to accept or reject a causal relationship between multiple immunizations and an increased risk of allergic disease, particularly asthma.** Meanwhile, the biological evidence that immunization might lead to infection, autoimmune disease, or allergy is more than only theoretical.

Research on the developing human immune system, especially in relation to vaccines, is limited. Studies of animal models are essential to advancing knowledge of the immune system, but those studies have limits because of important differences between humans and animals. Thus, the committee recommends continued research on the development of the human infant immune system.

Vaccine Safety Science IOM Reports

2004–2012

Year	Vaccines Reviewed	Conditions Studied	Literature **Supports** Causation	Literature **Rejects** Causation	Literature **Inadequate to Accept or Reject** Causation
2002	**Immunization Safety Review** Hepatitis B Vaccine and Demyelinating Neurological Disorders	Demyelinating neurological disorders (7)	0	1	**6**

Rejects Causation

The committee concludes that the evidence favors rejection of a causal relationship between hepatitis B vaccine administered to adults and multiple sclerosis.

Inadequate to Accept or Reject Causation

Transverse myelitis; first episode of a central nervous system demyelinating disorder; acute disseminated encephalomyelitis; optic neuritis; Guillain-Barré syndrome; brachial neuritis.

Year	Vaccines Reviewed	Conditions Studied	Literature **Inadequate to Accept or Reject** Causation
2004	**Immunization Safety Review** Influenza Vaccines and Neurological Complications	Demyelinating neurological disorders (4): Guillain-Barré syndrome, multiple sclerosis, optic neuritis and others	Guillain-Barré syndrome, multiple sclerosis, optic neuritis.

Supports Causation

Adult Swine Influenza vaccine 1976: Guillain Barré syndrome.

Inadequate to Accept or Reject Causation

Adult influenza vaccines after 1976: Guillain Barré syndrome, multiple sclerosis, optic neuritis, and other demyelinating neurological disorders.

Pediatric influenza vaccines (6-23 months): Demyelinating neurological disorders.

https://www.nap.edu/catalog/10393/ immunization-safety-review-hepatitis-b-vaccine-and-demyelinating-neurological-disorders

https://www.nap.edu/catalog/10822/immunization-safety-review-influenza-vaccines-and-neurological-complications

DT/Td: Diphtheria, Tetanus / DTwP/DTP: Diphtheria, Tetanus, whole-cell Pertussis / DTaP: Diphtheria, Tetanus, acellular Pertussis
MMR: Measles Mumps Rubella / OPV: Oral Polio Vaccine / IPV: Inactivated Polio Vaccine / Hib: *Haemophilus influenzae* type b

Year	Vaccines Reviewed	Condition Studied	Literature **Rejects** Causation
2004	**Immunization Safety Review** Vaccines and Autism Thimerosal Containing Vaccines (TCV) and MMR (Measles-Mumps-Rubella)	Autism	*No link between TCV/MMR and autism*

Rejects Causation

The committee concludes that the evidence favors rejection of a causal relationship between thimerosal-containing vaccines and autism. The committee concludes that the evidence favors rejection of a causal relationship between MMR vaccine and autism.

In the absence of experimental or human evidence that vaccination (either the MMR vaccine or the preservative thimerosal) affects metabolic, developmental, immune, or other physiological or molecular mechanisms that are causally related to the development of autism, the committee concludes that the hypotheses generated to date are theoretical only.

https://www.nap.edu/catalog/10997/immunization-safety-review-vaccines-and-autism

Year	Vaccines Reviewed	# of Conditions Studied	Literature **Supports** Causation	Literature **Rejects** Causation	Literature **Inadequate to Accept or Reject** Causation
2012	**Adverse Effects of Vaccines** Evidence and Causality DTaP, MMR, Varicella, Influenza (excluding H1N1 vaccines), HepB, HepA, HPV + Meningococcal	158*	18	5	**135**

Supports Causation

Anaphylaxis; oculorespiratory syndrome; transient arthralgia; febrile seizures; syncope; measles inclusion body encephalitis...

Inadequate to Accept or Reject Causation

Encephalopathy; transverse myelitis; stroke; multiple sclerosis; myocarditis; SIDS; fibromyalgia; chronic urticaria; Bell's Palsy; Guillain-Barré syndrome; immune thrombocytopenic purpura thrombocytopenia; chronic inflammatory disseminated polyneuropathy; opsoclonus myoclonus syndrome; optic neuritis; **autism (DTaP)**; ataxia; acute disseminated encephalomyelitis; seizures...

https://www.nap.edu/catalog/13164/adverse-effects-of-vaccines-evidence-and-causality

In total, 57 conditions were analyzed over 8 vaccines—see page 674 of IOM report for detailed summary table. Analysis is based on the evaluation of 12,000+ peer-reviewed articles—it was the largest study undertaken to date and the first comprehensive review since 1994.

Vaccine Safety Science IOM Report SIDS

In 2003, the IOM published a review dedicated to sudden infant death syndrome (SIDS)—the diagnosis most commonly given to infant deaths of uncertain cause. The committee reviewed epidemiologic evidence focusing on three outcomes: SIDS, all SUDI (sudden unexpected death in infancy), and neonatal death (infant death, whether sudden or not, during the first four weeks of life).

Based on this review, the committee concluded that the evidence favors rejection of a causal relationship between some vaccines and SIDS; **and that the evidence is inadequate to accept or reject a causal relationship between other vaccines and SIDS, SUDI, or neonatal death.**

The evidence regarding biological mechanisms is essentially theoretical, reflecting in large measure the lack of knowledge concerning the pathogenesis of SIDS.

Year of IOM Report	Outcomes	Vaccines Studied	Literature Supports Causation	Literature Rejects Causation	Literature **Inadequate to Accept or Reject** Causation
1991	SIDS	DTaP, DT, HepB, OPV, IPV,	-	DTaP, DT	**HepB, OPV, IPV**
2003	SIDS, sudden unexpected death (all SUDI) and neonatal death.	DTwP, DTaP, HepB, Hib, OPV, IPV,	-	DTwP	**DTaP, Hib, HepB, OPV, IPV, Multiple vaccines**

2003 Immunization Safety Review – Conclusions
*"The epidemiologic evidence regarding the relationship between SIDS and receipt of DTaP vaccine consists of **one uncontrolled observational study**. The committee concludes that the **evidence is inadequate to accept or reject a causal relationship between DTaP vaccine and SIDS.***

The committee concludes the evidence is inadequate to accept or reject causal relationships between SIDS and the individual vaccines Hib, HepB, OPV and IPV. Since the 1991 and 1994 IOM reports, no additional epidemiologic studies have been published. The limited data available are drawn from Vaccine Adverse Event Reporting System (VAERS) case reports, which alone are insufficient to establish any causal link.

The committee <u>does not recommend</u> a policy review of the recommended childhood vaccination schedule by any of the national or federal vaccine advisory bodies on the basis of concerns about SUDI.

The committee recommends continued research on the etiology and pathology of SIDS.
It notes that the National Institute of Child Health and Human Development (NICHD, 2001) is targeting five areas of research: (1) the brain and homeostatic control, (2) autonomic development and function, (3) infant care and the sleep environment, (4) infection and immunity and (5) genetics."

DT/Td: Diphtheria, Tetanus / DTwP/DTP: Diphtheria, Tetanus, whole-cell Pertussis / DTaP: Diphtheria, Tetanus, acellular Pertussis
MMR: Measles Mumps Rubella / OPV: Oral Polio Vaccine / IPV: Inactivated Polio Vaccine / Hib: *Haemophilus influenzae* type b

https://www.nap.edu/catalog/10649/immunization-safety-review-vaccinations-and-sudden-unexpected-death-in-infancy

Vaccine Safety Science IOM Report   2013

The Childhood Immunization Schedule and Safety
Stakeholder Concerns, Scientific Evidence, and Future Studies

*"Key elements of the immunization schedule—for example, the number, frequency, timing, order, and age at the time of administration of vaccines—**have not been systematically examined in research studies**."*

Chapter 7: Conclusions and Recommendations

"The committee supports the National Vaccine Advisory Committee Safety Working Group statement that 'the strongest study design, a prospective, randomized clinical trial that includes a study arm receiving no vaccine or vaccine not given according to the current recommended schedule, would be unethical and therefore cannot be done.'"—National Vaccine Advisory Committee (NVAC), 2009, p.38.

*In light of the ethical and feasibility requirements and the available evidence, the committee concludes that **new randomized controlled trials of the childhood immunization schedule are not justified at this time.***

Recommendation 6-2: The Department of Health and Human Services [HHS] should refrain from initiating randomized controlled trials of the childhood immunization schedule that compare safety outcomes in fully vaccinated children with those in unvaccinated children or those vaccinated by use of an alternative schedule.

The committee's literature searches and review were intended to identify health outcomes associated with some aspect of the childhood immunization schedule. Allergy and asthma, autoimmunity, autism, other neurodevelopmental disorders (e.g., learning disabilities, tics, behavioral disorders and intellectual disabilities), seizures, and epilepsy were included as search terms.

No studies have compared the differences in health outcomes that some stakeholders questioned between entirely unimmunized populations of children and fully. Experts who addressed the committee pointed not to a body of evidence that had been overlooked but rather to the fact that existing research has not been designed to test the entire immunization schedule.

The committee believes that although the available evidence is reassuring, studies designed to examine the long-term effects of the cumulative number of vaccines or other aspects of the immunization schedule have not been conducted."

Vaccine Research & Development
Duration of Clinical Trials

> "A typical vaccine development timeline takes 5 to 10 years, and sometimes longer, to assess whether the vaccine is safe and efficacious in clinical trials, complete the regulatory approval processes, and manufacture sufficient quantity of vaccine doses for widespread distribution."
> https://coronavirus.jhu.edu/vaccines/timeline

The development of a vaccine is a time-consuming and very expensive process, with total costs ranging from US$200 to 500 million. All products must undergo safety and efficacy experimentation in animals before being given to humans, and clinical trial data should cover at least two years. This time frame allows researchers to observe adverse events over time and identify any safety signals mid- to long-term.

Preclinical trials
Proof of Concept (1–10 years)

Preclinical testing of vaccine candidates typically starts with animal models, first in small mammals such as mice or rats and then non-human primates such as monkeys. Preclinical studies are important for eliminating potential vaccines that are either toxic or do not induce protective immune responses. Many vaccines that appear to be safe and induce protective immune responses in animals, fail in humans.

Only vaccine candidates that show promise in preclinical testing will initiate an Investigational New Drug (IND) application and move forward into phase I clinical trials on humans.

Phase I
Clinical Trials to Assess Safety, Dosing, and Immune Responses (1–2 years)

Phase I clinical trials are the first step in assessing vaccines in people. Typically involving one to several dozen healthy volunteers, phase I trials assess short-term safety (e.g., soreness at the site of injection, fever, muscle aches) and immune responses, often with different vaccine dosages.

Only if a vaccine candidate is shown to be safe in phase I will it normally move to phase II trials.

Phase II
Clinical Trials to Assess Safety and Immune Responses (2–3 years)

Accelerated: Phase II trials can be completed in three to four months, allowing for longer follow-up to better assess safety and immunogenicity. This timeline is shortened when phase I and phase II trials are layered on top of each other, as was the case for the COVID-19 vaccines.

Phase III
Clinical Trials to Assess Safety and Efficacy (2–4 years)

Accelerated: Assessment of safety and efficacy, particularly if conducted in areas with a high risk of infection, but with follow-up continuing for two years or more to assess long-term safety and efficacy. Analysis of the determined endpoints is key to the final report of this final clinical phase before potential licensure.

The COVID-19 biotechnological gene therapy products marketed as vaccines were launched with less than two months of safety data, and less than one year of clinical phase II and III trial data. Animal studies were inadequate to determine safety, and all trial phases were layered on top of each other, instead of waiting for the results from previous completed phases. This decision was justified by governments worldwide by the emergency use authorization status of these products.

Typical Timeline for Vaccine Research & Development

	Research	Development		Commercialization	
	Discovery Target identification + validation	**Phase I** Proof of concept	**Phase II** Early product development	**Phase III** Advanced development	**Phase IV** Pre-launch + post-licensure
KEY ACTIVITIES	• Hit discovery • Assay development + screening	• Assess safety, immunogenicity, and optimize dose schedule • Healthy volunteer studies • Dose escalation	• Expand safety, immunogenicity, and optimize dose schedule • Effects + adverse effects • Pre-regulatory data	• Assess safety • Effects + adverse events • Regulatory data	• Regulatory submission • Pre-marketing • Post-marketing surveillance
PEOPLE #	Animals	1-100	Hundreds	Thousands	Millions
DURATION	24–36 months	12–24 months	12–24 months	12–48 months	6–12 months *Post-marketing surveillance > 1 year*

Investigational New Drug (IND) application submitted to regulatory agency

Product License Application (PLA) accepted by regulatory agency

> "**Before vaccines are licensed, their efficacy has to be shown in clinical trials, which are generally not powered to evaluate safety.** Even phase III trials collect only limited safety data, mostly on common adverse events that occur shortly after vaccination such as local and systemic reactions related to the immunogenicity of the vaccine. As a result, when a new vaccine comes to market there is some uncertainty about its safety profile, specifically about rare events or those occurring a longer time after vaccination. Such effects cannot be detected until the vaccine is administered within large populations. That is the work of vaccine pharmacovigilance."
> https://www.jstor.org/stable/26963416

Phase IV
Scaling Up Vaccine Manufacturing
Post-Licensure Vaccine Safety Monitoring

Administered to millions of people with post-marketing surveillance initiated for all newly authorized biotechnological products. Safety signals have to be closely monitored, as rare adverse events are only observed once the product is administered to the mass population. Phase IV provides information about the safety and effectiveness of the vaccine in the general population, under normal conditions.

From start to finish—initial vaccine development to market authorization—the process of bringing a vaccine to market typically takes at least a decade.
The Regulatory Approval Process alone can take several years, except for accelerated medication deemed urgent and life-saving, which can take just months.

The licensing of vaccines should be a lengthy process because of the high safety and efficacy standards required when injecting healthy children and adults.

Considerable time is needed to acquire substantiating data from clinical trials—a process that is especially time-consuming for new vaccines.

Safety of Clinical Trials
Vaccine injury can be easily discounted and discarded during clinical trials by the principle investigator—events are dismissed as "unrelated" to vaccination, without much investigation. Negative trials do not have to be submitted, and only positive results based on limited endpoints, no true placebos and/or flawed safety design, are often presented to regulatory authorities.

193

Adjuvanted Vaccines Approval Timeline   1930–2022

Approval Year	Adjuvants	Formulation & Type	Vaccines
1930	**Alum**	Alum salts: aluminum hydroxide, aluminum phosphate, aluminum oxyhydroxide	DTP, hepatitis A & B, IPV, Hib, meningococcal, pneumococcal
1997	**MF59**	**Proprietary Chiron > Novartis > Seqiris** Squalene emulsion (with Tween 80) oil-in-water	Seasonal & pandemic influenza **Fluad®** **Focetria®** (2009)
1997	**Virosomes**	**Proprietary Berna Biotech** Liposomes: spherical vesicles made up of phospholipids, lecithin and cephalin.	**Epaxal®** hepatitis A **Inflexal-V®** influenza Solvay **Invivac®** influenza
2003	**RC-529**	**Proprietary Corixa Corp** Fully synthetic monosaccharide mimetic of MPL	Berna Biotech **Supervax®** *Only approved in Argentina*
2004	**Nano Alum AAHS**	**Proprietary Merck** Amorphous aluminum hydroxyphosphate sulfate	**Comvax®/Procomvax®** HepB-Hib **Recombivax-HB®** hepatitis B **Gardasil®** HPV **Vaxelis®** DTaP-IPV-HepB-Hib
2005	**AS04**	**Proprietary GSK** MPL adsorbed on alum adjuvant	**Fendrix®** hepatitis B **Cervarix®** HPV
2013	**AS03**	**Proprietary GSK** Squalene emulsion with polysorbate 80	Pre-pandemic H5N1
2015	**AS01**	**Proprietary GSK / Antigenics** MPL/QS21 Stimulon® in liposome base	**Mosquirix®** RTS,S malaria **Shingrix®** shingles **Abrysvo®/Arexvy®** RSV HIV vaccine in clinical trials
2017	**CpG 1018**	**Proprietary Dynavax** 22-mer oligonucleotide sequence	**Heplisav-B®** hepatitis B VLP* Valneva **VLA2001** COVID-19 (2022)
2021	**Allhydroxiquim-II**	**Proprietary ViroVax LLC** Alum adsorbed to TLR7/8 agonist	**COVAXIN®** COVID-19 *Approved in India*
2022	**Matrix-M™**	**Proprietary Novavax**	**Nuvaxovid®** COVID-19

*Virus-Like Particles

AS	Adjuvant System
MPL	3−O−desacly-monophosphoryl lipid A is a detoxified derivative of the cell wall lipopolysaccharide from *Salmonella minnesota* R595 strain—isolated from the surface of the bacteria. It stimulates activation of innate immunity via toll-like receptor 4 (TLR4).
QS21	A highly purified immunogenic naturally occuring substance derived from the Chilean soapbark tree *Quillaja saponaria Molina*, fraction 21 that is a key component of GSK's AS01. QS21-Stimulon® is licensed from Antigenics Inc., a wholly owned subsidiary of Agenus Inc.
CpG 1018	Cytosine phosphoguanine (CpG) motifs—a proprietary and synthetic form of DNA that mimics bacterial and viral genetic material, targeting toll-like receptor 9 (TLR9).
Matrix-M™	Consists of two individually formed 40 nm-sized particles, each with different and well-characterized saponin fraction (Fraction-A and Fraction-C). Matrix-A and -C particles are formed by formulating purified saponin from the Chilean soapbark tree *Quillaja saponaria Molina*, with cholesterol and phospholipid into nanoparticles.
Adjuvant 65	A proprietary peanut oil-based adjuvant made by Merck for use in its killed influenza vaccine in the 1960s, that was patented in 1964 and licensed in the UK in 1974. It was never approved in the USA, though it was still available—it was eventually abandoned due to purification issues.

It is important to note that adjuvants alone are not licensed, but each specific antigen-adjuvant formulation is evaluated in clinical trials before licensure.

"Saponins are natural glycosides which possess a wide range of pharmacological properties including cytotoxic activity. The name 'saponin' comes from the Latin word 'sapo,' which means 'soap' as saponins show the unique properties of foaming and emulsifying agents. *Quillaja saponin* is a natural effective emulsifier to form and stabilize oil/water emulsions with very small oil beads (ø<200 nm). They are stable in a wide range of environmental parameters (pH, ionic strength, temperature)."

"We now know that aluminium in adjuvants is dissolved and transported throughout the body, including the brain and we cannot discount the biological availability of this aluminium. It is a sobering thought that aluminium adjuvants have not had to pass any of the safety trials that would be expected of any drug or treatment.

Their application is historical and this should not necessarily be equated with their safety.

There is no consensus as to whether it is safe to introduce aluminium in prophylaxis or otherwise, and until the requisite research is carried out it is misleading to conclude that aluminium adjuvants are safe for all to use."

—Prof. Christopher Exley, PhD, FRSB

British biochemist and leading expert on human exposure to aluminum having spent over 35 years researching the subject, resulting in his moniker "Mr. Aluminum." He is the author of *Imagine You Are An Aluminum Atom* and of numerous scientific articles on the topic of aluminum.

The above quote is taken from *The Lancet* "Reflection and Reaction"– June 2004

https://www.thelancet.com/journals/laninf/article/PIIS1473-3099(04)01039-4/abstract

Types of Vaccines
Currently Available

2024

Below is a color-code presenting further details on the specific types of vaccines available.

Some vaccines are a mix between different types, like DTP, and other vaccines exist as various versions, like for the influenza vaccine (inactivated, live-attenuated, split-virion etc.) and are presented as a color gradient on the following pages.

STANDARD

Sv
SUBUNIT VIRAL

RECOMBINANT SUBUNIT VIRAL
Genetically engineered recombined genetic viral surface proteins, glycoproteins, or virus-like particles (VLP). Adjuvants may be required.

Iv
INACTIVATED VIRAL

INACTIVATED VIRAL
Killed or chemically-/heat- inactivated viral pathogen. Split-virion influenza vaccines are prepared based on the inactivated whole influenza virus that has been disrupted by a detergent. Adjuvants may be required.

Lv
LIVE VIRAL

LIVE-ATTENUATED VIRAL
Weakened live version of the viral pathogen—oral or injected. Adjuvants are not required.

Bt
BACTERIAL TOXOID

BACTERIAL TOXOID
Chemically altered toxin with reduced toxicity yet retained antigenicity—cannot provide immunity from disease as it cannot prevent infection or transmission. Adjuvants, multiple doses, and booster shots are required.

Sp
SUBUNIT PROTEIN

RECOMBINANT SUBUNIT PROTEIN
Inactivated and purified pieces of a pathogen and include polysaccharide* (bacteria encased in a layer of sugar), conjugate (bacteria protein attached to sugar), or surface protein with carrier protein or toxin. Adjuvants are required.

WP
WHOLE

WHOLE PATHOGEN
Live-attenuated mycobacteria (BCG-TB), live virus (smallpox), or inactivated whole pathogen (whole-cell pertussis).

Carbohydrates in the form of capsular polysaccharides and/or lipopolysaccharides are the major components on the surface of bacteria.

NUCLEIC ACID-BASED

MESSENGER RNA (mRNA or self-amplifying mRNA) or DNA (not currently licensed)
Genetically modified mRNA code for specific proteins, packaged in a positively charged lipid delivery system. DNA vaccines are more stable and do not require a lipid delivery system—instead they use electroporation.

VIRAL VECTOR-BASED
Pathogen genetic material is inserted into a "harmless" carrier microorganism that replicates after injection. Typical carriers include adenovirus, measles, influenza, or poxvirus.

Products using bioinformatics with genetic materials that enter cells and instruct them to produce proteins.
Marketed as vaccines, these novel biologics are actually experimental gene-based products.

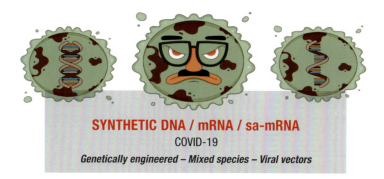

SYNTHETIC DNA / mRNA / sa-mRNA
COVID-19
Genetically engineered – Mixed species – Viral vectors

In 2020, the World Health Organization categorized vaccine types as the following:
live-attenuated, inactivated, subunit / recombinant and toxoid.

LIVE-ATTENUATED	**INACTIVATED**	**SUBUNIT / RECOMBINANT**	**TOXOID**
ORAL POLIO VACCINE (OPV)	WHOLE-CELL PERTUSSIS (WP)	ACELLULAR PERTUSSIS (AP)	DIPHTHERIA TOXOID
MEASLES	INACTIVATED POLIO VIRUS (IPV)	*HAEMOPHILUS INFLUENZAE* TYPE B (HIB)	TETANUS TOXOID (TT)
MUMPS	INFLUENZA	HUMAN PAPILLOMAVIRUS (HPV)	
RUBELLA	TICK-BORNE ENCEPHALITIS	HEPATITIS B	
ROTAVIRUS		MENINGOCOCCAL MEN ACWY	
INFLUENZA		PNEUMOCOCCAL (PCV-7, -10,-13)	
YELLOW FEVER		SHINGLES-SHINGRIX	
SHINGLES-ZOSTAVAX			
TUBERCULOSIS BCG			

This picto represents vaccines that are genetically engineered and are applied to many vaccines types, as most contain synthetic substances.

Swiss Vaccine Schedule History
From Cradle to Grave

Vaccines given in combination

DTP | Hi | IPV

Vaccines given as separate doses

DTP | Hi | IPV

1925

Sm SMALLPOX **D** DIPHTHERIA

1975

BCG TUBERCULOSIS **DTP** DTwP **OP** LIVE POLIO **M** MEASLES **Sm** SMALLPOX **Ru** RUBELLA **Flu** INFLUENZA

Newborn

2020

DURING EVERY PREGNANCY from 13 weeks / 2nd trimester

TDP Tdap **Flu** INFLUENZA

BASE

DTP DTaP **IPV** POLIO **Hi** HIB **HB** HEPATITIS B **MMR** MMR-II **Vz** VARICELLA **HP** HPV

COMPLEMENTARY

Mn MENINGOC **Pn** PNEUMOCC **Flu** INFLUENZA

EVERY YEAR

Abbreviations

DTwP	Diphtheria, tetanus, whole-cell pertussis (whooping cough)
DTaP	Children: diphtheria, tetanus, acellular pertussis – Infanrix®
Tdap	Adults: reduced diphtheria, tetanus, acellular pertussis – Boostrix®
BCG	Bacillus Calmette–Guérin
OPV	Oral live polio (Sabin)
IPV	Inactivated live polio (Salk)
MMR	Measles, mumps, rubella
Td	Tetanus, diphtheria
HepB	Hepatitis B
Hib	*Haemophilus influenzae* type b
Pn	Pneumococcal
Mn	Meningococcal
HPV	Human papillomavirus
Vz	Varicella (chickenpox)

2024

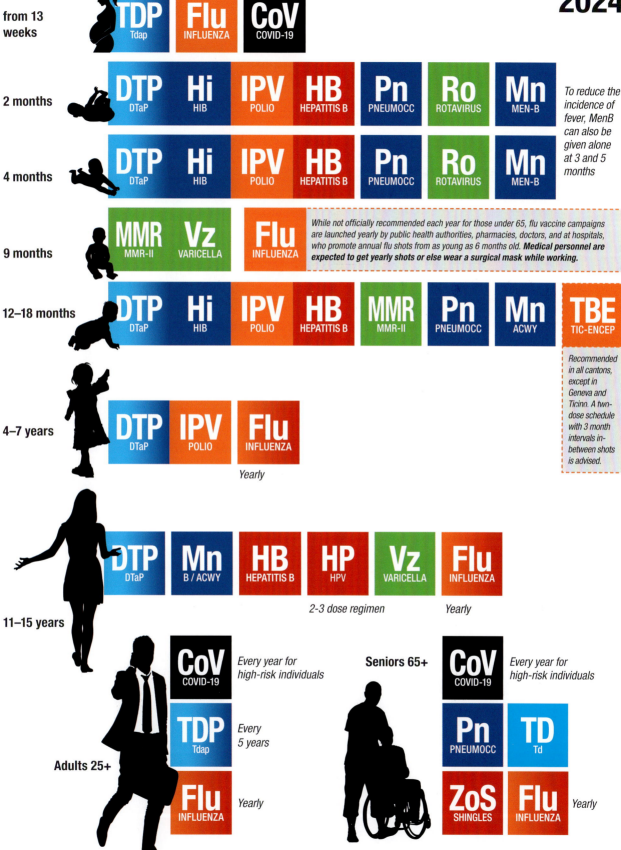

from 13 weeks

TDP Tdap | Flu INFLUENZA | CoV COVID-19

2 months

DTP DTaP | Hi HIB | IPV POLIO | HB HEPATITIS B | Pn PNEUMOCC | Ro ROTAVIRUS | Mn MEN-B

To reduce the incidence of fever, MenB can also be given alone at 3 and 5 months

4 months

DTP DTaP | Hi HIB | IPV POLIO | HB HEPATITIS B | Pn PNEUMOCC | Ro ROTAVIRUS | Mn MEN-B

9 months

MMR MMR-II | Vz VARICELLA | Flu INFLUENZA

*While not officially recommended each year for those under 65, flu vaccine campaigns are launched yearly by public health authorities, pharmacies, doctors, and at hospitals, who promote annual flu shots from as young as 6 months old. **Medical personnel are expected to get yearly shots or else wear a surgical mask while working.***

12–18 months

DTP DTaP | Hi HIB | IPV POLIO | HB HEPATITIS B | MMR MMR-II | Pn PNEUMOCC | Mn ACWY | TBE TIC-ENCEP

Recommended in all cantons, except in Geneva and Ticino. A two-dose schedule with 3 month intervals in-between shots is advised.

4–7 years

DTP DTaP | IPV POLIO | Flu INFLUENZA

Yearly

11–15 years

DTP DTaP | Mn B / ACWY | HB HEPATITIS B | HP HPV | Vz VARICELLA | Flu INFLUENZA

2-3 dose regimen — *Yearly*

Adults 25+

CoV COVID-19 — *Every year for high-risk individuals*

TDP Tdap — *Every 5 years*

Flu INFLUENZA — *Yearly*

Seniors 65+

CoV COVID-19 — *Every year for high-risk individuals*

Pn PNEUMOCC | TD Td

ZoS SHINGLES | Flu INFLUENZA — *Yearly*

USA Vaccine Schedule History
From Womb to Tomb

1983 — 11 vaccines for 7 diseases

Age	Vaccine
2 months	DTwP
	OPV
4 months	DTwP
	OPV
6 months	DTwP
15 months	MMR
18 months	DTwP
	OPV
4 years	DTwP
	OPV
14 years	Td

1994 — 18 vaccines for 9 diseases

Age	Vaccine
at birth	HepB *since 1991*
2 months	HepB
	DTP
	OPV
	Hib
4 months	DTP
	OPV
	Hib
6 months	DTP
	OPV
	Hib
	HepB
12 months	MMR
	Hib
15 months	DTaP/DTwP
4 years	DTaP/DTwP
	OPV
	MMR

Abbreviations

DTwP	Diphtheria, tetanus, whole-cell pertussis
DTaP/Tdap	Diphtheria, tetanus, acellular pertussis
OPV	Oral live polio (Sabin)
IPV	Inactivated live polio (Salk)
MMR	Measles, mumps, rubella
Td	Tetanus, diphtheria
HepB	Hepatitis B
Hib	*Haemophilus influenzae* type b
PCV/Pn	Pneumococcal
HPV	Human papillomavirus

2015 — 41 vaccines for 16 diseases by 16 years old

Age	Vaccine
pregnancy	DTaP
	Influenza
at birth	HepB
2 months	HepB
	DTP
	IPV
	Hib
	PCV
	Rotavirus
4 months	DTP
	IPV
	Hib
	PCV
	Rotavirus
6 months	DTP
	IPV
	Hib
	PCV
	Rotavirus
	HepB
	Influenza
12 months	MMR
	PCV
	Hib
	Varicella
	HepA
18 months	DTaP
	Influenza
	HepA
2 years	Influenza
3 years	Influenza
4 years	DTaP
	IPV
	MMR
	Varicella
5-18 years	Influenza *every year*
10 years	HPV
11 years	HPV
12 years	Tdap
	Meningococcal
16 years	Meningococcal

The combination of all these vaccines has never been adequately safety tested.

2024

Routine Vaccination Schedule

From 1st–2nd trimester

DTP DTaP | Flu INFLUENZA | CoV COVID-19 | RS RSV *Abrysvo®*

At birth

HB HEP-B | Mandated (in several US states) synthetic vitamin K injection is given to babies at birth that **contains aluminum.**

2 months

DTP DTaP | Hi HIB | IPV POLIO | HB HEP-B | Pn PNEUMOCC | Ro ROTAVIRUS

4 months

DTP DTaP | Hi HIB | IPV POLIO | Pn PNEUMOCC | Ro ROTAVIRUS

6 months

DTP DTaP | Hi HIB | IPV POLIO | HB HEP-B | Pn PNEUMOCC | Flu INFLUENZA | CoV COVID-19

Yearly — *From 6 months old 2 or 3 dose regimen*

12-15 months

Pn PNEUMOCC | MMR MMR-II | Vz VARICELLA | HA HEP-A

Hep A: 2-dose regimen from age 12-24 months

4-6 years

DTP DTaP | Hi HIB | IPV POLIO | MMR MMR-II | Vz VARICELLA | Pn PNEUMOCC | Flu INFLUENZA

Live-attenuated intranasal flu vaccine from 2–49 years old

Yearly

9-12 years

DTP DTaP | Mn MENINGOC | HP HPV | Flu INFLUENZA *Yearly*

2-3 dose regimen

16 years

Mn MENINGOC

The US federal government is the largest buyer of childhood vaccines.

Adults 18+

Flu INFLUENZA | CoV COVID-19 | DTP DTaP

Yearly | *Yearly* | *Every 5-10 years*

Seniors 60+

RS RSV *Arexvy®* | Pn PNEUMOCC | CoV COVID-19 | Flu INFLUENZA | ZoS SHINGLES

Yearly

A visual guide to vaccines
used in the routine immunisation schedule

July 2024

Vaccine trade name, manufacturer, abbreviation	Disease(s) protected against	Age due
Infanrix hexa (GSK) or **Vaxelis (Sanofi)** Hexavalent vaccine DTaP/IPV/Hib/HepB	Diphtheria, tetanus, pertussis (whooping cough), polio, *Haemophilus influenzae* type b (Hib) and hepatitis B	8,12,16 weeks
Rotarix (GSK) Rotavirus (oral)	Rotavirus gastroenteritis	8 weeks and 12 weeks
Bexsero (GSK) MenB	Meningococcal group B	8 weeks, 16 weeks and 1 year
Prevenar 13 (Pfizer) Pneumococcal conjugate vaccine (PCV)	Pneumococcal (13 serotypes)	12 weeks and 1 year
Menitorix (GSK) Hib/MenC	*Haemophilus influenzae* type b (Hib) and meningococcal group C	1 year
MMRvaxPro (MSD) or **Priorix (GSK)** MMR	Measles, mumps and rubella	1 year and 3 years 4 months
Fluenz Tetra (AstraZeneca) LAIV (nasal)	Influenza	Eligible paediatric age groups*
Boostrix-IPV (GSK) dTaP/IPV	Diphtheria, tetanus, pertussis and polio	3 years 4 months
Gardasil 9 (MSD) Human papillomavirus vaccine (HPV)	Cancers and genital warts caused by specific human papillomavirus (HPV) types	12 to 13 years (school Year 8)
Revaxis (Sanofi) Td/IPV	Tetanus, diphtheria and polio	14 years (school Year 9)
MenQuadfi (Sanofi) MenACWY	Meningococcal groups A, C, W and Y	14 years (school Year 9)
ADACEL (Sanofi) Tdap	Tetanus, diphtheria and pertussis	Pregnant women from 16 weeks gestation
Pneumovax 23 (MSD) Pneumococcal Polysaccharide Vaccine (PPV23)	Pneumococcal (23 serotypes)	65 years (and certain at-risk groups)**
Shingrix (GSK) Shingles	Shingles	From 1 September 2023 onwards: 50+ years and severely immunosuppressed; eligible age groups***; 70 years+ if Zostavax not available
Zostavax (MSD) whilst supplies available Shingles	Shingles	70+ years of age (immunocompetent)***

*See annual flu letter for eligible age groups: www.gov.uk/government/collections/annual-flu-programme
**See the Green Book: www.gov.uk/government/publications/pneumococcal-the-green-book-chapter-25
***See Shingles: guidance and vaccination programme: www.gov.uk/government/collections/shingles-vaccination-programme
Images of inactivated influenza vaccines not shown on this poster as several different products are available
Full details of the routine immunisation schedule available: www.gov.uk/government/collections/immunisation

Date of revision: July 2024. © Crown copyright 2024. VAPCO24 JUL 2024. UK Health Security Agency Gateway Number: 2024077. Produced by UK Health Security Agency

immunisation
The safest way to protect your health

The UK Health Agency, previously called Public Health England, provides official information on the vaccination schedule and supplies information—including the ingredients—on the brands of vaccines available in the country.

UK Health Security Agency

> **Vaccination during pregnancy**
> dTaP/Tdap: from 16 weeks
> Influenza/COVID-19:   can be given at any stage of pregnancy

The complete routine immunisation schedule — From July 2024

Age due	Diseases protected against	Vaccine given and trade name		Usual site[1]
Eight weeks old	Diphtheria, tetanus, pertussis (whooping cough), polio, *Haemophilus influenzae* type b (Hib) and hepatitis B	DTaP/IPV/Hib/HepB	Infanrix hexa or Vaxelis	Thigh
	Meningococcal group B (MenB)	MenB	Bexsero	Left thigh
	Rotavirus gastroenteritis	Rotavirus[2]	Rotarix[2]	By mouth
Twelve weeks old	Diphtheria, tetanus, pertussis, polio, Hib and hepatitis B	DTaP/IPV/Hib/HepB	Infanrix hexa or Vaxelis	Thigh
	Pneumococcal (13 serotypes)	Pneumococcal conjugate vaccine (PCV)	Prevenar 13	Thigh
	Rotavirus	Rotavirus[2]	Rotarix[2]	By mouth
Sixteen weeks old	Diphtheria, tetanus, pertussis, polio, Hib and hepatitis B	DTaP/IPV/Hib/HepB	Infanrix hexa or Vaxelis	Thigh
	MenB	MenB	Bexsero	Left thigh
One year old (on or after the child's first birthday)	Hib and MenC	Hib/MenC	Menitorix	Upper arm/thigh
	Pneumococcal	PCV booster	Prevenar 13	Upper arm/thigh
	Measles, mumps and rubella (German measles)	MMR	MMRvaxPro[3] or Priorix	Upper arm/thigh
	MenB	MenB booster	Bexsero	Left thigh
Eligible paediatric age groups[4]	Influenza (each year from September)	Live attenuated influenza vaccine LAIV[3,6]	Fluenz Tetra[3,6]	Both nostrils
Three years four months old or soon after	Diphtheria, tetanus, pertussis and polio	dTaP/IPV	Boostrix-IPV	Upper arm
	Measles, mumps and rubella	MMR (check first dose given)	MMRvaxPro[3] or Priorix	Upper arm
Boys and girls aged twelve to thirteen years	Cancers and genital warts caused by specific human papillomavirus (HPV) types	HPV[5]	Gardasil 9	Upper arm
Fourteen years old (school Year 9)	Tetanus, diphtheria and polio	Td/IPV (check MMR status)	Revaxis	Upper arm
	Meningococcal groups A, C, W and Y	MenACWY	MenQuadfi	Upper arm
65 years old	Pneumococcal (23 serotypes)	Pneumococcal Polysaccharide Vaccine (PPV23)	Pneumovax 23	Upper arm
65 years of age and older	Influenza (each year from September)	Inactivated influenza vaccine	Multiple	Upper arm
65 from September 2023[7]	Shingles	Shingles vaccine	Shingrix	Upper arm
70 to 79 years of age (plus eligible age groups and severely immunosuppressed)[7]	Shingles	Shingles vaccine	Zostavax[3,7] (or Shingrix if Zostavax contraindicated)	Upper arm

1. Intramuscular injection into deltoid muscle in upper arm or anterolateral aspect of the thigh.
2. Rotavirus vaccine should only be given after checking for SCID screening result.
3. Contains porcine gelatine.
4. See annual flu letter at: www.gov.uk/government/collections/annual-flu-programme
5. See Green Book HPV Chapter 18a for details on immunising immunocompromised young people who will need 3 doses.
6. If LAIV (live attenuated influenza vaccine) is contraindicated or otherwise unsuitable use inactivated flu vaccine (check Green Book Chapter 19 for details).
7. See Green Book Shingles Chapter 28a for details on eligible age groups including severely immunosuppressed individuals from age 50.

 For vaccine supply information for the routine immunisation schedule please visit portal.immform.phe.gov.uk and check Vaccine Update for all other vaccine supply information: www.gov.uk/government/collections/vaccine-update

 **i mmunisation** | The safest way to protect children and adults

How Safe Is the Vaccine Schedule
For the Developing Brain?

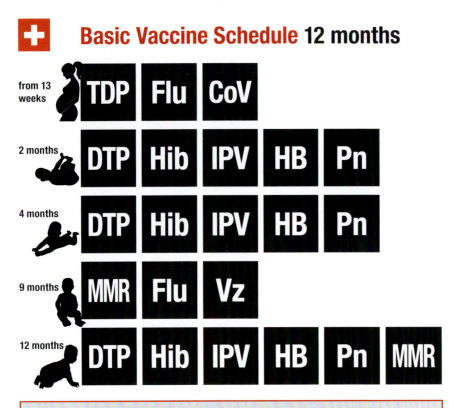

Basic Vaccine Schedule 12 months

from 13 weeks	TDP	Flu	CoV		
2 months	DTP	Hib	IPV	HB	Pn
4 months	DTP	Hib	IPV	HB	Pn
9 months	MMR	Flu	Vz		
12 months	DTP	Hib	IPV	HB	Pn	MMR

FDA "safety" limit for daily aluminum (Al³⁺) 0.85–1.25 mg/dose.
Based on oral ingestion by adult mice, observed for 28 days—evaluated on efficacy markers.
No adjustment is made for body weight.

The US vaccine schedule has even more vaccines from birth...

If a mother follows the recommended schedule (excluding the influenza shot) using Boostrix® during pregnancy, Infanrix Hexa®, Prevenar-13®* and NeisVac-C®, by the time her baby is 12 months old, it will have been exposed to over **3,800 mcg of aluminum,** by **injection** along with a chemical cocktail of several other synthetic ingredients...

...that have never been safety tested in combination.

Ingredients include known lung and skin irritants, neurotoxins like aluminum and foreign proteins derived from animals, humans, yeasts or genetically engineered.

TDP	Reduced-antigen tetanus, diphtheria, acellular pertussis (Tdap)
Flu	Influenza *recommended from 6 months old*
CoV	Coronavirus SARS-CoV-2
DTP	Diphtheria, tetanus, acellular pertussis (DTaP)
Hib	*Hemophilus Influenzae* type B
IPV	Inactivated polio
HB	Hepatitis B
Pn	Pneumococcal (PCV)
MMR	Measles, mumps, rubella
Vz	Varicella-chickenpox

*Prevenar-13® in the EU-CH-UK / Prevnar-13® in the US

Timeline of Early Brain Development
Vaccine Exposure Begins in the Womb

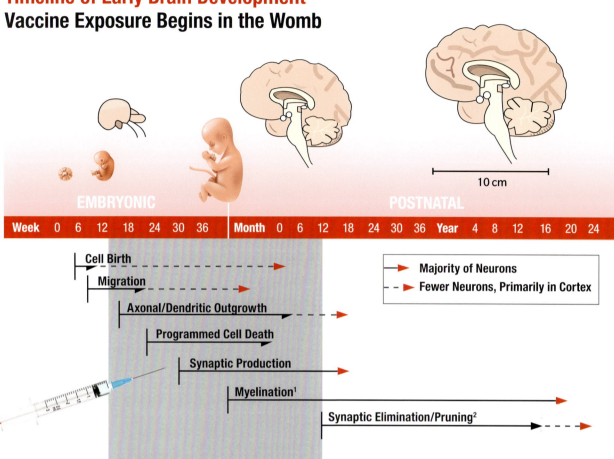

10 cm

EMBRYONIC

POSTNATAL

| Week | 0 | 6 | 12 | 18 | 24 | 30 | 36 | Month | 0 | 6 | 12 | 18 | 24 | 30 | 36 | Year | 4 | 8 | 12 | 16 | 20 | 24 |

Cell Birth

Migration

Axonal/Dendritic Outgrowth

Programmed Cell Death

Synaptic Production

Myelination[1]

Synaptic Elimination/Pruning[2]

→ **Majority of Neurons**

--→ **Fewer Neurons, Primarily in Cortex**

1. Myelination is the a process of creating the myelin sheath that surrounds the axons of the nerves, forming an electrically insulating layer (like the rubber coating that insulates copper wire).

2. Synaptic pruning first begins at 8 months in the visual cortex and at 24 months in the frontal cerebral cortex, removing unnecessary excitatory and inhibitory synaptic connections. Pruning also occurs in the brainstem and cerebellum.

By the age of 12 months old, the average child following the government recommended vaccine schedule will have been exposed to, <u>by injection:</u>

aluminum salts, multiple antigens like viruses and bacteria, polysorbate 80, and residual cellular debris containing DNA fragments from many species including fetal human cells, monkey cells, fetal avian cells, fetal bovine cells, insect cells, as well as genetically modified molecules, nanoparticles, and other undisclosed contaminants.

During this critical juncture of growth, repeated hyperstimulation of the immune system can trigger the cell danger response, shutting down functions like development and learning in favor of survival.

Tdap, Influenza and COVID-19 vaccines* are all recommended during pregnancy.
Some contain aluminum, genetic materials, and foreign proteins.
**The nervous, immune, digestive, respiratory, endocrine and skeletal systems
as well as filtering organs like the liver and kidneys, not to mention the brain and lungs,**
are all immature and in development during this critical phase of growth.

Since 2023, the newly licensed RSV vaccine is also recommended to pregnant women in the USA, during the RSV season from October to April. **205**

Timeline adapted from Andersen, S.L. (2003) "Trajectories of Brain Development: Point of Vulnerability or Window of Opportunity?" *Neuroscience and Biobehavioral Reviews, 27, 3-18.*

Vaccine Safety Concerns
Overview of Historical Publications 1885–1920

Below is an non-exhaustive list of historical publications written by esteemed doctors, surgeons, scholars, researchers, and scientists during the 35-year period between 1885 and 1920.

Vaccination failure has been observed since 1818 and anti-vaccine sentiment began shortly after compulsory smallpox vaccination laws were enforced in the mid-1800s. It was only in the late 20th century that adverse effects became officially registered through a centralized system in some, but not all, developed nations.

1885

The Story of A Great Delusion / 324 pages
William M. White

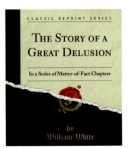

"The promise of vaccination, its absolute security and harmlessness, was speedily belied [disproved]. The vaccinated caught smallpox; they fell sick after the operation; they were afflicted with eruptions and swellings; they died. These mishaps were at first denied—stoutly denied; and when denial was no longer possible, it was attempted to explain them away."

https://archive.org/details/storyofgreatdelu00whitrich

1893

Leprosy and Vaccination / 124 pages
William Tebb

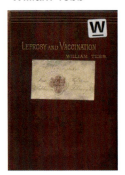

"Dr. John Murray, Inspector-General of Hospitals London, in a communication to Dr. P. S. Abraham, Secretary of the National Leprosy Fund, June 9th 1890, says: 'I consider that it [leprosy] is communicable from the sick to those that are well, probably through a broken surface, as an ulcer or wound, and that it may be communicated by inoculation.'"

https://archive.org/details/b21363432

1888

Encyclopaedia Britannica, 9th edition
Vol XXIV Charles Creighton, MA, MD

"The risks of vaccination may be divided into the risks inherent in the cowpox infection and the risks contingent to the puncture of the skin. Of the latter nothing special requires to be said; the former will be discussed under the five heads of (1) erysipelas, (2) jaundice, (3) skin eruptions, (4) vaccinal ulcers, and (5) so-called vaccinal syphilis."

https://archive.org/details/encyclopaedia-britannica-9ed-1875

1889

Jenner and Vaccinations – A Strange Chapter of Medical History / 376 pages
Charles Creighton, MA, MD

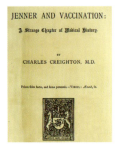

"When vaccination was passing through a storm of adverse criticism during the smallpox epidemic of 1805, Jenner wrote to one of his friends, that nothing of that kind ever shook his faith in cowpoxing."

https://archive.org/details/b21357067

1889

Vaccination – Proved Useless & Dangerous from Forty-Five Years of Registration Statistics / 54 pages

Alfred Russel Wallace, LLD, DCL, OXON, FRS

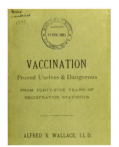

"On careful inspection it will be seen that on three separate occasions a considerable increase in vaccinations was followed by an increase of Small-pox. Let the reader look at the Diagram, and note that in 1863 there was a very great number of vaccinations, followed in 1864 by an increase in Small-pox mortality."

https://archive.org/details/b2136140x

1889

History and Pathology of Vaccination – Vol I & II

1000+ pages / Collection of essays edited by Edgar M. Crookshank, MB

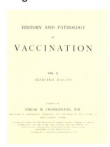

"I found that both official and unofficial vaccinators were completely occupied with the technique of vaccination, to the exclusion of any precise knowledge of the history and pathology of the diseases from which their lymph stocks had been obtained."

https://archive.org/details/historypathology02crooiala

1895

The Vaccination Question

Arthur Wollaston Hutton, MA

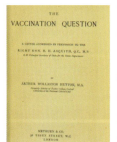

"Here is a question which even The Lancet admits to be 'difficult and momentous.' There have always been medical men disbelieving in vaccination, and their number to-day is rapidly increasing all the world over."

https://archive.org/details/b20398554

1898

A Century of Vaccination and What It Teaches

466 pages / Scott Tebb, MA, MD, DPH

"....the efficacy of this operation has been called in question by competent men, while its risks, so long denied, are now on all hands admitted. Can it be said that the Jennerian doctrine of vaccination has ever been placed on a truly scientific basis?"

https://archive.org/details/39002011125979.med.yale.edu

1898

Vaccination A Delusion – Its Penal Enforcement a Crime / 47 pages

Alfred Russel Wallace, LLD, DCL, OXON, FRS

"A large portion of the medical profession accepted, as proved, that vaccination protected against a subsequent inoculation of small-pox, when in reality there was no such proof, as the subsequent history of small-pox epidemics has shown."

https://archive.org/details/b21356336

1900

Compulsory Vaccination: A Curse and a Menace to Personal Identity / 344 pages

J.M. Peebles MD, PhD

"The vaccination practice, pushed to the front on all occasions by the medical profession, and through political connivance made compulsory by the state, has not only become the chief menace and gravest danger to the health of the rising generation, but likewise the crowning outrage upon the personal liberty of the American citizen."

https://archive.org/details/vaccinationcurse00peeb

Vaccine Safety Concerns
Overview of Historical Publications 1885–1920

1902

The Vaccination Superstition / 60 pages

John W. Hodge, MD

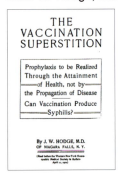

"The advocates of vaccination unhesitatingly assert that the vaccine disease protects its subjects from smallpox, but the facts, so far as we know them, do not warrant this assertion. Indeed, the theory which assumes to conserve health by propagating disease has always had a formidable array of facts to oppose it. From the days of Jenner to the present time, cases of smallpox have appeared among those who were supposed to be protected by vaccination, and these in no small numbers."

https://www.forgottenbooks.com/en/download/TheVaccinationSuperstition_10489841.pdf

1912

Leicester: Sanitation vs. Vaccination / 834 pages

J. T. Biggs

"It is significant to note that this official inquiry elicited the fact that these six little victims of vaccination had all been buried under misleading death certificates, no word as to vaccination appearing on any of those documents.

That practice of 'preserving vaccination from reproach' is believed to be common, and if the suspicion be well founded, it shows how our national vital statistics are considerably vitiated and the public deceived."

https://dn790007.ca.archive.org/0/items/leicestersanitat00biggrich/leicestersanitat00biggrich.pdf

1920

Horrors of Vaccination Exposed and Illustrated – Petition to the President / 244 pages

Chas M. Higgins

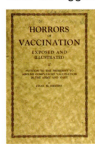

"Now the most barbarous and dangerous medical practice of today is the gross evil of compulsory vaccination, which is doubtless the greatest violation of common sense, medical propriety and the unalienable natural rights of the individual guaranteed in our basic American Charters, that any dogmatic, presumptuous profession or class of men has ever been guilty of."

https://ia800704.us.archive.org/28/items/39002086340891.med.yale.edu/39002086340891.med.yale.edu.pdf

That is just a selection of the earliest publications highlighting issues with the smallpox vaccine, from over a 100 years ago.

Since then, many vaccines have been added to the market, notably the triple DTP shot launched in 1947—recommended worldwide from the early 1950s. The 1990-2000s saw a dramatic increase in the number of vaccines added to the childhood vaccination schedule.

There have been thousands of books, medical and press articles, case studies, documentaries, and many lawsuits that highlight the numerous valid, serious safety concerns with vaccination—most of which are dismissed by authorities.

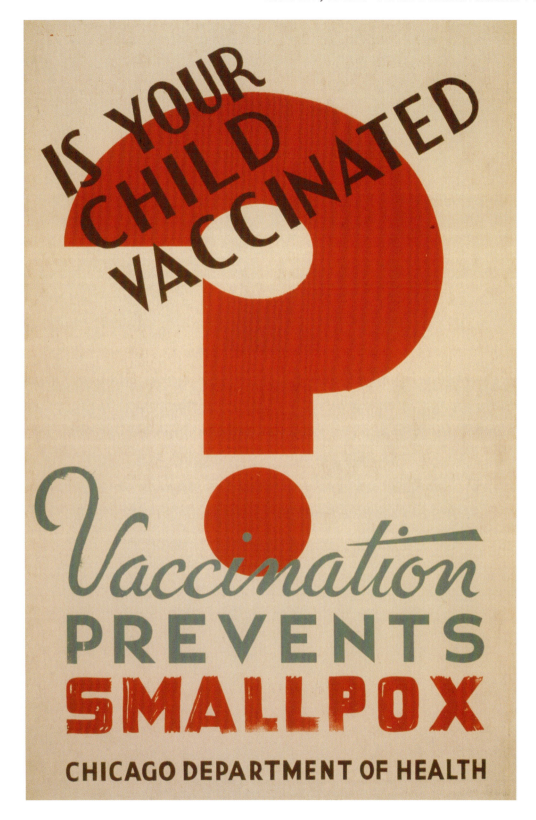

Chicago Department of Health, USA – Smallpox vaccine advertisement
1930s Public Domain Pictures / Library of Congress – Made by Illnois War Art Project Chicago

"Smallpox attained its maximum mortality after vaccination was introduced. The mean annual mortality for 10,000 population from 1850 to 1869 was at the rate of 2.04, whereas after compulsory vaccination, in 1871 the death rate was 10.24. In 1872 the death rate was 8.33 and this after the most laudable efforts to extend vaccination by legislative enactments."

—Dr. William Farr (1807–1883)

British physician who pioneered the quantitative study of morbidity (disease incidence) and mortality (death), helping establish the field of medical statistics. He was the compiler of Statistics of the Registrar General of London.

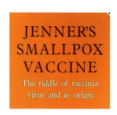

"Contrary to popular belief smallpox was not eradicated by mass vaccination. Though tried initially it proved difficult to implement in many countries and was abandoned in favour of surveillance-containment."

—Derrick Baxby, PhD (1940–2017)
British microbiologist and authority on orthopoxviruses.
Senior lecturer in medical microbiology, Liverpool University
and author of *Jenner's Smallpox Vaccine* (Heinemann, 1981)

Vaccine Injury A Brief Timeline

1721–2022

YEAR	EVENT
1721	Tuberculosis and syphilis are complications arising from variolation, a method that is banned in 1840. Variolated individuals could spread disease and there was a 2–3% fatality rate.
early 1800s	Smallpox following smallpox vaccination is observed. An initial single protective injection becomes a regular booster over the years.
mid-1800s	Deaths following smallpox vaccination are observed by medical staff and reported in medical journals and other media.
	Syphilis, post-vaccinal smallpox and skin necrosis "vaccinia" are recognized as direct consequences of vaccination.
1880s	Eminent British biologist Alfred Russel Wallace writes over several hundred pages on the dangers of vaccination.
1883	A. Lürman reports icterus (jaundice) following smallpox vaccine in Bremen, Germany—it is the first reported hepatitis B epidemic.
1887	English physician Charles Creighton publishes "History of Cowpox and Vaccinal Syphilis" highlighting the dangers of vaccines.
1890	The Royal Commission Report documents vaccine injury including witness testimonies.
1900	The International Classification of Disease includes death by smallpox vaccination.
1901	Twenty-two children in the USA die of tetanus after vaccination against smallpox and diphtheria.
1906	Austrian pediatrician Clemens von Pirquet introduces the term "allergy" for the hypersensitivity reactions he observes after cowpox vaccination.
1910	Post-vaccinal infantile paralysis is reported.
1923	Post-vaccinal encephalitis is reported.
1929	Tragic deaths of over 70 babies after BCG vaccine (against tuberculosis) in Lübeck, Germany.
1942	Post-vaccinal poliomyelitis after DTP vaccine, also called "provocation polio," is reported.
1944	Post-vaccinal poliomyelitis after smallpox vaccine is medically acknowledged.
1946	Tort moral payments are paid by government for neurological injury sustained following the Swiss smallpox vaccine mandate and national campaign: CHF 500–2,000.
1948	The first WHO report includes 15% of its pages dedicated to post-vaccinal encephalitis. While the issue is acknowledged it is downplayed as "rare."
1950	Neurological damage is reported after whole-cell pertussis vaccine containing both thimerosal and aluminum.

1951

January 1951 – Volume 7, p.29
"Smallpox vaccination with prolonged vaccinia."
J.A. Bigler & E.L. Slotkowski

The *American Journal of Pediatrics* reports the case of smallpox vaccine injury (necrosis) and subsequent death of a young Black girl.

YEAR	EVENT
1955	The Cutter Incident: 164 children are paralyzed and 10 die after inadequately inactivated polio virus vaccines are administered.
1955–1970s	Oncogenic simian virus SV40 contaminates cell cultures used in polio vaccines given to hundreds of millions of people worldwide.
1960s	Rubella vaccine is associated with arthritis.
1969	*Tinnerholm v. Parke-Davis.* After 10 years, parents are awarded US$651,783.52 in damages for their DTP-IPV brain injured three-month-old child.
	One of the first live measles vaccines is taken off the market in the UK following reports of death and brain damage. SSPE registry begins.
1963–1971	Neurological disorders are reported in the USA following measles vaccination, including 59 cases of encephalitis and one case of subacute sclerosing panencephalitis.
1970s	UK government reports 1 in 30,000 children suffer neurological damage after DTP shots.
1972	Cases of subacute sclerosing panencephalitis (SSPE) are observed after live measles vaccine in children in the UK.
1973	The UK Association of Parents of Vaccine Damaged Children is founded.
	Sudden infant death syndrome (SIDS) enters the ICD classification system (code 795.0).
1974	Adverse publicity on neurological injury and death after whole-cell pertussis vaccine (DTP) spreads over the UK.
1976	"Swine Flu" influenza vaccine disaster in the USA with thousands injured with neurological disease (500 x Guillain-Barré) with around 53 deaths in total. The vaccine was withdrawn after 26 deaths (~40M vaccinated in the US).
1977	Adverse events are officially registered by the FOPH and Swiss Centers of Pharmacovigilance.

DTwP: diphtheria, tetanus and whole-cell pertussis vaccine

The DHSS admits in classified documents that severe brain damage can follow vaccination.

1978–1979 The Tennessee SIDS Cluster: eleven infants die suddenly within eight days of their DTP vaccination, nine of whom had received the same lot of DTP vaccine from Wyeth Laboratories: lot 64201. Four of the eleven infants were dead in 24 hours.

1979 The UK Vaccine Damage Payments Act compensates those who have at least 80% disability (epilepsy without severe mental handicap would not be included) with a maximum £10,000 payment.

1980s Over 300 lawsuits are filed in the USA against vaccine manufacturers for brain damage following the DTP vaccine.

1986 President Reagan signs the National Childhood Vaccine Injury Act in the USA, shielding vaccine manufacturers from liability, to protect supply.

1990 The Vaccine Adverse Events Reporting System (VAERS) is officially set up in the USA as a passive surveillance system of vaccine injury and deaths. The Vaccine Safety Datalink (VSD) is also set up as a collaborative project between CDC's Immunization Safety Office and nine healthcare organizations to monitor vaccine safety.

1990s Links between hepatitis B vaccine and auto-immune disorders (including multiple sclerosis) are reported and lead to the temporary suspension of a national vaccination campaign in France.

1991 The Swiss registry "vaccinovigilance" is set up to record adverse events after vaccines—it is managed by Swissmedic from 2002.

The US Institute of Medicine concludes that scientific literature is **inadequate to accept or reject a causal relationship** between vaccines and the following diseases: aseptic meningitis; chronic neurologic damage; learning disabilities + attention deficit disorder; hemolytic anemia; juvenile diabetes; Guillain-Barré syndrome; erythema multiforme; peripheral mononeuropathy; radiculoneuritis + other neuropathies; thrombocytopenia; thrombocytopenia purpura.

Anthrax vaccine injury—Gulf War syndrome affects soldiers who never left the USA. Over 1,000,000 military men and women are injured by a mandatory vaccine campaign against anthrax and botulism. Over 35,000 are said to have died over the years.

1998 Gastrointestinal issues are linked to neuro-logical complications of the MMR vaccine.

1999 Merck's RotaShield® vaccine is withdrawn due to intussusception, after being licensed one year.

2000–2017 Over 49,000 cases of polio after live oral polio vaccine are reported in India.

2002 LYMErix™ Lyme disease vaccine is pulled from the market after slow sales and reports of arthritis and other injuries, which are all officially denied by the manufacturer.

2006 The preservative thimerosal is taken out of most vaccines in Switzerland after 60 years of use (and inadequate safety studies) due to concerns over possible cumulative neurotoxic effects.

2007 HPV campaign is suspended in Japan and Denmark after thousands of adverse events.

2009 H1N1 swine flu pandemic adjuvanted vaccines: cases of narcolepsy, cataplexy and increased preterm births in vaccinated pregnant women.

2013 Maximum tort moral payment award of CHF 70,000 for vaccine injury in Switzerland.

2017 Evidence of HPV vaccine harm in Japan, Colombia, Denmark, UK, Ireland, USA—injured girls suffer disabling autoimmune and neurological diseases, including reports of deaths.

2018 Since licensure of HPV vaccine Gardasil®, around 33 million people have been vaccinated in the USA, with an average of 411 VAERS reports listing 113 ER visits and 3 deaths per month. As of mid-August, there are 59,226 VAERS HPV reports; 8,682 serious; 16,380 emergency room visits; 432 deaths and 142 cases of cervical cancer.

2019 In the same year that the WHO declares vaccine hesitancy as a threat to public health, officials at the Global Vaccine Safety Summit admit to lacking safety science on vaccines—especially long-term effects.

2020 As of August 1, 458 HPV vaccine injury claims have been filed with the US Vaccine Injury Compensation Program (VICP). There have been 17 claims of deaths.

2021 COVID-19 injections worldwide cause unprecedented injuries, mostly neurological and cardiovascular. In the USA as of December 24, there are 21,002 deaths, 110,609 hospitalizations and 35,650 permanent disabilities reported to VAERS.

2022 Injuries and deaths continue to be recorded at astonishing rates, yet the COVID-19 vaccines remain on the market as the "benefits outweigh the risks." They continue to be recommended to babies, children and pregnant women.

FOPH: Federal Office for Public Health, Switzerland
DHSS: Department of Health and Social Security, UK

VAERS: Vaccine Adverse Events Reporting System, USA

Vaccine Injury Compensation Program
Awards Paid 1989–2024

US$5,256,660,556.32 in compensation to victims, including legal fees, has been paid out since the inception of VICP in 1988. The most expensive year to date is **2017—US$282,094,721 for 706 awards**—and by August, **2024** is the year with the most claims compensated—**1,019 awards for US$176,693,406.62**.

Fiscal Year	Number of Compensated Awards	Petitioners' Award Amount	Attorneys' Fees/Costs Payments	Number of Payments to Attorneys (Dismissed Cases)	Attorneys' Fees/Costs Payments (Dismissed Cases)	Number of Payments to Interim Attorneys'	Interim Attorneys' Fees/Costs Payments	Total Outlays
FY 1989	6	$1,317,654.78	$54,107.14	0	$0.00	0	$0.00	$1,371,761.92
FY 1990	88	$53,252,510.46	$1,379,005.79	4	$57,699.48	0	$0.00	$54,689,215.73
FY 1991	114	$95,980,493.16	$2,364,758.91	30	$496,809.21	0	$0.00	$98,842,061.28
FY 1992	130	$94,538,071.30	$3,001,927.97	118	$1,212,677.14	0	$0.00	$98,752,676.41
FY 1993	162	$119,693,267.87	$3,262,453.06	272	$2,447,273.05	0	$0.00	$125,402,993.98
FY 1994	158	$98,151,900.08	$3,571,179.67	335	$3,166,527.38	0	$0.00	$104,889,607.13
FY 1995	169	$104,085,265.72	$3,652,770.57	221	$2,276,136.32	0	$0.00	$110,014,172.61
FY 1996	163	$100,425,325.22	$3,096,231.96	216	$2,364,122.71	0	$0.00	$105,885,679.89
FY 1997	179	$113,620,171.68	$3,898,284.77	142	$1,879,418.14	0	$0.00	$119,397,874.59
FY 1998	165	$127,546,009.19	$4,002,278.55	121	$1,936,065.50	0	$0.00	$133,484,353.24
FY 1999	96	$95,917,680.51	$2,799,910.85	117	$2,306,957.40	0	$0.00	$101,024,548.76
FY 2000	136	$125,945,195.64	$4,112,369.02	80	$1,724,451.08	0	$0.00	$131,782,015.74
FY 2001	97	$105,878,632.57	$3,373,865.88	57	$2,066,224.67	0	$0.00	$111,318,723.12
FY 2002	80	$59,799,604.39	$2,653,598.89	50	$656,244.79	0	$0.00	$63,109,448.07
FY 2003	65	$82,816,240.07	$3,147,755.12	69	$1,545,654.87	0	$0.00	$87,509,650.06
FY 2004	57	$61,933,764.20	$3,079,328.55	69	$1,198,615.96	0	$0.00	$66,211,708.71
FY 2005	64	$55,065,797.01	$2,694,664.03	71	$1,790,587.29	0	$0.00	$59,551,048.33
FY 2006	68	$48,746,162.74	$2,441,199.02	54	$1,353,632.61	0	$0.00	$52,540,994.37
FY 2007	82	$91,449,433.89	$4,034,154.37	61	$1,692,020.25	0	$0.00	$97,175,608.51
FY 2008	141	$75,716,552.06	$5,191,770.83	74	$2,531,394.20	2	$117,265.31	$83,556,982.40
FY 2009	131	$74,142,490.58	$5,404,711.98	36	$1,557,139.53	28	$4,241,362.55	$85,345,704.64
FY 2010	173	$179,387,341.30	$5,961,744.40	59	$1,933,550.09	22	$1,978,803.88	$189,261,439.67
FY 2011	251	$216,319,428.47	$9,572,042.87	403	$5,589,417.19	28	$2,001,770.91	$233,482,659.44
FY 2012	249	$163,491,998.82	$9,241,427.33	1,020	$8,649,676.56	37	$5,420,257.99	$186,803,360.70
FY 2013	375	$254,666,326.70	$13,543,099.70	704	$7,012,615.42	50	$1,423,851.74	$276,645,893.56
FY 2014	365	$202,084,196.12	$12,161,422.64	508	$6,824,566.68	38	$2,493,460.73	$223,563,646.17
FY 2015	508	$204,137,880.22	$14,464,063.71	118	$3,546,785.14	50	$3,089,497.68	$225,238,226.75
FY 2016	689	$230,140,251.20	$16,298,140.59	100	$2,746,864.60	58	$3,398,557.26	$252,583,813.65
FY 2017	706	$252,245,932.78	$22,045,785.00	132	$4,454,379.49	52	$3,363,464.24	$282,109,561.51
FY 2018	521	$199,588,007.04	$16,689,908.68	113	$5,151,255.64	57	$4,999,766.30	$226,428,937.66
FY 2019	653	$196,217,707.64	$18,991,247.55	103	$5,292,700.23	65	$5,457,545.23	$225,959,200.65
FY 2020	733	$186,860,677.55	$20,165,188.43	114	$5,774,438.88	76	$5,090,482.24	$217,890,787.10
FY 2021	719	$208,258,401.31	$24,944,964.77	140	$6,920,048.74	53	$4,249,055.37	$244,372,470.19
FY 2022	927	$195,693,889.57	$22,992,062.07	102	$4,868,964.74	56	$6,329,886.09	$229,884,802.47
FY 2023	885	$123,810,693.81	$35,984,811.55	126	$6,760,733.64	61	$7,329,281.69	$173,885,520.69
FY 2024	1,019	$133,575,637.64	$31,676,261.02	107	$7,315,141.53	37	$4,126,366.43	$176,693,406.62
Total	**11,124**	**$4,732,500,593.29**	**$341,948,497.24**	**6,046**	**$117,100,790.15**	**770**	**$65,110,675.64**	**$5,256,660,556.32**

NOTE: Some previous fiscal year data has been updated as a result of the receipt and entry of data from documents issued by the Court and system updates which included petitioners' costs reimbursements in outlay totals.

"Compensated" are petitions that have been paid as a result of a settlement between parties or a decision made by the U.S. Court of Federal Claims (Court). The # of awards is the number of petitioner awards paid, including the attorneys' fees/costs payments, if made during a fiscal year. However, petitioners' awards and attorneys' fees/costs are not necessarily paid in the same fiscal year as when the petitions/petitions are determined compensable. "Dismissed" includes the # of payments to attorneys and the total amount of payments for attorneys' fees/costs per fiscal year. The VICP will pay attorneys' fees/costs related to the petition, whether or not the petition/petition is awarded compensation by the Court, if certain minimal requirements are met. "Total Outlays" are the total amount of funds expended for compensation and attorneys' fees/costs from the Vaccine Injury Compensation Trust Fund by fiscal year.

Since influenza vaccines (vaccines administered to large numbers of adults each year) were added to the VICP in 2005, many adult petitions related to that vaccine have been filed, thus changing the proportion of children to adults receiving compensation.

VICP is supposed to compensate victims under a flexible, no-fault system that avoids the antagonism and delay of traditional litigation. Claims are to be handled "quickly, easily and with certainty and generosity," states a House report accompanying the legislation in 1986. While authorities are quick to point out that payment is not an admission of causation, it is important to remember that the vast majority of injured victims never make it as far as court. According to the Government Accountability Office (GAO) in 2014, most cases last at least 2 years, with more than half taking over 5 years before being settled, some taking even 8 to 10 years! Among the many constraints of this program is the strict 3-year statute of limitations, beginning from the onset of vaccine-related symptoms.

https://www.hrsa.gov/vaccine-compensation/data/index.html

https://www.gao.gov/products/GAO-15-142

Petitions Filed, Compensated, and Dismissed by Alleged Vaccine
Since the Beginning of VICP – October 1, 1988 to August 1, 2024

Vaccines	Filed Injury	Filed Death	Filed Total	Compensated	Dismissed
DT	69	9	78	26	52
DTaP	501	89	590	261	283
DTaP-Hep B-IPV	105	42	147	50	73
DTaP-HIB	11	1	12	7	4
DTaP-IPV	18	0	18	8	5
DTaP-IPV-HIB	55	22	77	19	45
DTaP-IPV-HIB-HEPB	0	0	0	0	0
DTP	3,288	698	3,986	1,273	2,712
DTP-HIB	20	8	28	7	21
Hep A-Hep B	49	0	49	24	10
Hep B-HIB	10	0	10	5	3
Hepatitis A (Hep A)	152	8	160	79	45
Hepatitis B (Hep B)	767	62	829	311	459
HIB	50	3	53	24	22
HPV	912	23	935	180	559
Influenza	10,023	243	10,266	6,588	1,123
IPV	269	14	283	10	271
Measles	145	19	164	57	107
Meningococcal	140	3	143	73	31
MMR	1,064	64	1,128	437	621
MMR-Varicella	64	2	66	28	24
MR	15	0	15	6	9
Mumps	10	0	10	1	9
Nonqualified[1]	120	13	133	3	128
OPV	282	28	310	158	152
Pertussis	4	3	7	2	5
Pneumococcal Conjugate	373	26	399	164	105
Rotavirus	126	6	132	78	35
Rubella	190	4	194	71	123
Td	246	3	249	147	81
Tdap	1,351	9	1,360	808	167
Tetanus	213	3	216	114	52
Unspecified[2]	5,427	9	5,436	12	5,419
Varicella	124	11	135	75	41
Grand Total	26,193	1,425	27,618	11,106	12,796

[1] Nonqualified petitions are those filed for vaccines not covered under the VICP.
[2] Unspecified petitions are those submitted with insufficient information to make a determination.

https://www.hrsa.gov/sites/default/files/hrsa/vicp/vicp-stats-8-1-2024.pdf

International Classification of Death and Disease
By Vaccination 1900–2019

The first international edition, known as the International List of Causes of Death (ICD), was adopted by the International Statistical Institute in 1900.

Groundwork was done by early respected medical statisticians, Englishman William Farr (1807–1883) and Frenchman Jacques Bertillon (1851–1922), to create a uniform classification of causes of death.

The French government held the first International Conference for the Revision of the Bertillon, or International Classification of Causes of Death in August 1900. A detailed classification of causes of death consisting of 179 groups and an abridged classification of 35 groups was adopted on August 21, 1900. The next conference was held in 1909 and the French government called succeeding conferences in 1920, 1929 and 1938.

The World Health Organization was entrusted with the ICD at its creation in 1948, and published the sixth version, ICD-6.

Historically, ICD codes have been used to identify the multitude of human disease conditions and causes of death. Currently, the ICD coding system is used globally by hospitals to classify morbidity data from patient records for billing purposes.

For some reason, the codes associated with vaccine injury have been omitted from the WHO online ICD code application.
https://icd.who.int/browse10/2010/en#/T36-T50
ICD-10 – Clinical code database
https://www.icd10data.com/ICD10CM/Codes/S00-T88/T36-T50/T50-
ICD-11
https://icd.who.int/en

Websites accessed 10OCT2024

With ever-evolving codes that change every decade or two, tracking vaccine injury has become convoluted and complicated. Most clinicians are often not even aware of the codes related to vaccine-induced disease.

Only the ICD codes attributed to vaccination have been extracted and presented hereafter.

International Classification of Death
1900   Revision 1
01   Smallpox: vaccinated
04   Cowpox and other effects of vaccination

International Classification of Death
1929   Revision 4
44(1)   Vaccinia [cowpox]
44(2)   Other sequelae of vaccination

International Classification of Death
1938   Revision 5
195a(1)   Vaccinia
195a(2)   Other sequelae of vaccination
              against smallpox

International Classification of Diseases (ICD)
1948   Revision 6
(E940-E946) Complications due to non-therapeutic medical and surgical procedures.
E940   Generalized vaccinia following vaccination
E941   Post-vaccinal encephalitis
E942   Other complications of smallpox vaccination
E943   Post-immunization jaundice and hepatitis
E944   Other complications of prophylactic
              inoculation

1955 ICD revision 7
(Y40-Y49) Prophylactic inoculation and vaccination

Y40   Vaccination against smallpox
Y41   Inoculation against diphtheria
Y42   Inoculation against whooping cough
Y43   Inoculation against tuberculosis
Y44   Inoculation against influenza
Y45   Inoculation against tetanus
Y46   Inoculation against typhoid/paratyphoid fever
Y47   Inoculation against typhus fever
Y48   Inoculation against yellow fever
Y49   Inoculation against other infectious disease

1968 ICD Revision 8
(E930-E936) Surgical and medical complications and misadventures

E933   Complications and misadventures in prophylaxis with bacterial vaccines
E934   Complications and misadventures in prophylaxis with other vaccines

1977 ICD Revision 9
(978) Poisoning by bacterial vaccines

978.0   BCG
978.1   Typhoid and paratyphoid
978.2   Cholera
978.3   Plague
978.4   Tetanus
978.5   Diphtheria
978.6   Pertussis vaccine, including combinations with a pertussis component
978.8   Other and unspecified bacterial vaccines
978.9   Mixed bacterial vaccines, except combinations with a pertussis component

(979) Poisoning by other vaccines and biological substances

979.0   Smallpox vaccine
979.1   Rabies vaccine
979.2   Typhus vaccine
979.3   Yellow fever vaccine
979.4   Measles vaccine
979.5   Poliomyelitis vaccine
979.6   Other and unspecified viral and rickettsial vaccines
979.7   Mixed viral-rickettsial and bacterial vaccines, except combinations with pertussis component
979.8   Other and unspecified vaccines and biological substances

999.5   Serum sickness

1990 ICD Revision 10

T50   **Poisoning by diuretics and other and unspecified drugs, medicaments and biological substances**

T50.A   **Poisoning by, adverse effect of and underdosing of bacterial vaccines**

T50.A1   Poisoning by, adverse effect of and underdosing of pertussis vaccine, including combinations with a pertussis component

T50.A11   Poisoning by, adverse effect of and underdosing of pertussis vaccine, including combinations with a pertussis component, accidental (unintentional)
T50.A11A ...initial encounter
T50.A11D ...subsequent encounter
T50.A11S ...sequela

T50.A12   Poisoning by, adverse effect of and underdosing of pertussis vaccine, including combinations with a pertussis component, intentional self-harm
T50.A12A ...initial encounter
T50.A12D ...subsequent encounter
T50.A12S ...sequela

T50.A2   Poisoning by, adverse effect of and underdosing of mixed bacterial vaccines without a pertussis component

T50.A21   Poisoning by, adverse effect of and underdosing of mixed bacterial vaccine, without a pertussis component, accidental (unintentional)
T50.A21A ...initial encounter
T50.A21D ...subsequent encounter
T50.A21S ...sequela

T50.A22   Poisoning by, adverse effect of and underdosing of pertussis vaccine, including combinations with a pertussis component, intentional self-harm
T50.A22A ...initial encounter
T50.A22D ...subsequent encounter
T50.A22S ...sequela, etc....

T50.B   **Poisoning by, adverse effect of and underdosing of viral vaccines**

T50.Z   **Poisoning by, adverse effect of and underdosing of other vaccines and biological substances**

2019 ICD Revision 11

This revised edition has vaccine adverse events in an even more complicated structure with an all new coding system.

https://icd.who.int/ct11/icd11_mms/en/release

The Brighton Collaboration

Commissioned to Standardize Definitions and Diagnostic Criteria for Adverse Events Following Immunization (AEFI)

> According to the Brighton Collaboration, "Safety can only be inferred indirectly from the relative absence of multiple, likely adverse events following vaccination." https://www.ncbi.nlm.nih.gov/books/NBK20507/

The history of this nonprofit organization spans back to 1999, with its official beginnings in 2000, involving top government agencies such as the CDC and the WHO. It was created at the same time as the Bill & Melinda Gates Foundation and Gavi—institutions heavily invested in vaccine development and distribution. Its mission is to advance the science of vaccination safety or "vaccinovigilance."

According to the Brighton Collaboration, creating a globally acceptable and common vocabulary for AEFI to facilitate comparison of vaccine safety data, can achieve the "goal of maintaining trust in worldwide vaccination programs." Initially registered in Basel, Switzerland, the Brighton Collaboration was tasked with centralization of vaccine injury definitions, termed AEFI, as well as determining levels of diagnostic certainty for assessing causality. The Brighton Collaboration was awarded guardianship of the Vaccine Safety Datalink, the privacy-protected surveillance data of vaccinated individuals in nine US healthcare organizations—Kaiser Permanente represents most of the data.

In 2019, the Brighton Collaboration was absorbed by the US program the Task Force for Global Health, that has since joined forces with the Coalition for Epidemic Preparedness Innovations (CEPI) to launch the Safety Platform for Emergency vACcines (SPEAC) project. The Task Force works towards distributing vaccines more equitably across the planet, even though adequate safety surveillance systems are still lacking in many countries.

The Task Force for Global Health's Brighton Collaboration is today a group of more than 750 "independent" vaccine experts selected to protect and preserve public health by promoting vaccination safety. In partnership with CEPI—who, together with Gavi and the WHO were driving COVID-19 product development—the Brighton Collaboration convened global experts to draft lists of potential adverse events of special interest (AESI) for COVID-19 product candidates. These AESI were communicated in a CDC presentation in October 2020, but not reported by legacy media nor disclosed to the general public. See Appendices J and K on pages 290–291.

AEFI Case Definitions

AEFI-CDs are published in the journal *Vaccine*

| 2004 | Fever
Intussusception
Persistent crying
Nodule at injection site
Generalized convulsive seizure
Hypotonic-hyporesponsive episode (HHE) |
|------|------|
| 2023 | **Over 70 case definitions are established**, **including SIDS,** thrombocytopenia, acute disseminated encephalomyelitis, multiple sclerosis, arthritis, encephalitis, diarrhea, anaphylaxis, aseptic meningitis, Guillain Barré syndrome, etc. |

A Brief Timeline of The Brighton Collaboration

1999 Robert Chen, CDC medical officer, calls for improving the quality of vaccine safety data at a Vaccine Conference in Brighton, England. The steering committee; Harald Heijbel, Ulrich Heininger, Tom Jefferson and Elisabeth Loupi join to address this need.

2000 **The World Health Organization joins as an observer and funding agency.**

Official launch meeting of the Brighton Collaboration in Verona, Italy. Katrin Kohl (CDC) and Jan Bonhoeffer, University Children's Hospital Basel (UKBB) are appointed to build an international network and coordinate activities.

All parties concerned are heavily invested in pharmaceuticals with the CDC owning several patents and Basel hospital receiving money from the pharmaceutical industry.

2003 The Brighton Collaboration Foundation is created as a Swiss NPO with international scope, established with seed capital of the University Children's Hospital Basel (UKBB), and partnered with the CDC Foundation.

2004 The first scientific publications of the Brighton Collaboration includes the first six case definitions of adverse events following immunization (AEFI). Funding is provided by the CDC, WHO and EUSAFEVAC*.

2005 UN Council for International Organization of Medical Sciences (CIOMS) recommends the use of the Brighton Collaboration case definitions and guidelines.

2008 The European Medicines Agency (EMA) recommends the use of the Brighton Collaboration standards.

Computer-based classification of adverse events is automated. The ABC-Tool is available on the website.

The first method for evaluating case definitions is established and published.

2009 General guidelines for the collection, analysis, and publication of vaccine safety data in surveillance systems and clinical trials are published.

The Pilot of European Vaccine Safety Data Network is launched on request of ECDC.

The Brighton Collaboration contributes to WHO's Global Vaccine Safety Landscape Analysis.

2011 The Brighton Foundation US is registered in Boston, Massachusetts, as a public charity and associated partner to the Brighton Collaboration Foundation, Switzerland.

2012 Vaccine.GRID is established as an independent foundation and dedicated platform for the conduct of large-scale international observational vaccine safety and benefit-risk studies.

2015 The GAIA Network for Global Alignment of Immunization Safety Assessment in pregnancy is established.

2017 The GAIA project has been completed successfully and the Bill & Melinda Gates Foundation provides funding for a project extension (GAIA Phase II).

2019 The Brighton Collaboration is dissolved as a Partnership under Swiss law and reconstituted as a program of The Task Force for Global Health, www.taskforce.org

The Coalition for Epidemic Preparedness Innovations (CEPI) and the Brighton Collaboration launch the Safety Platform for Emergency vACcines (SPEAC) Project.

2020 In partnership with CEPI (Gavi-WHO) the Brighton Collaboration convenes global experts to draft lists of potential adverse events of special interest (AESI) for COVID-19 vaccine candidates, which are then reviewed by the WHO.

https://brightoncollaboration.us/history/

EUSAFEVAC is a Research Grant for Improved Vaccine Safety Surveillance from the European Commission.

The following pages present research compiled on pharmaceutical companies, and vaccine manufacturers' beginnings, subsequent mergers, and acquisitions. While every effort has been made to be as comprehensive as possible, this timeline is not intended to be an exhaustive and complete chronology.

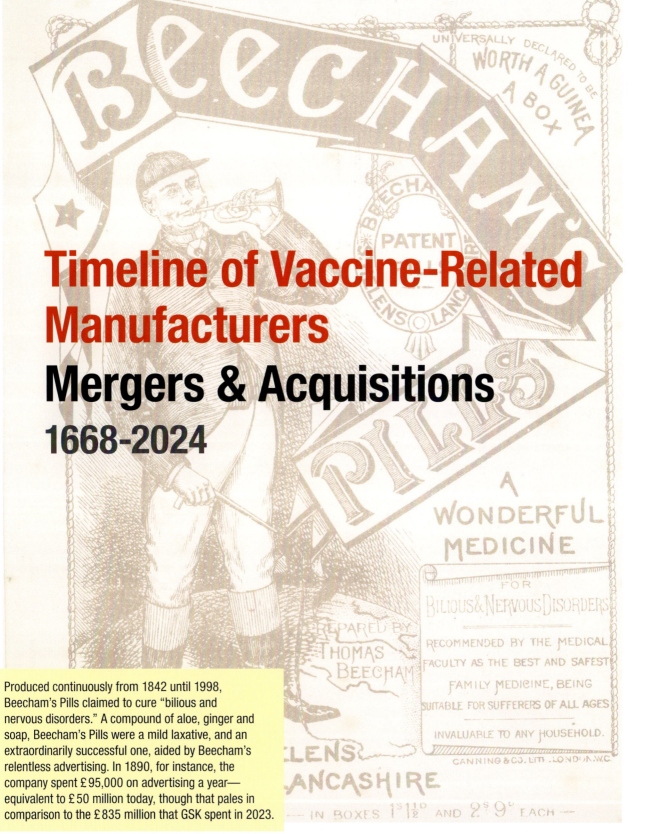

Wellcome Collection: "Beecham's Pills – A Wonderful Medicine" – Image circa 1899, in the public domain

Timeline of Vaccine-Related Manufacturers
Mergers & Acquisitions
1668-2024

Produced continuously from 1842 until 1998, Beecham's Pills claimed to cure "bilious and nervous disorders." A compound of aloe, ginger and soap, Beecham's Pills were a mild laxative, and an extraordinarily successful one, aided by Beecham's relentless advertising. In 1890, for instance, the company spent £95,000 on advertising a year—equivalent to £50 million today, though that pales in comparison to the £835 million that GSK spent in 2023.

Timeline of Vaccine-Related Manufacturers
Mergers & Acquisitions

The pharmaceutical industry has a complicated history of partnerships, mergers, and acquisitions.
Pfizer, GSK, Sanofi, and Merck currently dominate the global vaccine market, with Merck being the leader by 2023 revenue. What follows is a non-exhaustive and brief historical overview of vaccine manufacturers, biotech and pharmaceutical companies, major mergers, partnerships, and acquisitions.

1668 **Merck** is established when Friedrich Jacob Merck purchases an apothecary, Engel-Apotheke, in Darmstadt, Germany.

1715

Plough Court Pharmacy is established in London by Silvanus Bevan, which later becomes **Allen & Hanburys Ltd.**

1781 **Takeda** founder Chobei I, begins selling traditional Japanese and Chinese herbal medicines in Doshomachi, Japan.

1818–1827 Heinrich E. Merck takes over the family business and converts **E. Merck** from a pharmacy to a manufacturer of drugs and fine chemicals. The company becomes a major producer of alkaloids, including morphine, codeine, and cocaine, exporting its products to the US.

1828

William S. Merrell opens the Western Market Drug Store in downtown Cincinnati, Ohio, expanding into the wholesale drug business. Following his death in 1880 his sons form the **William S. Merrell Chemical Company.**

1830 The foundation of **Smith Kline** begins with the Smith & Gilbert drug house opening in Philadelphia, USA. After Gilbert withdraws, John K. Smith's younger brother George joins the company, which they build into a successful drug wholesale business.

1845 **Sharp & Dohme** begins as an apothecary shop in Baltimore, Maryland. By the 1920s, they build an extremely effective distribution network across the USA.

1848

Thomas Beecham, a former shepherd, is granted a license for his first shop in England. He develops the recipe for **Beecham's** laxative pills before building what was claimed to be the first factory in the UK with electricity in 1887.

1849

Charles Pfizer & Co. is established by German-American cousins Charles Pfizer and Charles Erhart.

1851

German pharmaceutical company **Schering** is founded by Ernst C. Friedrich Schering.

William R. Warner launches his own drug store in Philadelphia, Pennsylvania. In 1856 he goes on to invent a sugar-coated tablet process gaining him a place in the Smithsonian Institution.

St. Mary's Hospital is founded and opens its doors to patients. At St. Mary's, C.R. Alder Wright first synthesizes diamorphine (heroin) in 1874, and in 1902 Sir Almroth Edward Wright (no relation) starts a research department at St. Mary's Hospital Medical School in London, developing a system of anti-typhoid fever inoculation which is given to millions of soldiers during WWI.

1855

French, Richards & Co. is incorporated by Clayton French, William Campbell and William H. Richards.

The **Frederick Stearns Company** is founded by a pharmacist and will be bought by **Sterling Drugs Inc**. in 1944.

1857 The Swiss synthetic dye company that is to become **J.R. Geigy Ltd** in 1914 is founded by Johann Rudolf Geigy-Merian and Johann Muller-Pack.

1858 **SQUIBB**

The **Squibb** laboratory is founded by former Naval medical officer Edward Robinson Squibb in Brooklyn, New York, initially to supply the Army with reliable drugs. That same year the lab burns to the ground in an ether explosion. Despite severe burns, Squibb re-opens his laboratory a year later. By 1883, Squibb manufactures over 300 products and sells them globally. Squibb is known as an advocate of quality control and high purity standards early within the pharmaceutical industry.

1859 Europe's first tar dye company **CIBA** begins with Alexander Clavel taking up the production of fuchsine in his factory for silk-dyeing works in Basel, Switzerland. By 1873, the company is sold to Bindschedler and Busch, which is transformed in 1884 into a joint-stock company named "Gesellschaft für Chemische Industrie Basel" (**Company for Chemical Industry Basel**). The acronym, CIBA, is adopted as the chemical company's name in 1945.

1860 **Wyeth** is founded in Philadelphia, Pennsylvania, by pharmacist **John Wyeth** and his brother.

1863 **Solvay** is born out of a technological breakthrough, the ammonia-soda process developed by Ernest Solvay and a small circle of relatives including his brother Alfred.

Friedr. Bayer et Comp. is founded by dyestuffs merchant, Friedrich Bayer and cotton dyer Johann Friedrich Weskott.

Teerfarbenfabrik Meister, Lucius & Co. is founded in Germany and simplfies its name later to **Farbwerke Hoechst AG**.

1866

American drug maker **Parke, Davis & Co.** (formally incorporated in 1875) is founded in Detroit, Michigan, by Dr. Samuel P. Duffield. It begins as a business partnership between Dr. Duffield, Harvey Coke Parke and George S. Davis and is to become one of the largest pharmaceutical companies in the world. In 1897, they introduce the idea of physiological standardization and strict standards of purity to their products. In 1886, they initiate the practice of using lot numbers on the labels of all their products. The FDA does not require this until 1962.

1870 John and George Smith's nephew and Smith Kline manager Mahlon Smith, together with its bookkeeper Mahlon Kline—who becomes a partner—form **Smith, Kline & Co.**

1876

Family company **Eli Lilly & Co.,** is founded by and named after Colonel Eli Lilly, a pharmaceutical chemist and veteran of the American Civil War.

1880

American pharmacists Henry S. Wellcome and Silas Burroughs begin **Burroughs Wellcome & Co.,** a drug sales partnership in London. In 1884, BW develops a new compressed tablet for medicines, trade-marked as a "Tabloid" as an alternative to pills. Tabloid medications become so successful that BW opens its first large scale manufacturing facility in 1889. In 1894, it begins producing diphtheria antitoxin.

1882

Dr. H.W. Alexander of The Lancaster County Vaccine Farm begins making smallpox vaccine on his farm in Marietta, going from a single heifer—a female cow that has not had any offspring—to hundreds. Six years later, this **Marietta Vaccine Farm** is shipping internationally, providing vaccines to China, Canada, Mexico and South America. By 1894, the MVF is said to be the largest operation of its kind.

1885

Jordan Wheat Lambert, founder of the **Lambert Pharmacal Company** of St. Louis, purchases the famous formula for Listerine from Dr. Joseph Lawrence.

Boehringer Ingelheim is founded by Albert Boehringer in Ingelheim, Germany.

1886

The **Chemiefirma Kern und Sandoz** "Kern and Sandoz Chemistry Firm" is founded by Alfred Kern and Edouard Sandoz.

Warner gives up the retail pharmacy business and begins drug manufacture under the name **William R. Warner & Co.**

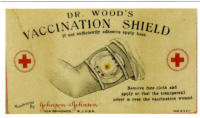

Johnson & Johnson is founded by brothers Robert, James, and Edward M. Johnson with the goal of manufacturing the first mass-produced antiseptic surgical dressings.

The pharmaceutical manufacturing firm **Upjohn Company** is founded in Michigan by Dr. William E. Upjohn. The company is originally formed to make friable pills, which could be "reduced to a powder under the thumb" and easily digested—a strong marketing argument at the time.

1887

William McLaren Bristol and John Ripley Myers purchase the **Clinton Pharmaceutical company** of Clinton, New York.

German drug company **E. Merck** opens its first sales office in New York, run by long-time employee Theodore Weicker.

1888

INSTITUT PASTEUR

The **Pasteur Institute** is founded in Paris.

1890

Vaccine and serum manufacturer **Parc Vaccinogène** is founded in Batavia, Dutch East Indies (present name Jakarta, Indonesia).

1891

Smith, Kline & Co. acquires **French, Richards & Co.** to form **Smith, Kline & French Co.**

Family member Georg Merck emigrates to the USA and—in partnership with Theodore Weicker—founds **Merck & Co.** in New York, as the US subsidiary of the German parent company **E. Merck.**

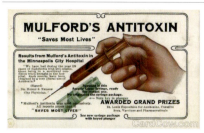

Pharmacist Henry Kendall Mulford and Milton Campbell incorporate **H.K. Mulford Company** in Philadelphia. It becomes the first to commercialize a diphtheria antitoxin and in the early 20th century, to produce both smallpox and rabies vaccines.

The private research institute of bacteriology and chemistry—later named the Blumenthal Institute for Bacteriology and Chemistry—is founded in the Russian Empire to develop vaccines. It is nationalized in 1919, and then in 1949 it is renamed the **Gamaleya National Center of Epidemiology and Microbiology**, part of the Health Ministry in Russia.

The **Lister Institute of Preventive Medicine,** informally known as the **Lister Institute**, is established as a research institute in London.

1892 E·R·SQUIBB & SONS

E.R. Squibb is joined by his two sons, Dr. Edward H. Squibb and Charles F. Squibb. The company is renamed **E.R. Squibb & Sons** and has capital assets of US$1.5 million.

1894 **Abbott Alkaloidal Company**, later renamed **Abbott Laboratories**, is founded by Chicago physician Wallace Calvin Abbott to formulate known drugs.

Wellcome Chemical Research Laboratories is established in the UK by Henry Wellcome to focus on biological experimentation including early forms of vaccines, notably diphtheria antitoxin.

1896

F. Hoffmann-La Roche AG is founded by Fritz Hoffmann-La Roche and known early on for producing various vitamin preparations and derivatives, becoming the first company to mass-produce synthetic vitamin C in 1934. Roche will become one of the largest pharmaceutical companies in the world.

1897
INSTITUT MERIEUX

Marcel Mérieux, former assistant to Louis Pasteur, creates the **Mérieux Biological Institute** in Lyon, France.

Pocono Laboratory in Swiftwater, Pennsylvania, is founded by Dr. Richard Slee and begins to produce Glycerinated Smallpox Vaccine.

Cutter Laboratories is a Berkeley-based and family-owned pharmaceutical company founded by Edward Ahern Cutter. Its early products include anthrax vaccine, hog cholera (swine fever) virus and anti-hog cholera serum—and a hog cholera vaccine.

1898 **Institut vaccinogène suisse de Lausanne** is established in Switzerland to produce smallpox vaccine using cows. It places itself freely under state control, who are its biggest client, but also supply other European countries, Central America, Africa and the Far East. The institute is run by directors Emile Félix and Jules Flück.

Bristol-Myers Co. BRISTOL-MYERS

Clinton Pharmaceutical company is renamed **Bristol, Myers and Company** and a year later is changed to **Bristol-Myers Corporation.**

Institut Sérothérapique et Vaccinal Suisse Berne
Sous le contrôle de l'Etat

L'institut bactério-thérapique Haefliger & Co. (Bern) and l'institut vaccinal suisse Charles Haccius (Lancy-Genève) merge to form **l'Institut bactério-thérapique et vaccinal Suisse,** under state control and the scientific direction of Prof. Doctor E. Tavel. The name is soon changed to **Swiss Serum and Vaccine Institute Berne** and is renamed **Berna Biotech** in 2001.

1899

Vaccinate Breeding Stock Now!

LITTLE PIGS of cholera-immune sows are themselves immune so long as they are suckling, but the litters of non-vaccinated sows share with their mothers the dangers of cholera infection. Vaccinate your breeding stock *now*, before breeding if possible—before farrowing certainly. Insure two generations at the trifling cost of one vaccination. Last year, before farrowing, thousands of valuable brood sows were immunized with

PITMAN-MOORE
Anti-Hog-Cholera Serum

Pitman-Moore is founded by Henry C. Pitman to manufacture products for vets and physicians. It becomes one of the world's largest producers of biological products to control diseases in cattle, horses, swine and pets. The company also produces vaccines for humans, including influenza, Salk polio and the Schwarz strain of measles vaccine.

1900 The Brazilian government responds to the bubonic plague pandemic by creating the **Federal Serum Therapy Institute**—headed by Baron Pedro Afonso, who owns and produces smallpox vaccines at the **Municipal Vaccine Institute** in Brazil.

1901 **Sterling Drug** is established with the name **Neuralgyline Co**. in Wheeling, West Virginia, by Albert H. Diebold and William E. Weiss, a pharmacist, with three local businessman. Its sole purpose is the manufacture and sale of a pain-relieving preparation.

Instituto Serumtherapico is founded in São Paulo, Brazil, to produce vaccine and serums.

1902 The **Danish Statens Serum Institute** is established to produce diphtheria antitoxin.

French pharmacist Henry Chibret founds **Chibret**, a company specializing in the manufacture and development of ophthalmological drugs.

Evans Sons, Lescher & Webb Ltd., is established in the UK and begins to produce sera and antitoxins against diphtheria, tetanus and meningitis. The company works in close collaboration with the Liverpool University Medical School.

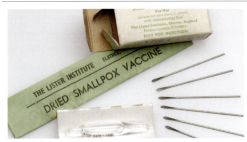

The Lister Institute of Preventive Medicine is renamed the **Jenner Institute of Preventive Medicine** until 1903, then the **Lister Institute** in honor of surgeon and medical pioneer, Dr. Joseph Lister. The institute is active in serum and vaccine lymph production and distribution from 1894–1950, notably making diphtheria antitoxin and smallpox vaccine. It also produces the Elstree smallpox strain that would be used during the WHO smallpox eradication program.

1904

BEHRING

Emil von Behring, the first recipient of the Nobel Prize in Physiology or Medicine, establishes **Behringwerke** in Marburg, Germany, for the purpose of working on disease prevention and for the manufacture of sera and vaccines, notably for diphtheria.

ISTITUTO SIEROTERAPICO E VACCINOGENO TOSCANO
"SCLAVO"

Istituto Sieroterapico e Vaccinogeno Toscano, or **Sclavo**, is founded in Siena, Italy, by Achille Sclavo, scientist and university professor of hygiene. Upon his death in 1930 the Institute becomes a cooperative and then in 1962 a private company.

Greer Laboratories is founded in the USA by R.T. Greer to collect source materials, such as roots, herbs, pollens.

For two years, **Sharp & Dohme** imports diphtheria, streptococcic, and tetanus antitoxin serum from the **Pasteur Institute** in Paris.

1905 The **Lister Institute** becomes a School of the University of London and continues to manufacture and sell vaccines.

A few years after Edward Squibb's death, his sons sell **E.R. Squibb & Sons** to former Merck-employee Theodore Weicker and Lowell M. Palmer. They incorporate the company, **E.R. Squibb & Sons Inc.**

The research department at London's **St. Mary's Hospital Medical School** is renamed the **Inoculation Department** and manufactures typhoid vaccines, diphtheria, and tetanus antitoxin. Upon Wright's death in 1947 it is renamed the **Wright Fleming Institute of Microbiology** which merges in 1988 with **Imperial College London.**

1907

American Cyanamid Company is founded by engineers Frank S. Washburn and Charles H. Baker in New York City to capitalize on a German patent they had licensed for the manufacture of nitrogen products for fertilizer. **Lederle Laboratories,** led by **Ernst Lederle former New York City Health Commissioner,** is the pharmaceutical and vaccine division.

1908 **Plough** is founded by Memphis-area entrepreneur, Abe Plough.

William R. Warner & Co. is acquired by Gustavus A. Pfeiffer & Company, a patent medicine company from St. Louis. Pfeiffer retains the **Warner** company name and moves the headquarters to New York, By the 1940s, some 50 companies have been acquired including Richard **Hudnut** Company in 1916.

1909 Sweden's **State Biological Laboratories** (SBL) is established to produce diphtheria antitoxin along with other vaccines and sera.

1911 **Pocono Laboratory** reopens as the **Slee Laboratories** after Dr. Richard Slee returns from his 1907–1911 role directing production at Lederle's Antitoxin Laboratory in NY. After a series of ownership transfers, Slee's manufacturing plant in Swiftwater, USA, eventually becomes a part of Sanofi in 2004.

The National Vaccine and Antitoxin Institute is established as a private company in Washington, DC, owned by Dwight T. Scott.

Pharmacia is founded in Stockholm, Sweden, by pharmacist Gustav Felix Grönfeldt. The company is named after the Greek word *φαρμακεία*, transliterated pharmakeia, which means "sorcery."

1913

The **Bayer Company Inc**. is established in New York. Bayer is Germany's third largest chemical company with 10,600 employees, some 7,900 of whom are based in Leverkusen. It holds 8,000 patents at home and abroad. Its international operations includes five foreign subsidiaries, 44 sales offices, and 123 agencies worldwide.

The Swedish international pharmaceutical company **Astra AB** is founded.

1914 CONNAUGHT ANTITOXIN LABORATORIES
UNIVERSITY OF TORONTO
Established for research investigation in Preventive Medicine and for the production and distribution of all Public Health Biological Products at Minimum Prices.

The **Connaught Medical Research Laboratories** is a non-commercial public health entity established by Dr. John G. FitzGerald, as part of the Canadian **University of Toronto** to produce the diphtheria antitoxin.

The **Kitasato Institute**—Japan's first private medical research facility—is established and will manufacture numerous vaccines.

1916

Hoechst AG is one of the co-founders of **IG Farben**, an advocacy group of Germany's chemicals industry to gain industrial power during and after World War I. It moves from an advocacy group into the well-known conglomerate in 1925.

Commonwealth Serum Laboratories (CSL) is established in Australia to provide vaccines for Australia and New Zealand. The first vaccine produced is in 1919 for the Spanish influenza pandemic.

Anchor Serum Company is incorporated in the USA, manufacturing and distributing animal health products, principally hog cholera virus vaccines.

Sandoz begins its pharmaceutical research.

1917

Neuralgyline Co. is renamed **Sterling Products.** Because supplies of drugs from Germany are cut off by the Allies, it sets up the **Winthrop Company** to manufacture active ingredients. In 1918, it will acquire the US assets of German company **Bayer AG** for US$ 5.3 million. From 1902 to 1986, Sterling Drug Inc. (named as such in 1942) acquires 130 companies, directly or indirectly.

Marietta / Lancaster County Vaccine Farm is purchased by veterinarian Dr. Samuel H. Gilliland and renamed **The Gilliland Laboratories**, and continues to produce smallpox vaccine.

1919

Following its expropriation under the Trading with the Enemy Act of 1917 at a US government auction, in partnership with Goldman Sachs and Lehman Brothers, George W. Merck buys back the company for US$ 3.5 million, but **Merck & Co.** remains a separate company from its former German parent **Merck** of Darmstadt.

China National Biotec Group Company Limited is founded as the **Central Office of Epidemic Prevention**—China's first institution dedicated to epidemiologic and bacteriological research and biological product manufacturing. Administered by the Ministry of Public Health of the People's Republic of China from 1949, it will become an important member of China National Pharmaceutical Group Co. Ltd., **Sinopharm**.

1923

Danish pharmaceutical company Nordisk Insulinlaboratorium, today **Novo Nordisk**, commercializes the production of insulin.

Organon International is founded in the Netherlands as a pioneer in contraception and fertility solutions. It first sells insulin.

1924

Joseph Nathan & Company chemist, Harry Jephcott, licenses a method to produce a Vitamin D supplement—Ostelin, which is its first pharmaceutical product.

Henry Wellcome brings together his company **Burroughs Wellcome & Co.**, with his philanthropic pursuits and research laboratories and museums to form the **Wellcome Foundation Ltd.**

Philip Hill, who made his money in real estate, acquires control of **Beecham** and buys up various companies to expand their portfolio in the next decades.

The company **Chobei Takeda & Co. Ltd.**, io incorporated with Choboi Takoda V as president. The company goes from being an individually owned business to a modern corporate organization integrating R&D, manufacturing, and marketing. The company changes its name to **Takeda Pharmaceutical Industries, Ltd.**, in 1943.

1925 Ayerst, McKenna & Harrison Ltd., later known as **Ayerst Laboratories**, is founded.

CRODA

Croda International is founded by British entrepreneur George W. Crowe and British chemist Henry J. Dawe in the UK. The company mainly produces lanolin, a natural fat in sheep's wool. One of the company's most popular products will be the aluminum hydroxide adjuvant Alhydrogel™.

1926

Imperial Chemical Industries plc (ICI) is founded from the merger of four companies: Brunner, Mond & Co., Nobel Explosives, the United Alkali Company, and British Dyestuffs Corporation. Competing with DuPont and IG Farben, ICI produces chemicals, explosives, fertilizers, insecticides, dyestuffs and paints.

American Home Products Corp is founded.

1928 **Pitman-Moore Company** becomes part of **Allied Laboratories** alongside Sioux Falls Serum Company, Sioux City Serum Company, Union Serum Company, and the Royal Serum Company in Kansas City, becoming the **Allied Consortium**.

UCB (Union Chimique Belge) Pharma Ltd. is founded by Belgian businessman Emmanuel Janssen in Belgium. UCB specializes in industrial chemicals and also has a small pharmaceutical division.

French chemical and pharmaceutical company **Rhône-Poulenc Rorer** is founded through the merger of Société Chimique des Usines du Rhône and Établissements Poulenc Frères. Very active in pharmaceuticals and fibers, the group diversifies after World War II to become a major producer of synthetic fibers (nylon, polyester), penicillin, and silicones.

1929 **H.K. Mulford Company** is acquired by and merges with Baltimore-based **Sharp & Dohme Corp.**

1930s **William S. Merrell Chemical Company** merges with a company started by Lunsford Richardson in 1898, Vicks Family Remedies (VapoRub), to become **Richardson-Merrell**.

1931 Los Angeles-based medical doctor Donald Baxter establishes **Baxter International**, as a manufacturer and distributor of intravenous therapy solution.

After John Wyeth's only son Stuart dies, controlling interest in **John Wyeth and Brothers** is left to Harvard University, who sells it to **American Home Products (AHP)** for US$ 2.9 million.

1934 The Research Foundation for Microbial Diseases of Osaka University, also referred to as the **BIKEN Foundation**, is established on an Osaka University campus by Japanese bacteriologist, Tenji Taniguchi. BIKEN manufactures vaccines and therapeutics.

C.L. Bencard Ltd., a company specializing in hay fever allergy prevention and treatment, including vaccines, is founded in Devon, UK.

1936

Swiss chemist Bernhard Joos establishes **Cilag**, acronym for **Chemical Industrial Laboratory AG**. Dr. Carl Nägeli, a well-known professor at the University of Zurich, is elected as president of the board.

Upon Henry Wellcome's death, ownership of the Foundation is transferred to the **Wellcome Trust.** Its first chairman is Sir Henry Dale, former director of Wellcome Physiological Research Laboratories.

Joseph Nathan & Co. pharmaceutical enterprise is established as a subsidiary company: **Glaxo Laboratories Ltd.**

1939 **Kern & Sandoz** becomes **Sandoz Ltd**. and its first steps are made into the agri-business.

1943 Beecham builds **Beecham Research Laboratories** in the UK. It also becomes the distributor of antigen products from the **Wright Fleming Institute.**

The Gilliland Laboratories, producer of smallpox vaccine for the US Army, tetanus vaccine, and gas gangrene antitoxin, is acquired by **American Home Products.**

Canadian Ayerst Laboratories, known for its estrogen products, is acquired by **American Home Products.**

1944 **Frederick Stearns Company** is acquired by **Sterling Drugs/Winthrop Company** and is merged to create **Winthrop-Stearns.**

1945 **Beecham** the company is renamed **Beecham Group Ltd.**

MEDICAL **EVANS** SUPPLIES

Evans Sons, Lescher & Webb Ltd is renamed **Evans Medical Supplies.**

Recherche et Industrie Thérapeutiques (RIT) is founded in Genval, Belgium, as a penicillin factory by Dr. Pieter De Somer and begins vaccine research and production in the 1950s.

1947 The pharmaceutical company **Labaz Group** is formed by Sociéte Belge de l'Azote et des Produits Chimiques du Marly.

Dr. Henry Foster, a young veterinarian, purchases one thousand rat cages and sets up a one-man laboratory in Boston overlooking the Charles River. He names the company **Charles River Breeding Laboratories** and will become a leading partner to the pharmaceutical industry for delivering solutions to accelerate the development of drugs, chemicals, medical devices and vaccines

1948

Degesch, a subsidiary of IG Farben, made the cyanide-based pesticide Zyklon B used in WWII concentration camps in Germany. **IG Farben** is seized by the allies in WWII and following the Nuremberg Trials (1947–1948), thirteen directors are convicted of war crimes, all of whom are released by the American High Commissioner in 1951. That same year, IG Farben is broken down into several companies: **Bayer, BASF, Hoechst,** and after several mergers the main successor companies become **Agfa, Bayer, BASF** and **Sanofi.**

1949 **Beecham Group** acquires **C.L. Bencard,** that specializes in therapeutic allergy vaccines.

1950

Dr. Frederik Paulsen and Eva Frandsen incorporate **"Nordiske Hormonlaboratorie"** later named **Ferring** in 1954.

Warner & Co. is renamed **Warner-Hudnut.**

1952 **Pfizer's Agriculture Division** opens a 732-acre research and development facility in Indiana, called the Vigo Plant—a former US Army biological weapons factory that produced both anthrax and botulinum toxin bombs. By 1988, the division is renamed **Pfizer Animal Health.**

Behringwerke becomes a wholly-owned subsidiary of **Farbwerke Hoechst AG.**

1953

Sharp & Dohme merges with **Merck & Co.**, also known as **Merck Sharp & Dohme** or **MSD**, outside the USA and Canada.

Litton Industries is founded and from 1951 will become a leading US defense contractor and a commercial biological research lab.

Dr. Paul Janssen establishes his own research laboratory within the **Richter-Eurpharma** company of his father, the name of the which is changed in 1956 to **NV Laboratoria Pharmaceutica C. Janssen.**

1954

Pitman-Moore is acquired by **Eli Lilly & Company** and is eventually taken over by **Dow Chemical Corporation** in 1959.

Sterling Drug makes its first move into vaccines and sera with the setup of the **Bayer Biological Institute** in former racing stables near Newmarket, England. Retired racehorses are used to produce antibodies for veterinary preparations. This and the rest of the UK veterinary business are eventually sold to **Pfizer** in 1962.

AHP becomes a leading US vaccine producer after supplying polio vaccine for Salk trials.

The Agro-Industrial division of **Eli Lilly & Company** is reorganized, launching **Elanco Products**, short for the parent company name that is then renamed in the 1990s to **Elanco Animal Health.**

1955

Burroughs Wellcome Fund is founded as an extension of the England-based **Wellcome Trust.**

Boehringer Ingelheim's Animal Health division is established as the company takes over **Pfizer's** veterinary program.

Indonesian vaccine manufacturer **Parc Vaccinogène** is renamed **Pasteur Institute.**

Warner-Hudnut and **Lambert Pharmacal Company** merge to form **Warner-Lambert.**

1958 *Glaxo*

Allen & Hanburys Ltd. is acquired by **Glaxo Laboratories Ltd**.

 成都生物制品研究所
CHENGDU INSTITUTE OF BIOLOGICAL PRODUCTS

The Chengdu Institute of Biological Products is founded as a prominent research and manufacturing base for biological products in China. It will become a part of **China National Biotec Corporation.**

P*h*RMA
RESEARCH • PROGRESS • HOPE

The **Pharmaceutical Manufacturers Association** is founded as a nonprofit scientific and trade group of over 100 companies that develop drugs and biologics. Currently known as the **Pharmaceutical Research and Manufacturers of America (PhRMA)**, it lobbies on behalf of drug makers, and is headquartered in Washington, DC.

1959 **Anchor Serum Company** is acquired by **North American Philips Corp**. and creates a subsidiary **Philips Roxane**, who go on to manufacture a short-lived measles vaccine and Rubelogen®, a rubella vaccine.

The **Allied Consortium**, including **Pitman-Moore** is acquired by **Dow Chemical Company.**

Cilag AG becomes a part of US giant **Johnson & Johnson.**

Evans Medical Supplies changes its name to **Evans Medical Ltd.**

1960 Pharmaceutical company **Immuno AG** is founded in Austria, specializing in blood and immunological products. The company will market the first ever tick-borne encephalitis vaccine in 1976, FSME*-Immun®.

1961 **NV Laboratoria Pharmaceutica C. Janssen** is purchased by **Johnson & Johnson** and renamed **Janssen Pharmaceutica.**

Glaxo acquires **Evans Medical Ltd.**

Mylan Pharmaceuticals is founded as a US-based drug distributor by Milan Puskar and Don Panoz.

1962

Biological E. Limited

Founded in India in 1953, **Biological E. Limited** is the first Indian private company to enter the vaccine manufacturing business.

Stallergenes is established in Lyon, France, by **Institut Mérieux**, to specialize in the treatment of allergies.

1963 French biotech company **BioMérieux** is founded by Marcel Mérieux's grandson and pharmacist Alain Mérieux, and specializes in diagnostic solutions. Family-owned **Institut Mérieux** remains the largest shareholder.

1966

Serum Institute of India is founded by commerce graduate student Cyrus Poonawalla as a part of the Cyrus Poonawalla Group. SII begins with the production of antitetanic serum made in horses from Poonawalla's stud farm, eventually becoming the world's largest vaccine manufacturer and major supplier to UNICEF and PAHO.

1968

Rhône-Poulenc acquires 51% majority interest in **Institut Mérieux.**

Based in Geneva, Switzerland, **The International Federation of Pharmaceutical Manufacturers & Associations (IFPMA)** is formed as a trade association representing pharmaceutical companies worldwide.

Litton Bionetics is created as a division of Litton Industries. It will become an important contractor to President Richard Nixon's cancer research facility at Fort Detrick, and a key supplier of monkeys and producer of viruses as well as other special biological materials. Litton Bionetics receives the then largest single contract awarded by the NIH in 1971 for US$6.8M.

SK&F

Belgian company **Recherche et Industries Thérapeutiques (RIT)** is acquired by **Smith, Kline & French** and renamed **SmithKline-RIT.**

1969 French ophthalmological drug company **Chibret** is acquired by the pharmaceutical company **Merck Sharp & Dohme**, who closes the Chibret headquarters in 2010.

Pitman-Moore's animal health division is acquired by **Johnson & Johnson.**

AkzoNobel

Through a series of acquisitions and mergers, **Organon** becomes AKZO, later **Akzo Nobel**. Organon is the human health care business unit of Akzo Nobel.

1970   **Warner-Lambert** acquires **Parke-Davis.**

Ciba-Geigy is formed by the merger of **J.R. Geigy Ltd**. and **CIBA**.

Synthélabo is founded through the merger of two French pharmaceutical laboratories, Laboratoires Dausse (founded in 1834) and Laboratoires Robert & Carrière (1899).

1971

One of the first biotechnology companies **Cetus Corporation** is founded and its chemist Kary Mullis, PhD, invents the Polymerase Chain Reaction (PCR) technique in the mid-1980s, for which he is awarded the Nobel Prize in 1993.

Bachem Feinchemikalien AG is founded in Switzerland, specializing in the development and manufacture of peptides and oligonucleotides as therapeutic agents.

 Schering-Plough

Schering Corporation merges with **Plough** to form **Schering-Plough.**

GC Biopharma

Initially established in 1967 in Korea as the **Sudo Microorganism Medical Supplies Co**., the name is changed to **Green Cross BioPharma Inc**., and becomes an important vaccine and biopharmaceutical manufacturer.

BiO-MeD

BioMed Ltd. is founded in India with the mission to produce "state of the art modern prophylactic measures." By 2024, the company produces nine vaccines.

1973   **sanofi**

Sanofi is founded as a subsidiary of French oil company Elf Aquitaine (subsequently acquired by Total in 2000), when Elf Aquitaine takes control of the **Labaz Group.**

1974

German pharmaceutical company **Bayer** acquires **Cutter Laboratories.**

Pasteur Institute creates **Pasteur Production**, a subsidiary specializing in manufacturing vaccines.

1976   ## Genentech

Genentech arises as the unlikely vision of two naive entrepreneurs: Herbert Boyer, a professor of microbiology at the University of California, San Francisco; and Robert Swanson, an unemployed venture capitalist. Its name, a contraction of *gen*etic *en*gineering *tech*nology, captures its agenda to pioneer the field of recombinant DNA.

Bio-Manguinhos is founded in Brazil to strengthen vaccine manufacturing capacity for the country's vaccination program.

Eli Lilly & Company gets out of the vaccine business and sells its vaccine product rights (notably its fractionated acellular pertussis vaccine) to **Wyeth**.

1977

Takeda first enters the US pharmaceutical market by developing a joint venture with **Abbott Laboratories** called **TAP Pharmaceuticals.** It is dissolved in 2008.

1978   **Richardson-Merrell National Laboratories** division—the largest producer of swine flu vaccine in the 1976 mass vaccination campaign—exits the vaccine business by donating its vaccine research and production facilities to the Salk Institute in San Diego. **Connaught Labs** acquires the vaccine factory at Swiftwater, Pennsylvania, from the Salk Institute.

Biogen Inc. is founded by four scientists as the first Europe-based biotech company (with close links to Harvard University). It is instrumental to the development of recombinant hepatitis B vaccine. Biogen licenses its hepatitis B technology to SmithKline Beecham and in 2003, merges with **Idec Pharmaceuticals** (founded in 1985) to become **Biogen Idec.**

1980

Israeli **Bio Technology General Corp** is established by Prof. Haim Aviv and other collaborators from the Weizmann Institute of Science. It is later to be renamed **Bio Technology Group (BTG).**

Dow Chemical acquires controlling interest of the Merrell pharmaceutical division of **Richardson-Merrell** company and becomes **Merrell Dow Pharmaceuticals.**

UK biotech company **Celltech** is founded.

1981

Boehringer Ingelheim expands its animal health operations in the United States with the acquisition of **Philips Roxane.**

CHIRON

American multinational biotechnology firm **Chiron Corporation** is founded.

1982   **Smith, Kline & French** acquires **Allergan**, an eye and skincare business for US$ 259 million, and merges with **Beckman Instruments, Inc.**, an American company specializing in diagnostics and measurement instruments. The company changes its name to **SmithKline Beckman.**

1983   Dr. David H. Smith resigns as chairman of the department of pediatrics at the University of Rochester School of Medicine in New York to co-found the company **Praxis Biologics** with two other Medical Center scientists—it goes public in 1987. By 1989, Praxis has the largest number of new vaccines in clinical trials and one of the finest manufacturing facilities in the USA.

Genetronics Biomedical Corporation is incorporated in the US, initially founded in Canada in 1979, under the name **Concord Energy Corp**. GBC is a technology leader in electroporation—applying pulsed rotating electric fields to open cell membrane pores to deliver pharmaceuticals or gene therapy.

Protein Sciences Corporation is founded in the US as a privately held biotech company. Its mission is to "improve health through the creation of innovative vaccines and biopharmaceuticals." Its breakthrough product will be a recombinant influenza vaccine FluBlok® (US) Supemtek® (EU), licensed several decades later.

1985   **Monsanto** acquires the pharmaceutical company **G.D. Searle & Company.**

Pasteur Production is acquired by **Institut Mérieux** and **Pasteur Vaccins** is created.

Dutch-German company **Rhein Biotech** is founded as a spin-off of the University of Dusseldorf. The company will produce vaccines for hepatitis, diphtheria, tetanus, typhoid, hemorrhagic fever, and influenza.

1986   The **Wellcome Trust** publicly sells shares of the Wellcome Foundation Ltd. to form **Wellcome plc.**

Danish **Novo Industri A/S** acquires the **Ferrosan Group,** now named **Novo Nordisk Pharmatech A/S.**

Chiron Corporation and **Ciba-Geigy** launch their joint vaccine venture company, **Biocine.**

British specialty biopharmaceutical company **Shire plc** is founded.

1987 **Pitman-Moore** is sold by **Johnson & Johnson** to **Animal Products Group of International Minerals and Chemical Corporation.**

Novavax is founded in the USA.

Biotech company **Vical Inc.**, is founded in San Diego to specialize in DNA delivery technology, including DNA vaccines for infectious diseases or cancer.

AHP merges its **Wyeth** and **Ayerst** divisions to unite its pharmaceutical businesses, forming **Wyeth-Ayerst Laboratories.**

1988 **Chiron** forms a joint venture company, **Mimesys Inc.** with **Johnson & Johnson** and the **Warner-Lambert Company** to develop the next generation of biotech drugs.

 MedImmune

Molecular Vaccines, Inc., renamed **MedImmune** a year later, is founded by Wayne T. Hockmeyer, David Mott and Dr. James Young.

To avoid a hostile takeover by Hoffman-La Roche, **Sterling Drugs** becomes a division of **Eastman Kodak** for the sum of US$ 5.1 billion.

1989 **Bristol-Myers Squibb**

Bristol-Myers and **Squibb** merge to become **Bristol-Myers Squibb**. While it has marketed few vaccines (DTwP and influenza), by 2018 it will have candidate cancer vaccines in development.

American Cyanamid Company acquires American company **Praxis Biologics Inc**., for around US$ 190 million and merges it with **Lederle Laboratories** to become **Lederle-Praxis Biologicals.**

Dow Chemical acquires 67% interest of **Marion Laboratories**, which is renamed **Marion Merrell Dow.**

The parent company of **Smith, Kline & French** Laboratories, SmithKline Corp., merges with the Beecham Group plc to form **SmithKline Beecham plc.**

North American Vaccine Inc. (NAVI) is incorporated in Montreal, Canada, and begins operations in the US a year later. It is one of only five companies licensed by the FDA to sell pediatric vaccines in the US.

VERTEX

Vertex Pharmaceuticals Inc., is founded by US chemist Joshua Boger and investor Kevin J. Kinsella to "transform the way serious diseases are treated." Vertex's beginnings are profiled by Barry Werth in his 1994 book *The Billion-Dollar Molecule.*

French **Institut Mérieux** acquires **Canadian Connaught Labs**, and the vaccine division is renamed **Pasteur Mérieux Sérums et Vaccins** until 1996.

1990 Florida-based **Berna Products Corp.** is established by Swiss **Berna Biotech.**

Predecessor to the international biotech firm **Mymetics,** the company **Hippocampe SA** is founded in France, conducting research on HIV surface proteins.

Cetus Corporation merges with **Chiron Corporation,** today a part of **Novartis.**

Hoffman-La Roche acquires a majority stake in **Genentech.**

1991 **Merck & Co., Inc.,** creates the **Merck Vaccine Division.**

UK biotech firm **Chiroscience** is founded.

Synthélabo acquires French Laboratories Delalande and Laboratoires Delagrange.

MCM Vaccine BV in Europe is established as a manufacturing and supply partnership between **MSD** and **Pasteur Mérieux Sérums et Vaccins** to co-develop a hexavalent pediatric vaccine—combining antigens developed by each company. The product of this partnership is the 6-in-1 vaccine Vaxelis®, first licensed in Europe in 2016, containing Merck's proprietary adjuvant AAHS.

UK-based **Medeva International plc**, buys the vaccine business of **Wellcome**. Medeva becomes the principal vaccine supplier to the National Health Service in the UK.

Sanofi and **Sterling Drug-Winthrop** form a strategic alliance.

1992 **Peptide Therapeutics Group plc** is founded in Cambridge, UK.

Hoffman-La Roche, more commonly known as **Roche**, acquires the patents from **Cetus Corporation** for the PCR technique for US$300M.

Sclavo S.p.A. is acquired by **Biocine**—the joint vaccine venture of **Chiron Corporation and Ciba-Geigy**—for US$64.4M

The California biopharmaceutical company **Aviron** is founded to develop cost-effective products against infections, mainly vaccines. One of its main products will be FluMist®, an intranasal influenza vaccine.

1993

IntroGene is established as a spin-off of Leiden University, Netherlands, with a mission to invent and develop products for gene therapy.

French **AxCell Biotechnologies** is founded, specializing in cell media for vaccinology.

The production department of Sweden's **State Biological Laboratories** (SBL) is separated from the Institute's other functions and is transformed into a private company called **SBL-Vaccin.**

Imperial Chemical Industries plc (ICI) demerges its bioscience business, splitting into two publicly listed companies: ICI and **Zeneca**—Zeneca would later go on to merge with **Astra AB** in 1999.

King Pharmaceuticals is founded in Tennessee, USA, and initially specializes in manufacturing drugs for other companies.

Biogenetech

Biogenetech is founded in Thailand and begins sales, marketing and distribution of biopharmaceutical products (including vaccines) and nuclear medicines.

British vaccine, drug, and diagnostics delivery company **PowderJect Pharmaceuticals** is founded, spun out of the University of Oxford. It develops a needle-free injection system for delivering medications and vaccines.

1994

German-Danish biotech company **Bavarian Nordic** is founded and becomes an established leader in biodefense.

A joint venture owned on a 50/50 basis by **Pasteur Mérieux Sérums et Vaccins** and **Merck Sharp & Dohme** (known as Merck Inc. in the US and Canada) is created to develop and commercialize combination vaccines originating from both companies' pipelines in Europe. This same year, PMSV becomes a wholly owned subsidiary of Rhône-Poulenc. In 1999, the joint venture will be named **Aventis Pasteur MSD** and in January 2005, **Sanofi Pasteur MSD**.

Federally-owned Australian vaccine producer **Commonwealth Serum Laboratories** is privatized, and renamed **CSL Ltd.**

Antigenics Inc., (renamed **Agenus** in 2011) is founded by American-Armenian scientist Garo H. Armen and American immunologist and physician Dr. Pramod K. Srivastava. Antigenics Inc. enters into a patent license agreement with the Mount Sinai School of Medicine, obtaining an exclusive worldwide license to patent rights relating to the heat shock protein technology for personalized cancer vaccines.

Behringwerke AG is split up into various spin-off companies for the production of blood products, vaccines and diagnostic equipment.

Sanofi purchases **Sterling Drug's** worldwide business for US$ 1.675 billion. **SmithKline Beecham** purchases Sterling's worldwide over-the-counter drug business for US$ 2.9 billion and promptly sells the US portion to **Bayer** for US$ 1 billion.

Pitman-Moore is sold to **Mallinckrodt Veterinary Corporation.**

Corixa Corporation is founded in Seattle, USA, as a biotech company developing immunotherapeutics to combat autoimmune and infectious diseases. It produces the novel vaccine adjuvant MPL.

American Home Products acquires **American Cyanamid Company** and its subsidiary **Lederle Laboratories / Lederle-Praxis Biologics** to become **Wyeth-Ayerst Lederle, Inc.**

1995 *GlaxoWellcome*

The **Wellcome Trust** sells its remaining shares to **Glaxo plc**, forming the world's largest pharmaceutical company **Glaxo Wellcome plc**.

US biotech company **VaxGen** is founded as a spin-off of **Genentech,** whose sole purpose is to develop AIDSVAX®, which ends in failure. Following the US-HHS withdrawal of an anthrax vaccine contract worth US$ 877.5 million, VaxGen stocks plummet. VaxGen merges with **diaDexus Inc.**, in 2010, who then files for bankruptcy in 2016. Its website now points to medical diagnostics company called **Diazyme.**

Hoechst AG buys a majority share of **Dow Chemical** for US$ 7.1 billion and renames it **Hoechst Marion Roussel Ltd.** The deal creates the world's second largest drug manufacturer behind Glaxo Wellcome and ahead of Merck.

American Home Products acquires the animal health division of **Solvay.**

Pharmacia &Upjohn

Swedish company **Pharmacia** merges with the American pharmaceutical company **Upjohn**, becoming **Pharmacia & Upjohn**.

Lucky Chemicals Co. Ltd., (founded in 1947) is renamed **LG Chemicals**, becoming the largest chemical company in Korea.

1996

Swiss companies **Ciba-Geigy** and **Sandoz** merge to form **Novartis.**

US biopharmaceutical company **Dynavax Technologies Corporation** is founded.

Indian microbiologist Dr. Krishna M. Ella and Mrs. Suchitra Ella set up **Bharat Biotech** in Genome Valley, Hyderabad, India. It becomes a leading vaccine manufacturer in India.

Russian pharmaceutical and vaccine company **Petrovax** is founded by scientist Arkady Nekrasov, PhD.

CHIRON BEHRING

Chiron Corporation obtains 49% of the human-vaccine business of **Behringwerke**, a subsidiary of the chemical and drug giant **Hoechst**, for US$ 118 million in cash. This acquisition creates **Chiron Behring Vaccines GmbH**, based in Marburg, Germany.

Pasteur Mérieux Sérums et Vaccins is renamed **Pasteur Mérieux Connaught**.

Dutch company **U-Bisys** is founded, specializing in the development of human antibodies and antibody-derivatives for therapeutic and diagnostic purposes.

1997

Veterinary vaccine company **Merial** is created through the merger of **Rhône Mérieux** (Rhône-Poulenc, France) and **MSD AgVet** (Merck & Co., USA)

AlphaVax is founded by American scientist Jonathan F. Smith and based in North Carolina. The company develops a range of vaccines for infectious diseases, biodefense, and cancer. It uses alphavaccine vector platform technology to generate a portfolio of vaccine candidates for infectious disease prevention and cancer immunotherapy.

1998

BioPort, formerly operated as state-owned **Michigan Biological Products**, is founded by Fuad El-Hibri, and is privatized as **Emergent BioSolutions** in 2004. Its main product is the anthrax vaccine, and will become one of BARDA's biggest biodefense partners.

Allergy Therapeutics Ltd. is founded as a fully integrated, specialty pharmaceuticals business focused on the treatment and prevention of allergies. The company is formed following the management buy-in of the **Bencard** allergy vaccine business (Pollinex®, Pollinex® Quattro®, Oralvac®) from **SmithKline Beecham**.

King Pharmaceuticals purchases the manufacturing site of **Warner-Lambert's Parke-Davis** division in Michigan, and creates the subsidiary **Parkedale Pharmaceuticals** for its sterile product line, including influenza vaccine Fluogen®.

Pasteur Institute in Indonesia becomes a state-owned company, and changes it name to **Bio Farma**.

Pitman-Moore is sold by Mallinckrodt Veterinary Corporation to **Schering-Plough.**

Chiron Corporation and **Hoechst Marion Roussel Limited (HMR)** form a vaccine-related joint venture known as **Chiron Behring Vaccines Private Ltd**., to be based in India. The joint venture is owned 51% by Chiron and 49% by HMR. Initially, it produces Rabipur®, a rabies vaccine, for distribution in India and other Asian countries.

Evans Medical Ltd. (and its holding company Evans Healthcare Ltd.) is acquired by **Medeva International plc** who changes its name to **Medeva Pharma Ltd.**

Intercell is formed as a spin-off of the Research Institute of Molecular Pathology in Vienna, Austria.

1999

Rhône-Poulenc merges with **Hoechst AG** to form **Aventis.**

Electrofect AS is founded in Oslo, Norway, as a biotech company specialized in "*in vivo* gene delivery and vaccine generation systems." Two years later, the name is changed to **Inovio AS.**

mΣDICAGO

Biopharmaceutical company **Medicago** is founded in Quebec, Canada, under Japanese majority parent company **Mitsubishi Tanabe Pharma Corporation**, which itself is under **Mitsubish Chemical Group.** Medicago's mission is to help improve public health using the power of plants, leveraging its plant-based therapeutics technology.

ISCONOVA AB

Isconova AB is founded in Sweden.

Pasteur Merieux and **Aventis** merge to become **Aventis Pasteur.**

Vivalis, a French biomanufacturing company is created, becoming one of the world's specialists in embryonic stem cells, in particular for vaccine production.

Celltech acquires **Chiroscience plc**, for US$ 527.7 million and is renamed **Celltech Chiroscience plc.**

AstraZeneca

AstraZeneca is founded through the merger of the Swedish **Astra AB** and the British **Zeneca Group** (itself formed by the demerger of the pharmaceutical operations of Imperial Chemical Industries in 1993).

Medeva Pharma Ltd. is acquired by **Celltech Chiroscience plc** for £ 563 million.

sanofi~synthelabo

Sanofi merges with **Synthélabo** to create **Sanofi-Synthélabo**; at the time of the merger Sanofi is the second largest pharmaceutical group in France in terms of sales and Synthélabo is the third largest.

 Agilent Technologies

Agilent Technologies, a life sciences, measurement, diagnostics, and applied chemical markets company, is founded as a spin-off of computing firm Hewlett-Packard.

Under an agreement worth US$400 million, **Wyeth-Lederle Vaccines**, a business unit of **Wyeth-Ayerst**, enters into a major collaboration with **Aviron** to market its intranasal influenza vaccine, FluMist®.

2000

Novartis and **AstraZeneca** combine their agrobusiness divisions to create a new company, **Syngenta**.

North American Vaccine, Inc. is purchased by **Baxter International** for US$257 million in stock.

German biopharmaceutical company **CureVac** is founded and enters into various collaborations with pharmaceutical companies and organizations, including agreements with **Boehringer Ingelheim, Sanofi Pasteur** (in 2011), **Johnson & Johnson, Genmab, the Bill & Melinda Gates Foundation, Eli Lilly & Company, GlaxoSmithKline, Coalition for Epidemic Preparedness Innovations (CEPI), the International AIDS Vaccine Initiative,** and the German government. CureVac develops therapies based on messenger RNA (mRNA) and is founded by Ingmar Hoerr, Steve Pascolo, Florian von der Mülbe, Günther Jung and Hans-Georg Rammensee.

Medeva Pharma Ltd. sells its assets related to vaccine manufacture to **Evans Vaccines Ltd.**

Dutch company **IntroGene** acquires **U-Bisys** to form **Crucell.**

PowderJect Pharmaceuticals acquires **Medeva's** vaccine business from **Celltech** under the name **Evans Vaccines Ltd.**

Glaxo Wellcome and **SmithKline Beecham** merge to form **GlaxoSmithKline plc**, better known as **GSK.**

Pfizer acquires **Warner-Lambert.**

Biotech company **Avidis** is launched as a spin-off from the **Medical Research Council** and **Cambridge University**, UK, specializing in proteomics and structural genomics.

Celltech Chiroscience and **Medeva** merge.

Peptide Therapeutics Group plc changes its name to **Acambis plc.**

Pharmacia & Upjohn completes a merger with **Monsanto** and **Searle** creating **Pharmacia Corporation.**

International biotech firm **Mymetics** is founded through a reverse merger of **ICHOR Corporation**, a dormant US company and French **Hippocampe SA**. Registered in the USA, its focus is on developing vaccines and therapies to combat AIDS and from 2003 is based in Switzerland.

2001

SINOVAC 科兴

Sinovac Biotech Co. Ltd is founded in the People's Republic of China, with its first products being China's first hepatitis A enzyme-linked diagnostic reagent and hepatitis A vaccine.

vbi

Variation Biotechnologies, or **VBI Vaccines** is founded in the USA, dedicated to the development and delivery of thermo-stable vaccines, and specializing in Virus-Like-Particle (VLP) platforms.

Northrop Grumman acquires **Litton Industries** for US$5.1 billion.

Walvax is founded in China to manufacture vaccines and immunotherapeutics.

ID Biomedical Corporation of Quebec is founded as an integrated biotechnology company dedicated to the research, development and manufacturing of vaccines. It receives a 10-year mandate from the Canadian government to assure "a state of readiness in the case of an influenza pandemic and provide influenza vaccine for all Canadians in such an event." It will supply the majority of the government's influenza vaccine purchases.

The Developing Countries Vaccine Manufacturers Network is officially established at the Partner's meeting of Gavi. Registered in Switzerland, the DCVMN begins with 10 members; 3 members with WHO prequalified (PQ) vaccines. Currently, there are 43 members—15 full members with WHO PQ.

Ablynx is founded in Belgium as a biotech company—a spin-off of the Vlaams Instituut voor Biotechnologie and the Free University of Brussels. It specializes in the development of novel therapeutic proteins called "Nanobodies"—single-domain antibodies from llamas and other camelid species.

Swiss Serum and Vaccine Institute is renamed **Berna Biotech Ltd.**

2002

Vaccine manufacturer **Chongqing Zhifei Biological Products Co., Ltd.** is founded in the People's Republic of China.

Berna Biotech acquires a 93% ownership of **Rhein Biotech.**

MedImmune acquires California biotech company **Aviron** for US$1.5 billion, adding the experimental nasal spray vaccine FluMist® to its portfolio. Its name is changed to **MedImmune Vaccines.**

Wyeth

American Home Products changes its name to **Wyeth.**

pevion

Swiss companies **Berna Biotech** and **Bachem** create a joint venture named **Pevion Biotech**, focusing on therapeutic and prophylactic peptide vaccines, notably for malaria.

2003   From 2003 to 2011, **Pfizer** acquires **Pharmacia Corporation,** Embrex Inc, Catapult Genetics, Bovigen, **Wyeth**, Fort Dodge Animal Health, Vetnex Animal Health Ltd, Synbiotics Corporation, Microtek, **King Pharmaceuticals**, and Alpharma— considerably expanding both its human and animal product portfolio.

US based **Berna Products Corp** is acquired by **Acambis plc.**

Biovac is created in South Africa as a partnership formed with the South African government to establish local vaccine manufacturing capabilities.

Novartis organizes all its generics businesses into one division and merges some of its subsidiaries into one company, reusing the predecessor brand name of **Sandoz.**

PowderJect Pharmaceuticals is acquired by **Chiron Corporation** for £542 million.

2004

UCB (Union Chimique Belge) Pharma acquires **Celltech Pharmaceuticals.**

Sanofi-Synthélabo merges with Aventis and is renamed the **Sanofi Aventis Group.** It becomes the world's third largest pharmaceutical company, behind Pfizer and GSK. **Aventis Pasteur,** the vaccine division of the Sanofi Aventis Group, changes its name to **Sanofi Pasteur** in January 2005.

ID Biomedical Corporation acquires all of **Shire Pharmaceuticals'** vaccine assets in a deal worth US$ 120 million.

CSL Behring
Biotherapies for Life™

Australian blood products group **CSL Limited** buys the plasma therapeutics business of France-based **Aventis** for US$ 925 million, combining it with ZLB Bioplasma to create **ZLB Behring**, renamed **CSL Behring** in 2007. CSL Behring operates **CSL Plasma**, one of the world's largest blood plasma collection networks.

2005

Israel-US **SciVac Therapeutics** is established with links to the Israeli company **Bio Technology General (BTG) Corp** for its recombinant hepatitis B Sci-B-Vac™.

GSK buys **Wyeth's** Marietta vaccine manufacturing facility for US$ 100 million, to develop cell culture-based influenza vaccines.

GSK acquires Canadian influenza vaccine manufacturer **ID Biomedical Corporation of Quebec.** The value of the transaction is approximately C$ 1.7 billion/US$ 1.4 billion.

Xiamen Innovax Biotech Co., Ltd., is founded as the key vaccine arm of **Beijing Wantai Biological Pharmacy Enterprise Co., Ltd**., a privately-owned high-tech company engaged in vaccine research, development and production in China.

Arabio is founded as a leading vaccine manufacturer in the Gulf Cooperation Council region (Bahrain, Kuwait, Oman, Qatar, Saudi Arabia, and the UAE).

Creative Biolabs is founded by a group of scientists to provide treatments and as a custom service in the field of vaccine development, preclinical assessment and GMP manufacturing, notably in cancer research.

inovio

US-based **Genetronics Biomedical Corp**., acquires Norwegian **Inovio AS** and becomes **Inovio Biomedical Corp**. It specializes in clinical-stage biotechnology for "developing DNA medicines [vaccines] to fight HPV-related diseases, cancer and infectious diseases." It develops SynCon™—"a proprietary computer algorithm to help engineer precisely designed DNA plasmids that work like software the body's cells can download to learn how to produce a target protein" and markets an electroporation-based delivery technology designed to optimally deliver plasmids into cells, Cellectra®.

BTG is sold by parent company **Savient Pharmaceuticals** to Swiss multinational biopharmaceutical firm **Ferring.**

GSK purchases biotech company **Corixa Corporation** for US$ 300 million.

2006

Dynavax Technologies acquires **Rhein Biotech** from Crucell for US$ 12 million.

Biopharmaceutical company **Imaxio,** is created through the merger between **Diagnogene** and **Avidis**, specializing in vaccines and genomics.

Dutch pharmaceutical company **Crucell** buys the Swiss **Berna Biotech** and acquires Swedish **SBL-Vaccin** for US$ 50 million. Together with US-based **Berna Products**, they join forces to become the sixth largest vaccine company worldwide.

Novartis acquires the California-based **Chiron Corporation** for US$ 7.5 billion. Chiron Vaccines and Chiron Blood Testing are merged into **Novartis Vaccines and Diagnostics.** The company **Chiron Behring Vaccines** keeps its name and continues to produce Purified Chick Embryo Cell rabies vaccine Rabipur® in India.

Schering Pharma AG is acquired by **Bayer AG** and merged to form **Bayer Schering Pharma AG**, which is renamed **Bayer HealthCare Pharmaceuticals** in 2011.

Harrisvaccines (formerly called Sirrah Bios) is founded by Dr. Harry Harris at Iowa State University Research Park, as a spin-out of his laboratory at ISU. Harrisvaccines becomes known for its RNA particle technology that uses pathogens collected from a farm to create autogenous vaccines for livestock. Specific genes are sequenced and inserted into RNA particles creating potent vaccines that provide herd-specific protection.

Merck KGaA acquires the Swiss family biotech company **Serono AG.**

Dutch biotechnology company **Crucell** acquires Florida-based **Berna Products Corp.**, from **Acambis plc.**

2007

PaxVax is founded as an American specialty vaccine company, with its headquarters in Redwood City, California.

Okairos AG, a biopharmaceutical company focused on the discovery and development of genetic T-cell vaccines, including hepatitis C virus, is founded as a spin-out from Merck Laboratories' Research Center in Rome. Led by Riccardo Cortese and Alfredo Nicosia, the company moves to Basel, Switzerland.

Canadian biotech **Tekmira Pharmaceuticals Corporation** is spun off of 1992-founded Inex Pharmaceuticals. Tekmira specializes in developing novel RNAi (RNA interference) therapeutics and providing its proprietary lipid nanoparticle and nucleic acid delivery technologies to pharmaceutical partners, including vaccine manufacturers.

MedImmune is acquired by **AstraZeneca** for US$ 15.6 billion—the MedImmune name, AstraZeneca biologics and research arm, will be retired in 2019.

2008

BioNTech is established based on research by Uğur Şahin, Özlem Türeci and Christoph Huber, with a seed investment of € 150 million, focusing on the development and production of technologies and drugs for individualized cancer immunotherapy.

Acambis is acquired for US$ 546 million and operates as a subsidiary of **Sanofi Pasteur SA.**

2009 **Merck & Co.** merges with **Schering-Plough**, keeping the name of Merck & Co.

Animal vaccine company **Merial** becomes a wholly-owned subsidiary of **Sanofi.**

Wyeth becomes a wholly owned subsidiary of **Pfizer** for US$ 68 billion.

Roche finalizes its purchase of **Genentech** for US$ 46.8 billion.

Canadian biotechnology company **Acuitas Therapeutics Inc**., is founded in Vancouver, British Columbia, specializing in the development of delivery systems for nucleic acid (DNA / RNA) therapeutics based on lipid nanoparticle (LNP) technology—one of the key components of mRNA vaccines against COVID-19. Acuitas enters into several agreements with pharmaceutical companies and institutions, like Pfizer-BioNTech, CureVac, and Imperial College London, to advance vaccine clinical trials.

Beijing Advaccine Biotechnology Co., Ltd., is founded in China to develop vaccines for hepatitis B, RSV, type 1 diabetes, and from 2020, works on a COVID-19 vaccine.

Virometix AG is founded in Switzerland—a spin-out from the University of Zurich—as a "clinical stage biotech company pioneering a new class of synthetic vaccines to tackle infectious diseases and cancer."

GSK and **Pfizer** launch **ViiV Healthcare,** a joint venture dedicated to delivering treatments and care for HIV communities.

2010 **Eubiologics** is founded in Korea and develops and supplies vaccines, adjuvants and carrier proteins (rCRM197).

Sinergium Biotech is founded in Argentina to develop vaccines and biopharmaceuticals.

Abbott completes its acquisition of Belgium-based **Solvay Pharmaceuticals** for US$ 6.2 billion, providing Abbott with an extensive portfolio of medicinal products.

ModeRNA Therapeutics is launched by scientist Derrick Rossi, Harvard University faculty member Timothy A. Springer, investors Kenneth R. Chien, Bob Langer, and Flagship Pioneering Venture Studios run by Noubar Afeyan. A year later Afeyan hires Stéphane Bancel, previously an executive at BioMérieux and Eli Lilly & Company. ModeRNA Therapeutics specializes in mRNA therapy and it will be 10 years before a product is brought to market.

King Pharmaceuticals is acquired by **Pfizer** for US$ 3.6 billion in cash.

2011

All pharmaceutical companies of **Johnson & Johnson** are united under a common logo; **Janssen**.

Dutch company **Crucell** is acquired by **Johnson & Johnson** for US$ 2.4 billion and the **Berna Biotech** plant in Berne operates under the name **Janssen Vaccines.**

Sanofi Aventis Group changes its name to **Sanofi.**

Nanolek is founded in Russia and develops and commercializes vaccines and chemical products to treat orphan diseases.

Elanco expands its footprint in Europe by acquiring **Janssen Animal Health.**

CODAGENIX

New York's Stony Brook University spinout **Codagenix** is founded as clinical-stage biotech company. It signs a 20-year licensing agreement with the university for "use of a software algorithm that weakens a virus by changing its DNA." This proprietary "synthetic attenuated virus engineering" technology allows a synthetic biology-driven approach to develop cancer and infectious disease products—notably a novel intranasal live-attenuated COVID-19 vaccine candidate called CoviLiv™.

Hookipa Pharma is founded in Austria as a clinical-stage biopharmaceutical company focused on developing novel immunotherapies to fight cancer and chronic infectious disease.

2012

Takeda

Takeda establishes its Global Vaccine Business Division. It is the largest pharmaceutical company in Asia, with a strong position in Brazil.

zoetis

Pfizer Animal Health is renamed **Zoetis**. It is to become the largest producer of vaccines and medicines for pets and livestock.

2013

ModeRNA Therapeutics is awarded up to US$ 25 million by **DARPA**. In the same year ModeRNA signs a five-year exclusive option agreement with **AstraZeneca** to discover, develop, and commercialize mRNA for treatments in the therapeutic areas of cardiovascular, metabolic, and renal diseases as well as selected targets for cancer. The agreement includes a US$ 240 million upfront payment to ModeRNA, "one of the largest ever initial payments in a pharmaceutical industry licensing deal that does not involve a drug already being tested in clinical trials."

 ARCTURUS therapeutics

Arcturus Therapeutics is founded in San Diego as a "global late-stage clinical mRNA medicines and vaccines company with enabling technologies: (i) LUNAR® lipid-mediated delivery, (ii) STARR® mRNA Technology (sa-mRNA) and (iii) mRNA drug substance along with drug product manufacturing expertise."

valneva

Valneva SE is founded through the merger of Austrian company **Intercell** and French company **Vivalis SA** worth US$ 174 million.

abbvie

Abbott Laboratories splits off its research-based pharmaceuticals business into **AbbVie.**

2014

Baxter announces that it is leaving the vaccines business—divesting its commercial vaccine portfolio to **Pfizer** and exploring options for its vaccines R&D program, including influenza.

Okairos is acquired by **GSK** for US$ 325M.

Affinivax is founded in Boston as a clinical-stage biotechnology company developing next generation vaccine technologies (notably pneumococcal vaccines) for both preventive and therapeutic purposes. The BMGF provides US$ 4 million in seed funding and takes 2 out of the 6 seats on the board.

 VIAQUA THERAPEUTICS

ViAqua Therapeutics is founded in Israel, a biotechnology company and developer of vaccines for aquaculture.

American biotech company **PaxVax** acquires **Crucell Switzerland AG**, taking over the former **Berna Biotech** factory and calling it **PaxVax Berna**. It markets its first vaccine—oral typhoid Vivotif®.

2015

Novartis agrees to acquire GSK's oncology pipeline while **GSK** will acquire Novartis' vaccine business (excluding flu vaccines) for US$ 7.1 billion.

Elanco Animal Health completes its acquisition of **Novartis Animal Health** for US$ 5.4 billion, which further positions Elanco as a global leader in animal health.

Novartis sells its flu vaccine business to **CSL** for US$ 275 million. CSL merges it into its BioCSL operation and then rebrands the combined business with Novartis Influenza Vaccines as **Seqirus**, creating the world's second largest influenza vaccine company.

The US company **DuPont** merges with **Dow Chemical** in a deal worth US$ 120 billion.

The pharmaceutical operations of **Rhône-Poulenc** become part of **Sanofi** and the chemicals divisions become part of **Solvay group** and **Bayer Crop Science.**

Tekmira Pharmaceuticals Corporation changes its name to **Arbutus Biopharma,** and will be instrumental in the development of LNP technologies for COVID-19 vaccines.

Harrisvaccines is acquired by and becomes a subsidiary of **Merck Animal Health.**

Merck KGaA acquires **SigmaAldrich.**

Cilag AG's market presence is associated with the image of the **Janssen** trademark.

Nature's Toolbox, Inc. (NTx) is founded in the US as a biotechnology company that provides a proprietary biological manufacturing (bioreactor systems) and development platform, to produce proteins (notably enzymes) and RNA.

Stallergenes and **Greer Laboratories** merge to become **Stallergenes Greer**, now based in Switzerland, creating a "world leader in allergy immunotherapy."

LimmaTech Biologics AG begins operations in Zurich, Switzerland, with the mission to develop "innovative vaccines against antimicrobial resistant pathogens," leveraging its proprietary self-adjuvanting and multi-antigen vaccine platform.

VBI VACCINES

Variation Biotechnologies is acquired by Israeli-US **SciVac Therapeutics** for US$ 92 million and keeps the name **VBI Vaccines** with its flagship Sci-B-Vac™ hepatitis B vaccine.

Gritstone Oncology Inc., later named **Gritstone Bio**, is founded in California as a "clinical-stage biotechnology company that aims to develop the world's most potent vaccines for cancer and infectious disease." The company files for bankruptcy in 2024.

Alpine Immune Sciences is founded in Seattle, USA as a "clinical stage bio-pharmaceutical company dedicated to discovering and developing innovative, protein-based immunotherapies" for inflammatory and autoimmune diseases and cancer.

2016 **Boehringer Ingelheim** announces an asset-swap deal with **Sanofi**, selling its consumer health division (valued at € 6.7 billion) and € 4.7 billion in cash, while acquiring **Merial** animal health division (valued at € 11.4 billion/US$ 12.4 billion). **Boehringer** becomes one of the global leaders in animal healthcare products.

Vaccitech plc is a biotechnology company, founded by Sarah Gilbert and Adrian Hill at The Jenner Institute, University of Oxford, as a University spin-off. It develops vaccines and immunotherapies for infectious diseases and cancer, such as hepatitis B, HPV, and prostate cancer. In 2020, Vaccitech co-invents a COVID-19 vaccine using the ChAdOx (chimpanzee-adenoviral vector) platform.

European joint venture **Sanofi Pasteur MSD** is terminated. **Sanofi Pasteur** acquires a range of assets including 36 vaccines and three pipelines products owned by SPMSD. Simultaneously, **MSD** will control the SPMSD entity as well as ten other vaccines and the tangible and intangible assets related to the distribution of those vaccines.

UK-based biotech company **Emergex Vaccines Ltd**. is established and will conduct its COVID vaccine trials in 2022 in collaboration with medical research center Unisanté in Lausanne, Switzerland.

Boehringer Ingelheim sells its US pet vaccines business and a manufacturing plant for US$885 million to **Elanco Animal Health**.

2017

SpyBiotech is founded by Jenner Institute vaccinologists Sumi Biswas, Simon Draper, and Jing Jin, with £5 million in seed funding from Oxford Sciences Innovation and GV (formerly Google Ventures). A clinical stage biotechnology company spun out from the University of Oxford, SpyBiotech has the exclusive rights to license the University's proprietary SpyTag/SpyCatcher technology—a protein superglue that binds antigens to vaccine delivery platforms.

ICOSAVAX

Icosavax Inc., is founded in Seattle by University of Washington researchers Neil King and David Baker, who design a general computational method to manufacture virus-like particles (VLPs) displaying complex antigens. The biotech company specializes in the production of self-assembling proteins and VLP platforms for vaccine delivery—primarily based on hepatitis B and HPV. This technology will be applied to the development of vaccines for respiratory syncytial virus (RSV), human metapneumovirus (hMPV), and SARS-CoV-2.

Sanofi acquires **Protein Sciences Corporation**, notably its recombinant FluBlok® Quadrivalent vaccine, in a deal worth up to US$750 million.

BIKEN

BIKEN foundation spins off its production division and form **BIKEN Co. Ltd**., a joint venture with the **Mitsubishi Tanabe Pharma Corporation**, one of Japan's oldest and biggest pharmaceutical corporations.

2018 **Sanofi** acquires **Ablynx** for US$4.8 billion who has considerable SARS patent holdings.

ModeRNA Therapeutics name changes to **Moderna** and has still no commercialized product. Until 2019, Moderna accumulates losses of US$1.5 billion, with a deficit of US$514 million in 2019 and are on the brink of bankruptcy prior to the COVID-19 pandemic.

SK bioscience is spun off from **SK Chemicals** as a subsidiary—a member of **SK Group**—and specializes in biopharmaceuticals and vaccines.

Bayer acquires **Monsanto** for US$60 billion.

Emergent BioSolutions acquires US-based company **PaxVax** for US$270 million. With this it obtains typhoid vaccine Vivotif®, a cholera vaccine Vaxchora® and an Adenovirus 4/7 vaccine candidate for the military under contract with the US DoD.

2019 **Elanco Animal Health** announces it has entered into an agreement with **Bayer AG** to acquire its animal health business in a transaction valued at US$7.6 billion.

Bavarian Nordic acquires two vaccines from the traveler portfolio of **GSK** Vaccines: Rabipur/RabAvert® and Encepur®.

Takeda completes its acquisition of **Shire plc** for US$ 62 billion, becoming a global R&D-driven biopharmaceutical leader with a presence in 80 countries and regions.

ViroVax, LLC is founded in Kansas, USA, as a biotechnology company specializing in research and development of vaccines, adjuvants (Allhydroxiquim), and therapeutics for infectious diseases, including Zika, West Nile, and Dengue.

Pfizer obtains cancer drugs through its US$ 11.4 billion acquisition of targeted oncology specialist **Array BioPharma.**

Chiron-Behring Vaccines is acquired by Indian **Bharat Biotech** from GSK's Asian subsidiary, becoming the world's biggest rabies vaccine producer. By 2024, Bharat Biotech manufactures fourteen different vaccines (Rotavac®, Biohib®, Biopolio®, Comvac®, HINVAC®, Indirab®, JENVAC®, Revac-B® and Typbar®) and is the first manufacturer to develop indigenous Indian COVID-19 vaccines, COVAXIN® and nasal drop vaccine iNCOVACC®.

GSK and **Pfizer** merge their consumer healthcare businesses, named Haleon in 2022, with GSK owning just over two thirds of the joint venture.

2020

Mylan merges with **Upjohn,** Pfizer's off-patent medicine division, to form **Viatris,** a new global healthcare company.

Pfizer partners with German company **BioNTech** to develop and manufacture mRNA-based COVID-19 biologicals.

SpyBiotech and **Serum Institute of India** sign an exclusive global licensing agreement for the development of a novel virus-like particle (VLP) vaccine targeting COVID-19.

VaxEquity is founded by Imperial College London Professor Robin Shattock, and specialist life sciences capital investor Morningside, to exploit VaxEquity's proprietary self-amplifying RNA (sa-mRNA) vaccine platform. In 2021, the company announces a strategic partnership with **AstraZeneca** worth US$ 95 million and in 2023, secures a £700 million grant from Innovate UK to fund a collaborative project advancing RNA vaccines and therapeutics.

 Serum Life Sciences

Serum Life Sciences Ltd., first known as Covicure Vaccines Private Ltd, is incorporated as a subsidiary of **Serum Institute of India** of the Cyrus Poonawalla group, to respond to the SARS-CoV-2 pandemic. Its main product is the COVID-19 vaccine Covishield®—manufactured in partnership with AstraZeneca. SLS also becomes one of the main UNICEF suppliers of hexavalent DTwP-HepB-IPV-Hib vaccines.

2021

Genvax Technologies, a US startup, is founded by Dr. Harry Harris and his son, Joel Harris—who sold their company Harrisvaccines to Merck Animal Health—bringing advances in self-amplifying mRNA (sa-mRNA) and nanoparticle vaccines for animal vaccines.

Novo Nordisk announces it will acquire **Dicerna Pharmaceuticals** and its RNAi therapeutics for US$ 3.3 billion.

2022   **HALEON**

Haleon is established as a corporate spin-off from **GSK** who own 13%, with **Pfizer** owning 32% of the consumer healthcare business.

Johnson & Johnson acquires biotech company **Momenta Pharmaceuticals**—that specializes in drugs for autoimmune disorders—for US$ 6.5 billion.

The Mount Sinai Health System launches **CastleVax, Inc**., a clinical-stage vaccine research and development company devoted to the commercial development of Newcastle disease viral vector mucosal vaccines against COVID-19, RSV + HMPV, and norovirus.

GSK purchases Boston-based biotech company **Affinivax** with an upfront payment of US$ 2.1 billion to obtain its pneumococcal multiple antigen presenting system (MAPS) vaccine technology.

Merck acquires immune-focused **Prometheus Biosciences** (Est. 1995) for US$ 10.8 billion, scoring five clinical and pre-clinical candidates for inflammatory bowel and immune-mediated diseases.

2023

VLP Therapeutics Japan, Inc., BIKEN Foundation and **Denka Company Limited** together announce the conclusion of a collaborative research agreement to develop a seasonal influenza vaccine using self-amplifying "replicon" RNA technology.

Pfizer acquires and merges with **Seagen**, a global biotechnology company that discovers, develops and commercializes transformative cancer medicines, for a total value of US$ 43 billion.

2024 Johnson&Johnson

In a deal worth approximately US$ 2 billion, **Johnson & Johnson** acquires **Ambrx Biopharma Inc**., a "clinical-stage bio-pharmaceutical company with a proprietary synthetic biology technology platform to design and develop next-generation antibody drug conjugates" for cancer treatments.

Johnson & Johnson pays over US$ 13 billion to acquire medical device company **Shockwave Medical**—expanding J&J's portfolio of coronary and peripheral artery disease treatments, "two fast growing segments of the healthcare market."

In a deal worth US$ 1.1 billion, **AstraZeneca** acquires the Seattle-based biotech company, **Icosavax**, that specializes in developing virus-like-particles (VLPs).

AbbVie finalizes its acquisition of **Cerevel Therapeutics** for approximately US$ 8.7 billion, in a move to expand its portfolio of novel therapies targeting mental illnesses and neurological disorders (such as epilepsy, Alzheimer's disease and dementia).

Vertex Pharmaceuticals acquires the Seattle-based biotech company, **Alpine Immune Sciences** for US$ 4.9 billion, gaining access to its treatment for an autoimmune disease of the kidney, and its "proprietary 'directed evolution' platform."

Information has been sourced from scientific articles, press releases, company websites, SEC filings, books, pharmaceutical trade journals, the Wayback Machine, and Wikipedia.
References and source materials are available on www.TheUltimateVaccineTimeline.com

The Big Four Vaccine Manufacturers
Summary of Markets & Mergers

The market information on the four pharmaceutical companies presented below is sourced from Fortune Business Insights.
https://www.fortunebusinessinsights.com/blog/top-5-companies-in-the-vaccines-market-2020-10344

GlaxoSmithKline (UK)

"GSK is the largest player in the global vaccines market, dominating the market with a 20% revenue share and generating **US$9.1 billion in sales in 2019.** The company is known for its vaccines against meningococcal disease, polio, hepatitis B, DTaP (diphtheria, tetanus, and acellular pertussis), and selling 701 million doses globally in 2019. GSK has partnered with Sanofi to develop a protein-based vaccine candidate, which completed its Phase 2 trials in 2019 and will enter Phase 3 trials towards 2020. It also has 15 pipeline candidates for Malaria, COPD, Shigella Diarrhea, and Respiratory Syncytial Virus. However, the coronavirus pandemic led to an 8% decline in its vaccine sales in the first nine months of 2020."

GSK Vaccines in 2023
RSV (Arexvy®); DTaP (Infanrix®); DTaP+IPV (Kinrix®); DTaP+Hepatitis B+IPV (Pediarix®); Hepatitis A (Havrix®); Hepatitis B (Engerix-B®); Hepatitis A+Hepatitis B (Twinrix®); Hib (Hiberix®); Hib; Influenza (Fluarix® Quadrivalent and FluLaval® Quadrivalent); Meningococcal-MCV4 (Menveo®); Meningococcal serogroup B vaccine (Bexsero®); Rabies (RabAvert®); Rotavirus (Rotarix®); Pneumococcal-PCV10 (Synflorix®/Streptorix®); Tdap (Boostrix®); Zoster vaccine recombinant adjuvanted (Shingrix®).

The brand names of GSK vaccines developed in Belgium always end with -RIX.

Pfizer (USA)

"This pharmaceutical behemoth holds a 14% share in the global vaccines market, with the company bringing **US$6.5 billion in sales revenues in 2019.** The company's mRNA-based COVID-19 vaccine, which is developed in collaboration with BioNTech SE, was the first to get approved by the UK's Medical and Pharmaceutical Regulatory Agency (MHRA). The first doses were administered in the UK on December 8, 2020. Pfizer also boasts late-stage vaccine candidates for Respiratory Syncytial Virus, Meningococcal, Pneumococcal, and *Clostridioides difficile.*"

Pfizer Vaccines in 2023
RSV (Abrysvo®); COVID-19 vaccine (Comirnaty®); Meningococcal serogroups A,B,C,W,Y vaccine (Penbraya®); Meningococcal serogroup B vaccine (Trumenba®); Pneumococcal-PCV13 (Prevnar®13); Pneumococcal-PCV20 (Prevnar®20); Tick-borne encephalitis vaccine (Ticovac®).

The Big Four Vaccine Manufacturers
Summary of Markets & Mergers

Merck Inc. (USA)

"Accounting for an 18% share, Merck & Co. is the second-largest player in the market, generating revenues worth **US$ 8.4 billion in 2019**. The company's vaccines against rotavirus, varicella, pneumococcal, and HPV dominated its sales in 2019, with the company selling 190 million vaccine doses globally. Merck has teamed up with the International AIDS Vaccine Initiative to develop a vaccine for COVID-19 that leverages the recombinant Vesicular Stomatitis Virus (rVSV) platform. The clinical trials for the vaccine were supposed to commence in the latter half of 2020. It also acquired Themis, a research-based pharmaceutical company, to formulate another coronavirus vaccine using a measles virus vector platform. Moreover, the company also features two late-stage vaccines against cytomegalovirus and pneumococcal."

logo for Merck KGaA

Not to be confused with German company, Merck KGaA (Germany) who has the right to the Merck name everywhere except Canada and the US, where Merck & Co., Inc., holds the rights to the Merck name.
Outside the US / Canada, Merck & Co is known as MSD, while Merck KGaA inside the US / Canada goes by EMD Group.

Merck Vaccines in 2023
Ebola Zaire Vaccine, Live (ERVEBO®); Hib (PedvaxHIB®); Hepatitis A (Vaqta®); Hepatitis B (Recombivax-HB®); HPV (Gardasil®9); Measles, Mumps and Rubella (MMR-II®); MMR+Varicella (ProQuad®); Pneumococcal-PCV15 (Vaxneuvance®); Pneumococcal-PPV23 (Pneumovax®23); Rotavirus (RotaTeq®); Varicella (Varivax®); Zoster (Zostavax®*); BCG Vaccine U.S.P. (TICE-BCG®).

Zostavax was discontinued in the USA in 2020, in the UK and Australia at the end of 2023. It remains available in the EU and Switzerland.

Sanofi (FR)

"Sanofi is a major player in the market, occupying the fourth position with a 13% share in the global market. The company has earned its reputation as a pharmaceutical bigwig due to its role as a leading supplier of vaccines against DTaP, Polio, *Haemophilus Influenzae* Type B, and Influenza. These also accounted for a large portion of the company's **US$ 6.3 billion vaccine sales in 2019**. Sanofi has joined forces with GSK to create an adjuvant recombinant protein-based vaccine for COVID-19. Additionally, the company is also collaborating with the US-based Translate Bio to develop an mRNA-based coronavirus vaccine."

Sanofi Vaccines in 2023
DTaP (Daptacel®); DTaP+Hib+IPV (Pentacel®); DTaP+IPV (Quadracel®); DT (pediatric); Hib (ActHIB®); Influenza (Fluzone® Quadrivalent); Influenza (Flublok® Quadrivalent); Influenza (Fluzone® High-Dose); Meningococcal-MCV4 (Menactra®); Poliovirus, inactivated (IPOL®); Rabies (Imovax®); Td (TENIVAC®); Tdap (Adacel®); Typhoid Vi, inactivated, injectable (TYPHIM Vi®); Yellow Fever (YF-Vax®).

The Big Four Vaccine Manufacturers
USA Violation Tracker 2000–2024

GSK
United Kingdom

Total fines paid in US dollars
US$ 9,572,803,406

Number of records
50

Top 5 Offense Groups	Penalty Total US$	No. of records
Financial offenses	3,400,000,000	1
Healthcare-related offenses	3,019,361,860	5
Safety-related offenses	1,782,642,648	9
Competition-related offenses	1,056,270,000	17
Government contracting-related offenses	313,656,216	11

Top 5 Primary Offense Types	Penalty Total US$	No. of records
Tax violations	3,400,000,000	1
Off-label/unapproved promotion of medical products	3,019,361,860	5
Drug or medical equipment safety violation	1,782,583,148	8
Price-fixing or anti-competitive practices	931,270,000	15
False Claims Act and related	313,656,216	11

Total Revenues 2023
US$ 38.4 billion

Data source accessed 2024-06-28
https://web.archive.org/web/20240628142145/https://violationtracker.goodjobsfirst.org/parent/glaxosmithkline

THESE COMPANIES ARE ENTRUSTED WITH THE DEVELOPMENT AND SAFETY OF VACCINES

No company executive has ever been jailed for the listed offenses

PFIZER
USA

Total fines paid in US dollars
US$ 11,190,883,623

Number of records
105

Top 5 Offense Groups	Penalty Total US$	No. of records
Safety-related offenses	5,637,021,255	16
Healthcare-related offenses	3,373,675,000	10
Government contracting-related offenses	1,148,191,225	20
Competition-related offenses	1,017,881,952	19
Employment-related offenses	6,335,164	6

Top 5 Primary Offense Types	Penalty Total US$	No. of records
Drug or medical equipment safety violation	5,636,840,000	9
Off-label / unapproved promotion of medical products	3,373,675,000	10
False Claims Act and related	1,148,191,225	20
Price-fixing or anti-competitive practices	922,965,384	13
Foreign Corrupt Practices Act	60,216,568	3

Total Revenues 2023
US$ 58.5 billion

Data source accessed 2024-10-25
violationtracker.goodjobsfirst.org/parent/pfizer

The Big Four Vaccine Manufacturers
USA Violation Tracker 2000–2024

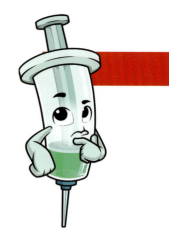

MERCK
USA

Total penalties paid in US dollars
US$10,710,400,031

Number of records
89

Top 5 Offense Groups	Penalty Total US$	No. of records
Safety-related offenses	5,427,585,000	8
Financial offenses	2,301,000,000	2
Healthcare-related offenses	1,391,000,000	6
Government contracting-related offenses	796,733,333	14
Competition-related offenses	538,000,000	7

Top 5 Primary Offense Types	Penalty Total US$	No. of records
Drug or medical equipment safety violation	5,427,450,000	5
Tax violation	2,300,000,000	1
Off-label / unapproved promotion of medical products	1,391,000,000	6
False Claims Act and related	796,733,333	14
Kickbacks and bribery	345,000,000	2

Total Revenues 2023
US$60.1 billion

Data source accessed 2024-10-25
violationtracker.goodjobsfirst.org/parent/merck

THESE COMPANIES ARE RELIED UPON FOR PROVIDING VACCINE SAFETY SCIENCE

Regulatory authorities only receive clinical trial data supplied by manufacturers

SANOFI
France

Total penalties paid in US dollars
US$ 1,348,822,885

Number of records
38

Top 5 Offense Groups	Penalty Total US$	No. of records
Safety-related offenses	513,794,039	4
Government contracting-related offenses	489,458,415	15
Competition-related offenses	328,497,768	14
Employment-related offenses	15,360,000	1
Consumer protection-related offenses	1,600,000	1

Top 5 Primary Offense Types	Penalty Total US$	No. of records
False Claims Act and related	489,458,415	15
Product safety violation	458,006,000	1
Price fixing or anti-competitive practices	302,680,768	12
Drug or medical equipment safety violation	55,781,039	2
Foreign Corrupt Practices Act	25,200,000	1

Total Revenues 2023
US$ 46.6 billion

Data source accessed 2024-10-25
violationtracker.goodjobsfirst.org/parent/sanofi

Top 10 Pharmaceutical Companies by Revenue
2022

		US$
1.	Pfizer	100,330,000,000
2.	Johnson&Johnson	94,940,000,000
3.	Roche	66,260,000,000
4.	MERCK	59,280,000,000
5.	abbvie	58,050,000,000
6.	NOVARTIS	50,540,000,000
7.	Bristol-Myers Squibb	46,160,000,000
8.	sanofi	45,220,000,000
9.	AstraZeneca	44,350,000,000
10.	GSK	36,150,000,000

https://www.fiercepharma.com/pharma/top-20-pharma-companies-2022-revenue

PFIZER IS THE FIRST PHARMACEUTICAL COMPANY TO SURPASS US$100 BILLION IN ANNUAL SALES

SERUM INSTITUTE OF INDIA

Although Serum Institute of India is not in the top ten pharmaceutical companies by revenue, it is important to note that SII is the largest manufacturer of vaccines by doses distributed globally.

Since the mid-1960s, SII has developed many vaccines including tetanus toxoid (TT), diphtheria-tetanus toxoid (DT), tetanus reduced dose diphtheria toxoid (Td), diphtheria-tetanus-pertussis (DTP), measles, rubella, measles-rubella (MR), measles-mumps-rubella (MMR), bacille Calmette-Guérin (BCG), rabies, hepatitis B (HepB), *Haemophilus influenzae* type B (Hib), DTP-HepB, DTP-Hib, DTP-HepB-Hib, meningococcal A conjugate (MenAfriVac), pandemic influenza H1N1, trivalent live attenuated seasonal influenza, rotavirus, inactivated polio (IPV), trivalent oral polio (tOPV), and bivalent oral polio (bOPV).

SII produced the very first prequalified vaccine by any manufacturer in a developing country. Its measles vaccine was prequalified by the WHO in 1992, followed by its DTP vaccine in the subsequent year. Since then, a total of twenty-seven vaccines by SII—the maximum by any manufacturer—have been prequalified by the WHO, including TT, DT, Td, DTP, HepB, Hib, DTP-HepB, DTP-Hib, DTP-HepB-Hib, measles, rubella, MR, MMR, BCG, pandemic H1N1, trivalent seasonal influenza, MenAfriVac, IPV, OPV and malaria.

Since 1992, SII has been providing large quantities of vaccines to UNICEF and PAHO (since 2000 purchased by Gavi) for distribution in their respective regions, which cover Eastern Europe, Asia, Africa and South America. **By 2024, SII is the largest vaccine supplier for UNICEF.**

Every year in India, there are about 30 million pregnant women with around 27 million babies born. With its current population of 1.44 billion people and the most births, India is the largest vaccine market.
Take note that most DTP vaccines given in LMICs still contain the whole-cell pertussis toxin, which in Western countries has been withdrawn from the market and replaced with acellular pertussis following legitimate safety concerns. However, the WHO still supports using cheaper whole-cell pertussis in Africa, Asia, the Middle East and South America. Thimerosal is also still being used as a preservative in "trace amounts" in single-dose vaccines given to babies and pregnant women despite being phased out in the West.

Global Vaccine Industry Revenues
Impressive Growth Since 1982

1982
The US federal price paid to vaccinate each child is under US$70

1982
US$170,000,000

1982

Vaccines on the market
Diphtheria
Tetanus
Pertussis
Polio
Measles
Mumps
Rubella
Influenza

2020
The US federal price paid
to vaccinate each child
is over US$2,000

2000
US$5,000,000,000

2020
US$60,000,000,000

Expected to rise to US$150 billion by 2030

REVENUES GENERATED IN 2021
From COVID-19 vaccines:
- **Pfizer-BioNTech's Comirnaty®** BNT162b2 generated US$81.3 billion
- **Moderna's Spikevax®** mRNA-1273 generated US$12.2 billion

REVENUES GENERATED IN 2019
In terms of revenue, the five leading vaccine products were:
- **Pfizer's Prevnar-13®** pneumococcal conjugate vaccine generated US$5.95 billion
- **Merck's Gardasil®** HPV (human papillomavirus) vaccine generated US$3.7 billion
- **GSK's Shingrix®** shingles vaccine generated US$2.3 billion
- **Sanofi's Pentacel®** diphtheria, tetanus, pertussis, polio and *Haemophilus influenzae* type B, 5-in-1 vaccine generated US$2.2 billion
- **Sanofi's Fluzone®** influenza vaccine generated US$2.1 billion

"Vaccination traverses and tramples upon all these safeguards and wisdoms; it goes direct to the blood, or, still worse, to the lymph, and not with food; it puts poison, introduced by puncture, and that has no test applicable to it, and can have no character given to it but that it is fivefold animal and human poison, at a blow in the very centre, thus otherwise guarded by nature in the providence of God. This is blood assassination, and like a murderer's life. The point, however, here is that this amazing act is the homicidal insanity of a whole profession."

—Dr. James Garth Wilkinson (1812–1899)

English medical doctor and homeopathic practitioner and author of
On Human Science, Good and Evil, and Its Works – Lippincott, Philadelphia, 1876.

"Vaccination is the most outrageous insult that can be offered to any pureminded man or woman. It is the boldest and most impious attempt to mar the works of God that has been attempted for ages. The stupid blunder of doctorcraft has wrought all the evil that it ought, and it is time that free American citizens arise in their might and blot out the whole blood poisoning business."

—Dr. James Martin Peebles, MD, MA, PhD (1822–1922)

American physician and author of *Vaccination: A Curse and Menace to Personal Liberty,* among many other books.

*"Vaccination is a gigantic delusion. It has never saved a single life.
It has been the cause of so much disease, so many deaths,
such a vast amount of utterly needless and altogether undeserved suffering,
that it will be classed by the coming generation among the greatest errors
of an ignorant and prejudiced age, and its penal enforcement
the foulest blot."*

—Alfred Russel Wallace (1823–1913)

British naturalist, explorer, geographer, anthropologist, biologist, and illustrator, who is best known for independently conceiving the theory of evolution through natural selection; his paper on the subject was jointly published with some of Charles Darwin's writings in 1858.

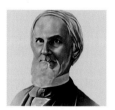

*"There is among profounder thinkers and observers growing conviction that
vaccination, so far from being benefit to mankind, is itself utterly useless
as preventive, irrational and unscientific in theory, and actually
the means of disseminating disease afresh where it is performed."*

—Dr. Alexander Wilder, MD (1823–1908)

American physician, professor of pathology in the US Medical College of New York,
author of the 1899 *Vaccination: A Medical Fallacy,* and medical editor
of *The New York Medical Tribune* in 1879.

"I fear that in some instances wholesale vaccinations and revaccinations at the commencement of an epidemic have spread small-pox among those who remain unvaccinated. At least it happened curiously enough that, in the best vaccinated districts of Trinidad, there was the most small-pox."

—Dr. Robert H. Bakewell, MD, MRCS (1831–1908)

Member of the Royal College of Surgeons and fellow of the Royal Medical and Chirurgical Society, vaccinator general and health officer of shipping for colony of Trinidad, and medical attendant of the Small-pox Hospital, Port of Spain.

"We are so constituted that we can never receive other proteins into the blood than those that have been modified by digestive juices. Every time alien protein penetrates by effraction, the organism suffers and becomes resistant..."

—Dr. Charles Richet (1850–1935)

French physiologist known for his pioneering work in immunology. He was awarded the Nobel Prize in Physiology or Medicine in 1913 for his work on anaphylaxis—allergic reactions.

"During the last considerable epidemic at the turn of the century, I was a member of the Health Committee of London Borough Council, and I learned how the credit of vaccination is kept up statistically by diagnosing all the revaccinated cases (of smallpox) as pustular eczema, varioloid or what not—except smallpox."

—George Bernard Shaw (1856–1950)

Irish comic dramatist, literary critic, political activist, and socialist propagandist. He was the winner of the Nobel Prize for Literature in 1925, which he initially refused stating, "I can forgive Nobel for inventing dynamite, but only a fiend in human form could have invented the Nobel Prize." He would later accept the award itself, but not the prize money.

"Vaccination is a barbarous practice and one of the most fatal of all the delusions current in our time. Conscientious objectors to vaccination should stand alone, if need be, against the whole world, in defense of their conviction. The fact of the matter is that it is only the self-interest of doctors that stands in the way of the abolition of this inhuman practice, for the fear of losing the large incomes that they at present derive from this source blinds them to the countless evils which it brings."

—Mahatma Gandhi (1869–1948)

Indian lawyer, anti-colonial nationalist, and political ethicist who employed nonviolent resistance to lead the successful campaign for India's independence from British rule and in turn inspired movements for civil rights and freedom.

"Dr. Thomas Francis did not mention in his key evaluation of the 1954 Salk field trials that those who contracted polio after their first inoculation and before their second inoculation were placed in the 'not inoculated'* list."

—Maurice Beddow Bayly (1887–1961)

English physician, he was a member of the National Anti-Vaccination League, the Animal Defense and Anti-Vivisection Society, and the English section of the Theosophical Society, as well as the author of several publications including *The Story Of The Salk Anti-Poliomyelitis Vaccine* in 1956.

*This is the same technique that has been used with COVID-19 injections. People injected and diagnosed with COVID-19 within 14 days of their first or second shot have been labeled "unvaccinated."

"Without propaganda there can, of course, be no large scale immunization, but how perilous it is to mix up propaganda with scientific fact. If we baldly told the whole truth it is doubtful whether the public would submit to immunization."

—Dr. Charles C. Okell, MRCP (1888–1939)

British professor of bacteriology at University College Hospital Medical School and editor of the *Journal of Hygiene* and assistant editor of *The Journal of Pathology and Bacteriology.* He was awarded the Military Cross following his wartime positions in France, Palestine, and Egypt. He worked for some time on the staff of the Wellcome Physiological Research Laboratories at Beckenham and became head of the bacteriological department in 1924.

"I have come to the conclusion that no vaccine or antiserum can be regarded as completely safe. Some are very much safer than others, but no vaccine or antiserum that has been used has been free from complications or accidents of one sort or another."

—Sir Graham S. Wilson (1895–1987)

British medical physican and co-writer of *Topley and Wilson's Principles of Bacteriology and Immunity.* He was a devout supporter of vaccination despite being aware of the risks involved.

The quote above is taken from a 1966 lecture he delivered at the London School of Hygiene & Tropical Medicine on *The Hazards of Immunization* that was expanded upon and released as a book with the same title.

"There is a great deal of evidence to prove that immunization of children does more harm than good."

—Dr. J. Anthony Morris (1919-2014)

US Chief Vaccine Control Officer, bacteriologist and research virologist at the National Institutes for Health and Federal Drug Administration 1940–1976.

"It is sad, but true, that the vaccination temple guardians and the industrial lobby supporting them do not accept criticism, refuse to carry out necessary investigations (e.g. on long-term vaccine safety), persecute researchers in vaccine safety science, and push media to make black and white dichotomy between 'normal' pro-vaccination and 'dangerous/irresponsible' anti-vaccination positions."

—Prof. Romain K. Gherardi

Former head of the Expert Neuromuscular Pathology Center, Assistance Publique-Hôpitaux de Paris, Henri Mondor Paris-Est University hospitals, Créteil, France.

"The best vaccination is to get infected yourself."

—Dr. Anthony Fauci

American physician-scientist and immunologist who served as the director of the National Institute of Allergy and Infectious Diseases (NIAID) from 1984–2022 and Chief Medical Advisor to the President (Trump, Biden). He was the highest paid civil servant and due to his position at the National Institutes of Health (NIH) was instrumental in the allocation of billions of dollars in government grant money to scientific research.

"Studies are increasingly pointing to the conclusion that vaccines represent a dangerous assault to the immune system leading to autoimmune diseases like multiple sclerosis, lupus, juvenile onset diabetes, fibromyalgia, and cystic fibrosis, as well as previously rare disorders like brain cancer, SIDS (sudden infant death syndrome), childhood leukemia, autism, and asthma."

—Dr. Zoltan P. Rona, MD, MSc

Canadian medical doctor who practices complementary and integrative medicine and is an expert in nutritional biochemistry and clinical nutrition.

"It should be of concern that the effect of routine vaccinations on all-cause mortality was not tested in randomized trials. All currently available evidence suggests that DTP vaccine may kill more children from other causes than it saves from diphtheria, tetanus or pertussis. Though a vaccine protects children against the target disease it may simultaneously increase susceptibility to unrelated infections."

—Dr. Peter Aaby

Danish anthropologist with a doctoral degree in medicine, he was highly involved in vaccination campaigns in Africa. In 1978, he established the Bandim Health Project, a health and demographic surveillance system site in Guinea-Bissau in West Africa.

"In the same way that during my Microsoft career I talked about the magic of software, I now spend my time talking about the magic of vaccines.

They are the most effective and cost-effective health tool ever invented. I like to say vaccines are a miracle. Just a few doses of vaccine can protect a child from debilitating and deadly diseases for a lifetime.

The benefits of widespread vaccination are mostly explained in terms of the lives vaccines save, and based on that measure alone, vaccines are the best investment to improve the human condition."

—William "Bill" H. Gates II

American business tycoon and founder of Microsoft.

Through the Bill & Melinda Gates Foundation, he has invested over US$10 billion in vaccines from 2000–2020. He announced at the World Economic Forum in 2019 that his investment in vaccines has brought an over 20-to-1 return, yielding some US$200 billion and "saving millions of lives."

"You know, 200,000 Americans needlessly perished because they refused the COVID vaccine, so that this anti-science—and they were victims, basically, of this kind of anti-science aggression. Now, it's a killing force. And that's why we need to care about it—because if we're health care providers or biomedical scientists, you know, it used to be enough just to want to save lives. And now—an added burden is now trying to figure out a way to combat the anti-vaccine, anti-science aggression, because it becomes such a killing force."

—Dr. Peter Hotez

American physician and vaccine developer, he is a highly published Yale and Rockefeller University alumnus, the dean of National School of Tropical Medicine and professor of pediatrics and molecular virology and microbiology at the Baylor College of Medicine, where he is also the co-director of the Texas Children's Center for Vaccine Development (CVD).

"We are also seeing another epidemic—a dangerous epidemic of misinformation. Vaccine is trust. First trust in science."

—Dr. Antonio Guterres

Portuguese socialist politician, diplomat, former Portuguese Prime Minister from 1995–2002, and Secretary-General of the United Nations 2017–2025. Quoted from an official televised statement made in April 2020.

Appendices

Placebo Studies
Prelicensure Clinical Trials

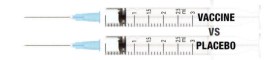

VACCINE
VS
PLACEBO

Placebo comes from the Latin "I shall please."
The first observed "pleasing" medical benefit of the placebo effect was pain relief. While reported in medical literature since the 18th century, it was not until the mid-20th century that the use of placebos in randomized clinical trials (RCTs) became accepted and expected in the scientific community.

Randomized placebo-controlled double-blind clinical trials are considered the gold standard in medical science to ensure the safety of new pharmaceutical products. A placebo is defined as an inert and inactive substance, a "sugar pill" or saline solution. Saline controls are necessary in clinical trials for market approval of drugs. Inert placebos provide a baseline for determining efficacy and evaluating safety and identifying common side effects.

Placebo-controlled clinical trials are not, however, required for vaccine approval. "Controls" can be active substances: other vaccines, adjuvant alone, or all excipients and preservatives without the antigen ("vaccine-sans-antigen").

The exact composition of a control is often not specified, and medical journals do not require disclosure of this information, which may remain proprietary and confidential. There are no policies from regulators governing public availability of details on all the components of the placebo control. This "raises concerns about the reliability and transparency of clinical trials, and undermines the ability to make informed clinical decisions."

The main reasons for not giving inert placebos
in clinical trials for pediatric and adult vaccine are:
- It is **unethical** to withhold a proven therapeutic
- It may be **impractical** in the blinding process

Public health authorities assert that the highest standards of safety are applied to vaccines, classed as biologicals, because they are given to healthy people, mostly babies and children. However, for injected pediatric vaccines currently on the US and European market, there are very few placebo-controlled prelicensure studies using inactive substances to establish a baseline for safety—even for completely new vaccines like Gardasil® and Cervarix®. Though a "saline placebo" is mentioned in the Gardasil® clinical data, the placebo consisted of amorphous aluminum hydroxyphosphate sulfate (AAHS) or vaccine-sans-antigen—including polysorbate 80 and L-histidine. Some influenza vaccine trials report having used saline.

The first highly publicized placebo-controlled trial was for the inactivated polio virus vaccine by Jonas Salk in the mid-1950s. The largest field trial at the time involved millions of children in the USA receiving either the thimerosal-preserved vaccine or "placebo," though the placebo was actually Medium 199; the synthetic nutrient culture used to cultivate the polio virus.

Public health authorities, regulatory agencies, the scientific community, and pharmaceutical companies confidently assert the safety of pediatric vaccines even though they have never been tested against an inert substance. For true safety profiles to be established when comparing results and looking at the statistical significance between two groups, a reliable baseline for adverse reactions is absolutely necessary.

The following pages present a summary of the various vaccines marketed to children with information on the control used during the pivotal prelicensure clinical trials. The data was collected from several sources; manufacturer package inserts, medical literature, ICAN, Dr. Sherry Tenpenny's "Vaccine Bootcamp" course, and the excellent books *Vaccines: A Reappraisal (2017), Ideological Constructs of Vaccination (2018),* and *Turtles All the Way Down: Vaccine Science and Myth (2022).*

Routine Childhood Vaccines in the USA & Europe – Placebo-Controlled Prelicensure Clinical Trials

Diseases targeted	Vaccine	Manufacturer	Control Group	Placebo
Influenza (flu) A/B TIV trivalent inactivated QIV quadrivalent inactivated LAIV live-attenuated	**Afluria® TIV**	Seqirus	Fluzone® TIV	NO
	Afluria® QIV	Seqirus	Fluarix® QIV / Fluzone® QIV	NO
	Fluarix® QIV*	GSK	Controls receive either the mono-, di-, or trivalent versions of Sanofi Pasteur's flu vaccine	NO
	FluLaval® Quadrivalent	ID Biomedical Corporation	Fluzone® QIV—adult study unspecified placebo	NO
	FluLaval® Trivalent	GSK	Fluzone® TIV	NO
	Fluzone® QIV*	Sanofi Pasteur	TIV / LAIV	NO
	Fluzone® TIV*	Sanofi Pasteur	Physiologic saline	YES
	FluMist® LAIV	MedImmune	Saline	YES
	Vaxigrip® Tetra*	Sanofi Aventis	Mutagrip trivalent	NO
	Flucelvax® Quad QIV	Seqirus	Flucelvax® (IIV3) or an investigational vaccine	NO
COVID-19	**Comirnaty®***	Pfizer-BioNTech	Saline placebo group vaccinated after 6 months in Phase III	YES
	Spikevax®*	Moderna	Saline placebo group vaccinated after 6 months in Phase III	YES
Hepatitis A	**Havrix®**	GSK	Engerix-B® vaccine	NO
	Vaqta®	Merck	AAHS and thimerosal	NO
Hepatitis B	**Heptavax-B®**[1]	Merck	Aluminum hydroxide or phosphate, thimerosal and other undisclosed ingredients	NO
	Engerix-B®	GSK	Heptavax-B® plasma-based vaccine	NO
	HB VaxPro® Recombivax-HB®	Merck	No control group	NO
Hemophilus influenzae **type b**	**b-CAPSA-1®**	Praxis	Meningococcal vaccine	NO
	HIB-Immune®	Lederle	Meningococcal vaccine	NO
	HIB-Vax®	Pasteur Mérieux Connaught	Meningococcal vaccine	NO
	ActHIB®	Sanofi	Daptacel®, IPOL® & PCV	NO
	Hiberix®	GSK	ActHIB® + DTaP, HepB, IPV	NO
	PedvaxHIB®	Merck	Lyophilized PedvaxHIB®	NO

also recommended to pregnant women

1) Heptavax-B® – Though not routinely given to children, it was tested on children and recommended to newborns of hepatitis-B+ mothers. It would also serve as the control for Engerix-B® which would become routine for all newborns.

Routine Childhood Vaccines in the USA & Europe – Placebo-Controlled Prelicensure Clinical Trials

Diseases targeted	Vaccine	Manufacturer	Control Group	Placebo
Human papillomavirus HPV	Gardasil®	Merck	AAHS aluminum adjuvant and vaccine-sans-antigen: sodium borate, L-histidine, polysorbate 80	NO
	Gardasil®9	Merck	Gardasil® or saline placebo to subjects previously injected with 3 doses of Gardasil	YES[2]
	Cervarix®	GSK	Aluminum hydroxide adjuvanted hepatitis A vaccine	NO
Meningococcal ACWY	Menactra®	Sanofi	Menomune®	NO
	Menomune®	Sanofi	Menactra®	NO
	Menveo®	GSK	Menomune®, Boostrix®, Menactra® or Mencevax®	NO
Meningococcal B	Bexsero®	GSK	Men-ACWY vaccine, Japanese encephalitis or aluminum placebo	NO
	Trumenba®	Pfizer	Vaccine-sans-antigen: 0.78 mg L-histidine and 50 mcg polysorbate-80	NO
Pneumococcal	Pneumovax-14®	Merck	Aluminum / phenol 0.25%	NO
	Pneumovax-23®	Merck	Phenol 0.25%	NO
	Prevnar 7®	Wyeth	Experimental meningococcal polysaccharide vaccine	NO
	Prevnar 13®	Pfizer	Prevnar®7	NO
Polio	IPV-Salk	Eli Lilly / Parke-Davis	Medium 199	NO
	OPV-Sabin	Pfizer	No control group	NO
	IPOL®	Sanofi	DTP	NO
Respiratory Syncytial Virus RSV	Abrysvo®*	Pfizer	Vaccine-sans-antigen	NO
Rotavirus	RotaShield®	Wyeth-Lederle	Uninfected tissue-culture fluid	NO
	Rotarix®	GSK	Vaccine-sans-antigen	NO
	RotaTeq®	Merck	Probably vaccine-sans-antigen, as placebo components were redacted from FDA documents	NO
Varicella / Chickenpox	Varivax®	Merck	Vaccine-sans-antigen: stabilizer and 45 mg neomycin	NO

2) Saline placebo was only given to those previously vaccinated with Gardasil®. It is important to note that Merck actually wrote "saline placebo" in the 796-page published original Gardasil® trial clinical data, however the "saline" was actually the vaccine-sans-antigen.

Routine Childhood Vaccines in the USA & Europe – Placebo-Controlled Prelicensure Clinical Trials

Diseases targeted	Vaccine	Manufacturer	Control Group	Placebo
COMBINATION VACCINES				
DTwP/DTP	Legacy vaccine that has never been tested in a RCT with placebo control, though it is still administered in LMICs to babies and even pregnant women.			**NO**
Tdap	**Boostrix®***	GSK	Decavac® or Adacel®	**NO**
	Adacel®*	Sanofi	Td (for adult use)	**NO**
DTaP	**Infanrix®**	GSK	DTP	**NO**
	Daptacel®	Sanofi	DT or DTP	**NO**
DTaP-IPV-HepB-Hib	**Hexyon®**	Sanofi Pasteur	Infanrix Hexa®	**NO**
	Vaxelis®	Sanofi / Merck	Infanrix Hexa®, Pentacel®, Recombivax-HB®	**NO**
	Infanrix Hexa®	GSK	Individual vaccines (Kinrix®, Engerix-B®)	**NO**
DTaP-IPV-HepB	**Pediarix®**	GSK	ActHIB®, Engerix-B®, Infanrix®, IPV and OPV	**NO**
DTaP-IPV-Hib	**Pentacel®**	Sanofi	HCPDT[3)], Poliovax®, ActHIB®, Daptacel® and IPOL®	**NO**
DTaP-IPV	**Quadracel®**	Sanofi	Daptacel® and IPOL®	**NO**
	Kinrix®	GSK	Infanrix® and IPOL®	**NO**
Hepatitis A + B	**Twinrix®**	GSK	Individual hepatitis A and hepatitis B vaccines	**NO**
Measles Mumps Rubella MMR	**MMR-I®** **MMR-II®**	Merck	Individual vaccines, Attenuvax®, Mumpsvax®, Meruvax®	**NO**
	M-M-R VaxPro®	Merck	MMR-II® / ProQuad®	**NO**
	Priorix®	GSK	Individual vaccines	**NO**
Measles Mumps Rubella Varicella MMRV	**ProQuad®**	Merck	MMR and Varicella	**NO**

also recommended to pregnant women

3) This study included HCPDT (hybrid component pertussis vaccine combined with diphtheria and tetanus toxoids adsorbed—not licensed in the USA), a vaccine made of the same components as Daptacel, but containing twice the amount of detoxified pertussis toxin (PT) and four times the amount of filamentous hemagglutinin (FHA)—20 mcg detoxified PT and 20 mcg FHA. "Hypotonic-hyporesponsive episode (HHE) was observed following 29 (0.047%) of 61,220 doses of HCPDT; 16 (0.026%) of 61,219 doses of an acellular pertussis vaccine made by another manufacturer; and 34 (0.056%) of 60,792 doses of a whole-cell pertussis DTP vaccine. There were 4 additional cases of HHE in other studies using HCPDT vaccine for an overall rate of 33 (0.047%) in 69,525 doses."

Childhood Vaccination Policies

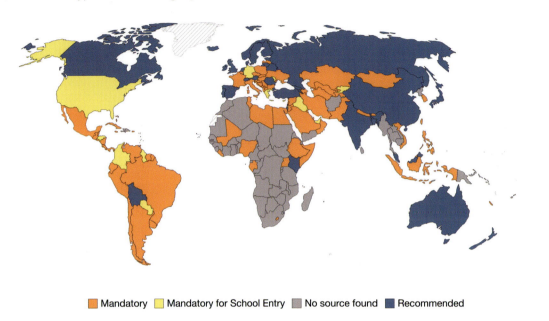

Which countries have mandatory childhood vaccination policies

Countries are mapped based on having requirements or recommendations for at least one vaccine in 2019.

■ Mandatory ■ Mandatory for School Entry ■ No source found ■ Recommended

Data source: Vanderslott & Marks (2021). Charting mandatory childhood vaccination policies worldwide. Vaccine.
Note: Policies can vary at the state level in some countries.
OurWorldInData.org/vaccination | CC BY

There are currently mandatory childhood vaccination programs in Europe:
in France, Italy, Slovakia, Greece, Croatia, Bulgaria, Czech Republic, Hungary, Latvia, Poland,
with polio vaccine being the only mandatory vaccine in Belgium.

https://ourworldindata.org/grapher/mandatory-childhood-vaccination

The information is extracted from a guest post on *Our World In Data* by Tatjana Marks and Samantha
Vanderslott, from the Oxford Vaccine Group and Oxford Martin School, University of Oxford,
Centre for Clinical Vaccinology and Tropical Medicine.

They charted "mandatory childhood vaccine policies worldwide as they are becoming an increasingly important
policy intervention for governments trying to address low vaccination rates."

"The term 'mandatory' and 'mandates' are taken to mean quite different things across countries.
Whilst the term is commonly used, it is poorly defined. Mandates require vaccination for a certain purpose, most
commonly related to school entry for children. While definitional disagreements still persist, it remains important to
better understand what policies are in place across countries and the reasons driving changes in policy over time."

https://ourworldindata.org/childhood-vaccination-policies

Plotkin's Vaccines – Vaccine Pricing Evolution

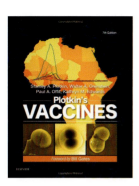

2018
7th Edition

Stanley A. Plotkin
Walter A. Orenstein
Paul A. Offit
Kathryn M. Edwards

Foreword by Bill Gates

According to the world's leading authority on vaccination, Dr. Stanley A. Plotkin, there are over 50 diseases for which vaccines have been developed and used around the globe. To delve deep into the intricate technical details of each vaccine, *Plotkin's Vaccines* is the academic reference bible. While this information is from the 7th edition, an 8th edition was released in 2023.

1. Adenovirus
2. Anthrax
3. Biodefense and Special Pathogens
4. Cancer
5. Cholera
6. Cytomegalovirus
7. Dengue
8. Bacterial Diarrhea
9. Diphtheria Toxoid
10. Ebola
11. Enterovirus 71
12. Epstein-Barr Virus
13. *Haemophilus influenzae* Type b
14. Hepatitis A
15. Hepatitis B
16. Hepatitis C
17. Hepatitis E
18. Herpes Simplex Virus
19. Human Immunodeficiency Virus (HIV)
20. Human Papillomavirus (HPV)
21. Inactived Influenza
22. Live Influenza
23. Japanese Encephalitis
24. Lyme Disease
25. Malaria
26. Measles
27. Meningococcal Capsular Group A, C, W and Y Conjugate
28. Meningococcal Capsular Group B
29. Mumps
30. Norovirus
31. Parasitic Disease
32. Pertussis
33. Plague
34. Pneumococcal Conjugate and Pneumococcal Common Protein
35. Pneumoccocal Polysaccharide
36. Inactivated Poliovirus
37. Live Poliovirus
38. Rabies
39. Respiratory Syncytial Virus (RSV)
40. Rotavirus
41. Rubella
42. Smallpox and Vaccinia
43. *Staphylococcus aureus*
44. Streptococcus Group A
45. Streptococcus Group B
46. Tetanus Toxoid
47. Tickborne Encephalitis
48. Tuberculosis
49. Typhoid Fever
50. Varicella
51. Yellow Fever
52. Zika Virus
53. Zoster

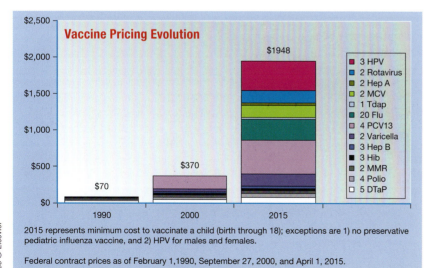

2015 represents minimum cost to vaccinate a child (birth through 18); exceptions are 1) no preservative pediatric influenza vaccine, and 2) HPV for males and females.

Federal contract prices as of February 1,1990, September 27, 2000, and April 1, 2015.

As is clearly shown in the graph taken from the 2018 edition of *Plotkin's Vaccines*, on page 1426, **from 2000 to 2015, the cost to vaccinate each child rose significantly from** US$70 to US$1948.

Figure 73.3. Sums of the prices of the Centers for Disease Control and Prevention (CDC) contract prices for vaccines needed to fully vaccinate a child from birth through 18 years of age. The vaccine contract prices are from February 1, 1990, September 27, 2000, and April 1, 2015. DTaP, diphtheria, pertussis, and acellular pertussis; Hep A, hepatitis A; Hep B, hepatitis B; Hib, *Haemophilus influenzae* type b; HPV, human papillomavirus; MCV, meningococcal conjugate vaccine; MMR, measles, mumps, rubella; PCV, pneumococcal conjugate vaccine; Tdap, adolescent-adult formulation tetanus, diphtheria, and acellular pertussis vaccine.

APPENDIX D

Vaccine Composition

Antigens – Active ingredient
The substance used to induce antibody production, traditionally whole (or parts of) inactivated or live-attenuated viruses, proteins, bacteria, or bacterial toxoids. More recently genetic engineering is used to create antigen-encoding mRNA/DNA.

Adjuvants
Metallic salts like aluminum hydroxide, aluminum phosphate, and Merck's proprietary AAHS and oil-in-water based emulsions, are used to enhance the immune response. LNPs are also recognized as having adjuvant-like inflammatory qualities.

Antibiotics
Antimicrobial agents (neomycin, streptomycin, gentamicin, and sulfates) are used in the production process to prevent contamination by microorganisms in the vaccine.

Excipients
Other "inactive" ingredients that include stabilizers (like sugars, gelatin or polysorbate 80), preservatives, fillers, and buffering agents like sodium chloride or sodium phosphate to balance pH levels.

Cell debris and contaminants
Vaccine antigens are often grown in cell cultures. Remnants of the growth medium, like foreign proteins and residual genetic material, cannot be completely filtered out of the final product.

Water H_2O
The main component of the sterile liquid suspension (the solvent) that serves as a carrier or binder for the vaccine formulation—or the reconstituent diluent for freeze-dried vaccines.

Typical vaccine dose: 0.5 mL–1.0 mL

Trace amounts: ≤ 0.5 mcg per dose

In addition to disease antigens (such as weakened, killed, or parts of viruses or bacteria—or more recently, genetic code), vaccines contain very small amounts of other ingredients, excipients.

Some excipients are added to a vaccine for a specific purpose. These include:
- **Preservatives**, to prevent contamination. *For example, thimerosal/ethylmercury, 2-phenoxyethanol, EDTA.*
- **Adjuvants**, to help stimulate a stronger immune response. *For example, aluminum salts like aluminum phosphate and aluminum hydroxide or Merck's proprietary amorphous aluminum hydroxide sulfate.*
- **Stabilizers**, to keep the vaccine potent during transportation and storage. *For example, sugars, MSG, BSA*, gelatin.*
- **Buffers**, to optimize the pH levels. *For example, phosphate, sodium chloride, potassium chloride, trometamol.*
- **Surfactant**, to keep the viscosity optimal for absorption. *For example, polysorbate 80, sodium deoxycholate.*

Others are residual trace amounts of materials that were used during the manufacturing process, and can include:
- **Cell culture materials**, used to grow the vaccine antigens. *For example, egg protein, kidney cells, fetal bovine serum, various culture media, MRC-5 (cell line derived from an aborted human fetus).*
- **Inactivating ingredients**, used to kill viruses or inactivate toxins. *For example, formaldehyde, ß-propiolactane, glutaraldehyde.*
- **Antibiotics**, used to prevent contamination by bacteria. *For example, neomycin, gentamicin, polymyxin B.*
- **Other manufacturing residue**, that is impractical to remove 100% from vaccine solution.

Excipient text adapted from https://www.cdc.gov/vaccines/pubs/pinkbook/downloads/appendices/B/excipient-table-2.pdf

BSA: bovine serum albumin. Due to potential allergic reactions, the WHO has set a guidance of ≤ 50 ng residual BSA per vaccine dose.

Injected Substances

 Corrosive

 Acute Toxic

 Irritant

 Health Hazard

 Environmental Hazard

Aluminum phosphate/hydroxide *adjuvant*
Known to be inflammatory and neurotoxic, this adjuvant is associated with Alzheimer's disease, dementia, seizures, autoimmune issues and cancer. It disturbs calcium metabolism and other biochemical pathways, increasing oxidative stress. This toxin accumulates in tissues and can pass the blood-brain barrier causing neurological injury. Damage increases with each dose injected, bypassing the kidneys which eliminate ingested aluminum.

Thimerosal / Merthiolate *preservative*
Sodium ethylmercury thiosalicylate
Can cause damage to the brain, nervous system, gut, liver, bone marrow and kidneys, and is linked to autoimmune disorders and neurological conditions. Used in vaccines since the 1930s and removed between 1999 and 2006, it is still in multidose influenza and Td vaccines and in "trace amounts" in meningococcal and JE* vaccines. Thimerosal remains in routine DTP and hepatitis B vaccines given to babies in LMICs.

 JE: Japanese encephalitis

Human and animal cells, and foreign proteins *residue*
Residual genetic material from aborted human fetuses, pigs, sheep, rabbit brains, dog and monkey kidneys, cows, calf serum and insects may be included in vaccines. The current permitted limit for residual genetic materials in vaccines is **<10 ng per dose,** though levels up to 10x higher have been found by independent researchers in Italy and the USA.

Monosodium Glutamate MSG *stabilizer*
Generally recognized as safe (GRAS) by the FDA, this excitotoxin has been linked to migraines, obesity, cardiac issues, developmental delays and infertility, as well as neurodegenerative disorders. Often used in Asian cooking and processed foods, it artificially enhances food flavor by exciting taste neurons, and may be neurotoxic.

Genetically modified yeast, animal, bacterial and viral DNA as well as cellular debris *manufacturing residue*
Can be incorporated into the recipient's DNA via insertional mutagenesis** and unknown genetic mutations may be carried down generations. Such residues may also deleteriously affect the genetic expression of proteins.

Beta-Propiolactone ßPL *inactivating agent*
An alkylating agent that acts as a superior sterilizing agent for biologicals because of its capacity to inactivate a wide variety of bacteria, fungi and viruses. It is a suspected gastrointestinal, liver, nervous, respiratory and skin poison, and may also be a cancer-causing agent.

Glutaraldehyde *inactivating agent*
Is a transparent oily liquid with a pungent odor, used as a disinfectant, preservative, and fixative. Exposure can cause lung irritation, asthma and breathing problems, and is toxic if ingested. It can cause birth defects in animals.

Phenol *preservative*
Chemically known as carbolic acid, phenol has largely replaced thimerosal to prevent microbial contamination in vaccines. It is highly caustic to tissues and toxic if swallowed, inhaled or touched, and may damage organs.

Formaldehyde (formalin) *inactivating agent*
Known to cause cancer in humans. Probable gastro-intestinal, liver, respiratory, immune, nerve and reproductive system poison. In January 2016, formaldehyde was classified as a carcinogenic, mutagenic and reprotoxic chemical (CMR) substance by the EU, a substance that can cause cancer, genetic mutations, or birth defects.

Polysorbate 80 / Tween™ *surfactant/emulsifier*
GRAS as a food additive, this emulsifying agent (surfactant) used to stabilize aqueous formulations, has been linked to cancer in animals and to autoimmune issues as well as infertility. It can bypass the blood-brain barrier and may help facilitate the passage of other ingredients into the brain.

Other: Proprietary ingredients and other undisclosed residual contaminants from the manufacturing process.

Safety standards for thimerosal (ethylmercury) by injection have never been established.
The only standards set by authorities apply to ingested methylmercury, which is chemically similar. Ethylmercury is considered to be less toxic, though adequate science is still lacking and safety limits remain absent, despite thimerosal being still used in childhood vaccines in LMICs.

---

*insertional mutagenesis: when foreign genetic material can incorporate itself into the host cell genome.

https://pubchem.ncbi.nlm.nih.gov

Frequency of Adverse Events Listed in Package Inserts

The following tabulated list is the numeric classification of adverse events (AE) as reported in vaccine package insert leaflets provided by the manufacturer.

Classification of AE frequency	People affected	Per 1,000,000 people vaccinated, the number of probable victims
Very common ≥ 1/10 **≥ 10%**	1 or more in 10	**≥ 100,000**
Common ≥ 1/100 to < 1/10 **1-10%**	1 or more in 100 to less than 1 per 10	**10,000 to 100,000**
Uncommon ≥ 1/1,000 to < 1/100 **0.1-1%**	1 or more in 1,000 to less than 1 per 100	**1,000 to 10,000**
Rare ≥ 1/10,000 to < 1/1,000 **0.01-0.1%**	1 or more in 10,000 to less than 1 per 1,000	**100 to 1,000**
Very rare < 1/10,000 **< 0.01%**	less than 1 in 10,000	**< 100** HOW MUCH COLLATERAL DAMAGE IS ACCEPTABLE?

The American Institute of Medicine (IOM) committees published reports in 1991, 1994a, 1994b, and 2012 and found that the following are causally related to vaccination:

- Acute encephalopathy (brain inflammation)
- Anaphylaxis (whole-body allergic reaction)
- Febrile seizures (convulsions with fever)
- Guillain-Barré syndrome (peripheral nerve inflammation)
- Brachial neuritis (arm nerve inflammation)
- Deltoid bursitis (shoulder inflammation)
- Acute & chronic arthritis (joint inflammation)
- Hypotonic / hyporesponsive episodes (shock and "unusual" shock-like state)
- Protracted, inconsolable crying and screaming
- Vaccine strain infection (smallpox, live polio, measles, varicella zoster vaccines)
- **Death (smallpox, live polio, measles vaccines).**

https://www.nvic.org/nvic-archives/institutemedicine.aspx

Under federal law, "only those adverse events for which there is some basis to believe there is a causal relationship between the drug and the occurrence of the adverse event" should be included. *21 CFR 201.57: format of labeling for prescription drug products and biological products.*

BOTH VACCINES CONTAIN ALUMINUM AND ARE MANUFACTURED BY GSK

DTaP-IPV-Hib-HepB Infanrix Hexa® *from 2 months old*	Tdap Boostrix® *for pregnant women and from 4 years old*	
Irritability, abnormal crying, restlessness, pain, redness, local swelling at the injection site (≤ 50 mm), fever ≥ 38°C, fatigue, loss of appetite	Swelling, pain, and redness at injection site, irritability, insomnia, headaches, fatigue	VERY COMMON
Pruritus*, vomiting, diarrhea, nervousness, local swelling at the injection site (> 50 mm)*, fever > 39.5°C, injection site reactions, including induration	Vomiting, diarrhea, gastrointestinal issues, nausea, anorexia, vertigo	COMMON
Diffuse swelling of the injected limb, sometimes involving the adjacent joint**, upper respiratory tract infection, somnolence, cough*	Convulsions, seizures, upper respiratory infections, lymphadenopathy (inflamed and enlarged lymph nodes), brain fog, conjunctivitis, fainting, cough, rash, muscular and joint pains, fever > 39°C, ulcer, flu-like symptoms	UNCOMMON
Extensive swelling reactions, swelling of the entire injected limb**, vesicles at injection site, angioneurotic edema* (fluid retention related to an allergic reaction), apnea*, collapse or shock-like state (hypotonic hyporesponsive episode)***, allergic reactions (including anaphylactic and anaphylactoid reactions), lymphadenopathy, thrombocytopenia, rash	Allergies, hypotonic episodes (sudden onset of poor muscle tone, reduced consciousness, and pale or bluish skin), convulsions, urticaria	RARE
Dermatitis, urticaria*, hemiconvulsion-hemiplegia epilepsy***	Neurological disorders, Guillain-Barré syndrome, paralysis, brachial neuritis (nerve damage)	VERY RARE

*Observed only with other GSK DTaP-containing vaccines

**Children primed with acellular pertussis vaccines are more likely to experience swelling reactions after booster administration in comparison with children primed with whole cell vaccines. These reactions resolve over an average of 4 days.

***Analysis of postmarketing reporting rates suggests a potential increased risk of convulsions (with or without fever) and hemi-convulsion-hemiplegia epilepsy syndrome when comparing groups which reported use of Infanrix Hexa® with Prevenar®13, to those which reported use of Infanrix Hexa alone.

There is no active surveillance of pregnant women who are vaccinated with Boostrix®. There are no human fertility studies to validate the safety of Boostrix® on babies and its effects on neurological and immunological development in the critical first year of life.

Adverse Events (AE) Terminology

MedDRA® is an international subscription*-based, clinically validated medical dictionary used by regulatory authorities and biopharmaceutical companies to classify a diverse range of product-related adverse events for safety data collection. MedDRA® is used "for registration, documentation and safety monitoring of medical products [pharmaceuticals, vaccines and drug-device combinations] both before and after a product has been authorized for sale." It is designed to be a "rich and highly specific standardized medical terminology to facilitate sharing of regulatory information internationally for medical products used by humans."

First published in 1999, MedDRA® largely replaced the WHO's Adverse Reaction Terminology (ART) that was created in 1968 and has been discontinued since 2015. The WHO-ART is the coding system applied to WHO's VigiBase adverse event database, managed since 1978 by the Uppsala Monitoring Center in Sweden.

MedDRA has a five-level hierarchical structure. The top level is the SOC, representing 27 broad classes grouped by etiology and manifestation site. The most frequently used level of terms is Preferred Terms which are ~18,000 distinct and unambiguous descriptors. The MHRA in the UK has also adopted a subset of MedDRA® Lowest Level Terms—a list of 1,491 Patient-Friendly Terms for its reporting scheme, YellowCard, that collects, collates and investigates suspected adverse drug reactions.

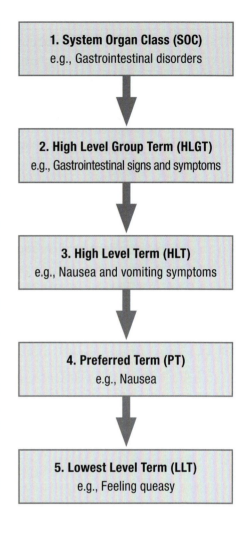

1. **System Organ Class (SOC)**
e.g., Gastrointestinal disorders

2. **High Level Group Term (HLGT)**
e.g., Gastrointestinal signs and symptoms

3. **High Level Term (HLT)**
e.g., Nausea and vomiting symptoms

4. **Preferred Term (PT)**
e.g., Nausea

5. **Lowest Level Term (LLT)**
e.g., Feeling queasy

Patient-Friendly Term (PFT)
Used in MHRA-UK YellowCard

The International Conference on Harmonisation of Technical Requirements for Registration of Pharmaceuticals for Human Use (ICH) was created in 1990 at a WHO conference on Drug Regulatory Authorities in Paris. MedDRA® is an initiative of ICH, and is a trademark registered by the International Federation of Pharmaceutical Manufacturers and Associations (IFPMA) on behalf of ICH.

"In November 1998, the ICH awarded a maintenance contract for MedDRA® to a US information technology firm called BDM Technologies. BDM was later acquired by the system integration company called TRW, and TRW was subsequently acquired by Northrop Grumman Corporation."

*MedDRA® is distributed by subscription on a sliding scale fee based on company revenues, from US$0 for nonprofit entities and regulatory agencies, to US$63,800 per year for companies with annual revenues up to US$20 billion.
https://www.meddra.org/subscription-rates

"MedDRA: The Tale of a Terminology: Side Effects of Drugs Essay"
https://doi.org/10.1016/S0378-6080(09)03160-2

AE System Organ Class & Grades

SOC, the highest level of the MedDRA® hierarchy, is identified by anatomical or physiological system or etiology.

Within each SOC, AEs are listed and accompanied by descriptions of severity (Grade).

Adverse Events by System Organ Class (SOC)

- Blood and lymphatic system disorders
- Cardiac disorders
- Congenital, familial, and genetic disorders
- Ear and labyrinth disorders
- Endocrine disorders
- Eye disorders
- Gastrointestinal disorders
- General disorders and administration site conditions
- Hepatobiliary disorders
- Immune system disorders
- Infections and infestations
- Injury, poisoning, and procedural complications
- Metabolism and nutrition disorders
- Musculoskeletal and connective tissue disorders
- Neoplasms [cancer] benign, malignant and unspecified (incl. cysts and polyps)
- Nervous system disorders
- Pregnancy, puerperium, and perinatal conditions
- Psychiatric disorders
- Renal and urinary disorders
- Reproductive system and breast disorders
- Respiratory, thoracic, and mediastinal disorders
- Skin and subcutaneous tissue disorders
- Social circumstances
- Surgical and medical procedures
- Vascular disorders

Death is classed under "general disorders"

Grades

Grade refers to the severity of the AE.

Grades 1 through 5 consist of unique clinical descriptions of severity for each AE based on this general guideline:

- **Grade 1**
  Mild; asymptomatic or mild symptoms; clinical or diagnostic observations only; intervention not indicated.

- **Grade 2**
  Moderate; minimal, local or noninvasive intervention indicated; limiting age-appropriate instrumental **ADL***.

- **Grade 3**
  Severe or medically significant but not immediately life-threatening; hospitalization or prolongation of hospitalization indicated; disabling; limiting self care **ADL****.

- **Grade 4**
  Life-threatening consequences; urgent intervention indicated.

- **Grade 5**
  Death related to AE.

Activities of Daily Living (ADL)
Instrumental ADL refer to preparing meals, shopping for groceries or clothes, using the telephone, managing money, etc.
**Self care ADL refer to bathing, dressing and undressing, feeding self, using the toilet, taking medications, and not bedridden.*

https://www.meddra.org/how-to-use/basics/hierarchy

MedDRA® has thousands of medical codes for adverse events after vaccination that are classed by organ system. All systems of the body are being affected by vaccines, especially the nervous system and the blood.

Adverse Events Following Immunization (AEFI)

General definition AEFI

Any untoward medical occurrence which follows immunization and which does not necessarily have a causal relationship with the usage of the vaccine. The adverse event may be any unfavorable or unintended sign, abnormal laboratory finding, symptom or disease. An AE is a term that is a unique representation of a specific event used for medical documentation and scientific analyses.

Cause-specific definitions

1. **Vaccine product-related reaction**

   An AEFI that is caused or precipitated by a vaccine due to one or more of the inherent properties of the vaccine product, whether related to the active component or one of the other components (e.g., adjuvant, preservative, or stabilizer).

2. **Vaccine quality defect-related reaction**

   An AEFI that is caused or precipitated by a vaccine that is due to one or more quality defects of the vaccine product, including its administration device, as provided by the manufacturer.

3. **Immunization error-related reaction**

   An AEFI that is caused by inappropriate vaccine handling, prescribing, or administration and thus, by its nature, is preventable.

4. **Immunization anxiety-related reaction**

   An AEFI arising from anxiety about the immunization.

5. **Coincidental event**

   An AEFI that is caused by something other than the vaccine product, immunization error, or immunization anxiety.

Adapted from the WHO 2012 *Global Vaccine Safety Blueprint – Immunization, Vaccines, and Biologicals.*

"The Department of Immunization, Vaccines and Biologicals thanks the Bill and Melinda Gates Foundation whose financial support has made the production of this document possible."

https://iris.who.int/bitstream/handle/10665/70919/WHO_IVB_12.07_eng.pdf?sequence=1&isAllowed=y

According to the FDA, serious adverse events are defined as:

- death
- a life-threatening adverse event
- inpatient hospitalization or prolongation of existing hospitalization
- a persistent or significant incapacity or substantial disruption of the ability to conduct normal life functions
- a congenital anomaly/birth defect
- a medical or surgical intervention to prevent death, a life-threatening event, hospitalization, disability, or congenital anomaly.

AEFI Causality Assessment / Step 1–4

Annex 1: Worksheet for AEFI causality assessment

Step 1 (Eligibility)

Patient ID/Name :	DoB/Age:	Sex: Male/Female
Name one of the vaccines administered before this event	What is the Valid Diagnosis?	Does the diagnosis meet a case definition?

Create your question on causality here

Has the _____ vaccine / vaccination caused _____ (The event for review in step 2 - valid diagnosis)

Is this case eligible for causality assessment?   Yes/No; If, "Yes", proceed to step 2

Step 2 (Event Checklist) ✓ (check) all boxes that apply

	Y	N	UK	NA	Remarks
I. Is there strong evidence for other causes?					
1. In this patient, does the medical history, clinical examination and/or investigations, confirm another cause for the event?	☐	☐	☐	☐	
II. Is there a known causal association with the vaccine or vaccination?					
Vaccine product					
1. Is there evidence in published peer reviewed literature that this vaccine may cause such an event if administered correctly?	☐	☐	☐	☐	
2. Is there a biological plausibility that this vaccine could cause such an event?	☐	☐	☐	☐	
3. In this patient, did a specific test demonstrate the causal role of the vaccine ?	☐	☐	☐	☐	
Vaccine quality					
4. Could the vaccine given to this patient have a quality defect or is substandard or falsified?	☐	☐	☐	☐	
Immunization error					
5. In this patient, was there an error in prescribing or non-adherence to recommendations for use of the vaccine (e.g. use beyond the expiry date, wrong recipient etc.)?	☐	☐	☐	☐	
6. In this patient, was the vaccine (or diluent) administered in an unsterile manner?	☐	☐	☐	☐	
7. In this patient, was the vaccine's physical condition (e.g. colour, turbidity, presence of foreign substances etc.) abnormal when administered?	☐	☐	☐	☐	
8. When this patient was vaccinated, was there an error in vaccine constitution/ preparation by the vaccinator (e.g. wrong product, wrong diluent, improper mixing, improper syringe filling etc.)?	☐	☐	☐	☐	
9. In this patient, was there an error in vaccine handling (e.g. a break in the cold chain during transport, storage and/or immunization session etc.)?	☐	☐	☐	☐	
10. In this patient, was the vaccine administered incorrectly (e.g. wrong dose, site or route of administration; wrong needle size etc.)?	☐	☐	☐	☐	
Immunization anxiety (Immunization stress related responses - ISRR)					
11. In this patient, could this event be a stress response triggered by immunization (e.g. acute stress response, vasovagal reaction, hyperventilation, dissociative neurological symptom reaction etc)?	☐	☐	☐	☐	
II (time): Was the event in section II within the time window of increased risk (i.e. 'Yes' response to questions from II 1 to II 11 above)					
12. In this patient, did the event occur within a plausible time window after vaccine administration?	☐	☐	☐	☐	
III. Is there strong evidence against a causal association?					
1. Is there a body of published evidence (systematic reviews, GACVS reviews, Cochrane reviews etc.) **against** a causal association between the vaccine and the event?	☐	☐	☐	☐	
IV. Other qualifying factors for classification					
1. In this patient, did such an event occur in the past after administration of a similar vaccine?	☐	☐	☐	☐	
2. In this patient, did such an event occur in the past independent of vaccination?	☐	☐	☐	☐	
3. Could the current event have occurred in this patient without vaccination (background rate)?	☐	☐	☐	☐	
4. Did this patient have an illness, pre-existing condition or risk factor that could have contributed to the event?	☐	☐	☐	☐	
5. Was this patient taking any medication prior to the vaccination?	☐	☐	☐	☐	
6. Was this patient exposed to a potential factor (other than vaccine) prior to the event (e.g. allergen, drug, herbal product etc.)?	☐	☐	☐	☐	

Note: Y: Yes; N: No; UK: Unknown; NA: Not applicable.

"Causality assessment is the systematic review of data about an AEFI case; it aims to determine the likelihood of a causal association between the event and the vaccine(s) received."

The 2018 revision brings "greater clarity to the categorization of 'AEFI cases ineligible for classification' and 'unclassifiable cases'; A broader consideration on a spectrum of stress responses to immunization when assessing causality for immunization anxiety related AEFI; Attention to 'falsified vaccines' during AEFI causality assessment [reports have been received of falsified vaccines for yellow fever, meningitis, rabies and pentavalent vaccines]; Use of clearer language and semantics in the checklist questions."

"Causality assessment of AEFI should ideally be performed by a reviewing team or committee of reviewers whose areas of expertise could include pediatrics, neurology, general medicine, forensic medicine, pathology, microbiology, immunology and epidemiology. Other external medical experts should be invited for the review of specific events."

All quotes are extracted from WHO document
https://www.who.int/publications/i/item/9789241516990

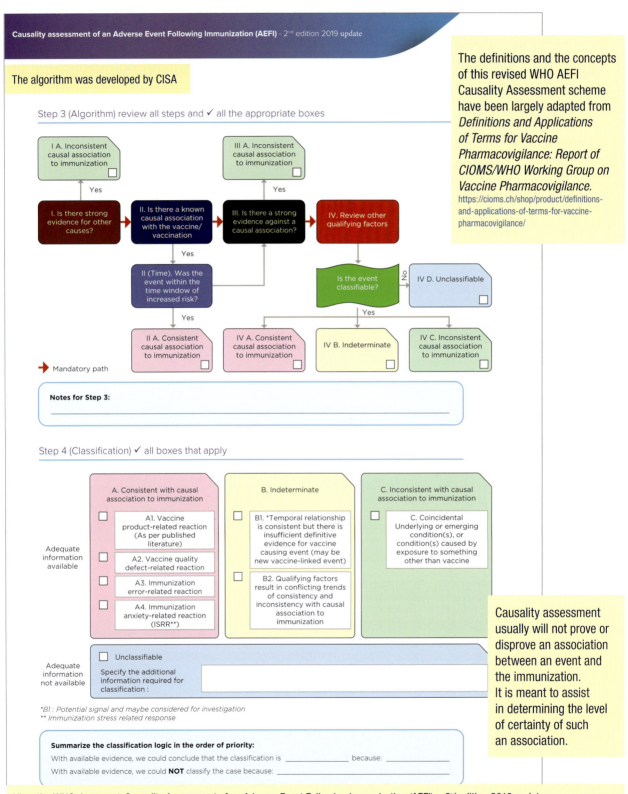

Causality assessment of an Adverse Event Following Immunization (AEFI) - 2nd edition 2019 update

The algorithm was developed by CISA

The definitions and the concepts of this revised WHO AEFI Causality Assessment scheme have been largely adapted from *Definitions and Applications of Terms for Vaccine Pharmacovigilance: Report of CIOMS/WHO Working Group on Vaccine Pharmacovigilance.*
https://cioms.ch/shop/product/definitions-and-applications-of-terms-for-vaccine-pharmacovigilance/

Step 3 (Algorithm) review all steps and ✓ all the appropriate boxes

I A. Inconsistent causal association to immunization ☐

III A. Inconsistent causal association to immunization ☐

I. Is there strong evidence for other causes? — Yes →

II. Is there a known causal association with the vaccine/vaccination →

III. Is there a strong evidence against a causal association? — Yes →

IV. Review other qualifying factors

II (Time). Was the event within the time window of increased risk? — Yes

Is the event classifiable? — No → IV D. Unclassifiable ☐

II A. Consistent causal association to immunization ☐

IV A. Consistent causal association to immunization ☐

IV B. Indeterminate ☐

IV C. Inconsistent causal association to immunization ☐

→ Mandatory path

Notes for Step 3:
_____

Step 4 (Classification) ✓ all boxes that apply

A. Consistent with causal association to immunization

Adequate information available

☐ A1. Vaccine product-related reaction (As per published literature)

☐ A2. Vaccine quality defect-related reaction

☐ A3. Immunization error-related reaction

☐ A4. Immunization anxiety-related reaction (ISRR**)

B. Indeterminate

☐ B1. *Temporal relationship is consistent but there is insufficient definitive evidence for vaccine causing event (may be new vaccine-linked event)

☐ B2. Qualifying factors result in conflicting trends of consistency and inconsistency with causal association to immunization

C. Inconsistent with causal association to immunization

☐ C. Coincidental Underlying or emerging condition(s), or condition(s) caused by exposure to something other than vaccine

Causality assessment usually will not prove or disprove an association between an event and the immunization.
It is meant to assist in determining the level of certainty of such an association.

Adequate information not available

☐ Unclassifiable
Specify the additional information required for classification : _____

*B1 : Potential signal and maybe considered for investigation
** Immunization stress related response

Summarize the classification logic in the order of priority:
With available evidence, we could conclude that the classification is _____ because: _____
With available evidence, we could **NOT** classify the case because: _____

View the WHO document *Causality Assessment of an Adverse Event Following Immunization (AEFI) – 2nd edition 2019 update*
https://www.who.int/publications/i/item/9789241516990

AEFI Reported in Package Inserts

HPV vaccine – Gardasil®9
https://www.fda.gov/media/90064/download?attachment

Additionally, the following postmarketing adverse experiences have been spontaneously reported for GARDASIL:

Blood and lymphatic system disorders: Autoimmune hemolytic anemia, idiopathic thrombocytopenic purpura, lymphadenopathy.

Respiratory, thoracic and mediastinal disorders: Pulmonary embolus.

Gastrointestinal disorders: Pancreatitis.

General disorders and administration site conditions: Asthenia, chills, death, malaise.

Immune system disorders: Autoimmune diseases, hypersensitivity reactions including anaphylactic/anaphylactoid reactions, bronchospasm.

Musculoskeletal and connective tissue disorders: Arthralgia, myalgia.

Nervous system disorders: Acute disseminated encephalomyelitis, Guillain-Barré syndrome, motor neuron disease, paralysis, seizures, transverse myelitis.

Infections and infestations: Cellulitis.

Vascular disorders: Deep venous thrombosis.

Adverse events observed during clinical trials and in postmarketing experience are listed in vaccine product inserts.
The fact that these adverse events are listed does not imply they are causally linked to the vaccine, however under US federal law (21 CFR 201.57) "only those adverse events for which there is some basis to believe there is a causal relationship between the drug and the occurrence of the adverse event" are included. All US vaccine package inserts are available via the link below.
https://www.fda.gov/vaccines-blood-biologics/vaccines/vaccines-licensed-use-united-states

---

DTaP vaccine – Infanrix®
https://www.fda.gov/media/75157/download?attachment

6.2   Postmarketing Experience

In addition to reports in clinical trials, worldwide voluntary reports of adverse events received for INFANRIX since market introduction are listed below. This list includes serious events and events which have a plausible causal connection to INFANRIX. These adverse events were reported voluntarily from a population of uncertain size; therefore, it is not always possible to reliably estimate their frequency or establish a causal relationship to vaccination.

Infections and Infestations: Bronchitis, cellulitis, respiratory tract infection.

Blood and Lymphatic System Disorders: Lymphadenopathy, thrombocytopenia.

Immune System Disorders: Anaphylactic reaction, hypersensitivity.

Nervous System Disorders: Encephalopathy, headache, hypotonia, syncope.

Ear and Labyrinth Disorders: Ear pain.

Cardiac Disorders: Cyanosis.

Respiratory, Thoracic, and Mediastinal Disorders: Apnea, cough.

Skin and Subcutaneous Tissue Disorders: Angioedema, erythema, pruritus, rash, urticaria.

General Disorders and Administration Site Conditions: Fatigue, injection site induration, injection site reaction, Sudden Infant Death Syndrome.

The Brighton Collaboration – SPEAC Report
Priority List of Adverse Events of Special Interest: COVID-19

V2.0. 25-05-2020   Safety Platform for Emergency vACcines (SPEAC) Project

The Brighton Collaboration is responsible for determining case defininitions of AEFIs and AESIs and for establishing the necessary criteria to establish levels of diagnostic certainty.

4. Results

Table 1 lists AESIs considered potentially applicable to COVID-19 vaccines based on known association with vaccination in general. The rationale for including the AESI is further delineated in the last column of table 1.

Adverse events of special interest applicable to COVID-19 vaccines

TABLE 1. AESI RELEVANT TO VACCINATION IN GENERAL (EVENTS LISTED IN RED HAVE EXISTING BC CASE DEFINITIONS) IN THE TOOLBOX.)

BODY SYSTEM	AESI TYPE	RATIONALE FOR INCLUSION AS AN AESI (SEE FOOTNOTE)
Neurologic	Generalized convulsion	1, 2, 4
	Guillain-Barré Syndrome (GBS)	2
	Acute disseminated encephalomyelitis (ADEM)	3
Hematologic	Thrombocytopenia	1, 2
Immunologic	Anaphylaxis	1, 2
	Vasculitides	3, 4
Other	Serious local/systemic AEFI	1, 2

Levels of diagnostic certainty

1. Proven association with immunization encompassing several different vaccines
2. Proven association with vaccine that could theoretically be true for CEPI vaccines under development
3. Theoretical concern based on immunopathogenesis.
4. Theoretical concern related to viral replication during wild type disease.
5. Theoretical concern because it has been demonstrated in an animal model with one or more candidate vaccine platforms.

Table 2 focuses on AESIs relevant to particular vaccine platforms that are being considered in the COVID-19 vaccine development programs.

TABLE 2. AESI RELEVANT TO SPECIFIC VACCINE PLATFORMS FOR COVID-19 VACCINES

BODY SYSTEM	VACCINE PLATFORM SPECIFIC AESIs*	KNOWN/POSSIBLE ASSOCIATION WITH
Neurologic	Aseptic meningitis Encephalitis / Encephalomyelitis	Live viral vaccines including measles
Immunologic	Arthritis	r-VSV platform
Other	Myocarditis	MVA platform

*Review of nucleic acid platforms, and protein platforms has not been conducted since these are novel **(i.e., new).**

https://brightoncollaboration.org/priority-list-of-adverse-events-of-special-interest-covid-19/

National Center for Immunization & Respiratory Diseases

https://www.fda.gov/media/143530/download

CDC Post-Authorization / Post-Licensure Safety
Monitoring of COVID-19 Vaccines

Tom Shimabukuro, MD, MPH, MBA
CDC COVID-19 Vaccine Task Force
Vaccine Safety Team

October 22, 2020

List of Vaccine Adverse Events Reporting System
(VAERS) of adverse events (AE)

Preliminary list of VAERS AEs of special interest

- COVID-19 disease
- Death
- Vaccination during pregnancy and adverse pregnancy outcomes
- Guillain-Barré syndrome (GBS)
- Other clinically serious neurologic AEs (group AE)
  - Acute disseminated encephalomyelitis (ADEM)
  - Transverse myelitis (TM)
  - Multiple sclerosis (MS)
  - Optic neuritis (ON)
  - Chronic inflammatory demyelinating polyneuropathy (CIDP)
  - Encephalitis
  - Myelitis
  - Encephalomyelitis
  - Meningoencephalitis
  - Meningitis
  - Encephalopathy
  - Ataxia

- Seizures / convulsions
- Stroke
- Narcolepsy / cataplexy
- Autoimmune disease
- Anaphylaxis
- Non-anaphylactic allergic reactions
- Acute myocardial infarction
- Myocarditis / pericarditis
- Thrombocytopenia
- Disseminated intravascular coagulation (DIC)
- Venous thromboembolism (VTE)
- Arthritis and arthralgia (not osteoarthritis or traumatic arthritis)
- Kawasaki disease
- Multisystem Inflammatory Syndrome (MIS-C, MIS-A)

Preliminary list of VSD pre-specified outcomes for RCA

Near real-time sequential monitoring
(Rapid Cycle Analysis)

- Acute disseminated encephalomyelitis (ADEM)
- Acute myocardial infarction (AMI)
- Anaphylaxis
- Acute respiratory distress syndrome (ARDS)
- Arthritis and arthralgia / joint pain
- Convulsions / seizures
- Disseminated intravascular coagulation (DIC)
- Encephalitis / myelitis / encephalomyelitis / meningoencephalitis / meningitis / encephalopathy (not ADEM or TM)
- Guillain-Barré syndrome (GBS)
- Immune thrombocytopenia (ITP)
- Kawasaki disease (KD)
- Multisystem Inflammatory Syndrome in Children (MIS-C)
- Myocarditis / pericarditis
- Narcolepsy / cataplexy
- Stroke
- Transverse myelitis (TM)
- Venous thromboembolism (VTE)

Artificial intelligence-driven Rapid Cycle Analysis (RCA) methods in Vaccine Safety Datalink were developed to allow population-based monitoring of potential outcomes associated with a vaccine in near real-time by examining outcome rates in recent vaccinees during risk intervals in relation to rates during comparison intervals. The associations produced by this approach are considered statistical signals that indicate the need for additional analytic investigation.

Who Is Accountable?
Responsibility and Liability

Below is the list of key agencies responsible for the implementation of vaccines and surveillance of adverse events following their administration.

World / Europe

WHO	**World Health Organization** *global authority on public health and vaccine policy*
SAGE	**Strategic Advisory Group of Experts on Immunization**
GACVS	**WHO Global Advisory Committee on Vaccine Safety**
VigiAccess	**WHO International Database of Reported Possible Vaccine Adverse Events**
UNICEF	**The UN Children's Fund** *a major supplier of vaccines in the developing nations*
EMA	**European Medicines Agency** *regulatory agency granting authorization*
EudraVigilance	**EU Passive Surveillance System for Vaccine Injury Reports**

USA

1986 National Childhood Vaccine Injury Act *max. payout for death US$ 250,000*

HHS	**Department of Health and Human Services**
NIH	**National Institutes of Health**
NIAID	**National Institute of Allergy and Infectious Diseases**
CDC	**Centers for Disease Control and Prevention**
FDA	**Food and Drug Administration** *regulatory agency granting authorization*
CBER	**Center for Biologics Evaluation and Research** *performs regulatory oversight and licensure*
VRBPAC	**Vaccines and Related Biological Products Advisory Committee** *FDA's expert advisors*
ACIP	**Advisory Committee on Immunization Practices** *recommends vaccine schedule*
VICP	**The National Vaccine Injury Compensation Program**
CISA	**Clinical Immunization Safety Assessment Project** *investigates biologic risks of AEFI*
VAERS	**Vaccine Adverse Events Reporting System managed by CDC/FDA**

Switzerland

2013 Law on Epidemics *"Tort moral" max. payout CHF 70,000*

FOPH	**Federal Office of Public Health**
Swissmedic	**Swiss Agency for Therapeutic Products** *regulatory agency granting authorization*
FCV	**Federal Commission for Vaccination** *recommends vaccine schedule*
Vaccinovigilance	**Passive Surveillance System for Vaccine Injury Reports**

United Kingdom

1979 Vaccine Injury Act *max. payout for death-disability £ 120,000*

DHSC	**Department of Health and Social Care**
MHRA	**The Medicines and Healthcare products Regulatory Agency**
JCVI	**Joint Committee on Vaccination and Immunisation** *recommends vaccine schedule*
YellowCard	**Passive Surveillance System for Vaccine Injury Reports**

HHS/CDC Poster from the 1980s

Adults: Get vaccinated to stay healthy

- Measles
- Mumps
- Rubella
- Hepatitis B
- Tetanus
- Diphtheria
- Influenza
- Pneumococcal Disease

Don't become an endangered species

U.S. DEPARTMENT OF HEALTH & HUMAN SERVICES
Public Health Service
CDC

Vaccines are the only product where manufacturers have:
– immunity from liability
– legal mandates enforcing their use
– promotion using taxpayer $$$
– government-funded advertising
– guaranteed sales in a global market

Vaccines are the only medical product that does not have to:
– study pharmacokinetics
– prove long-term safety
– use a saline placebo control in trials
– be studied for carcinogenic properties
– be studied for genotoxic properties

- **National regulatory agencies are responsible for granting authorizations.**

- **National public health agencies and their expert committees are responsible for making official recommmendations on vaccines and schedules.**

- **Local medical authorities and associations are responsible for implementing, promoting, communicating, and enforcing vaccine policy in their territories.**

CDC / FDA – VAERS Vaccine Adverse Events Reporting System

If a medical professional suspects an adverse event following a vaccination, they are legally obliged to report it to the CDC. Falsifying a VAERS report is considered fraud. This passive surveillance system has an estimated underreporting factor of at least 31 to 41. Even the authorities admit that less than 10% of vaccine injuries are ever reported. **A VAERS report takes about 30 minutes to fill out with time-limited pages to complete.**

NIAID Task Force Report on Safer Childhood Vaccines

According to the 1986 NCVIA, the Task Force on Safer Childhood Vaccines (TFSCV) was required to prepare a report and recommendations for the Secretary, HHS, in consultation with the Advisory Commission on Childhood Vaccines, an external advisory group charged with providing advice to the Secretary on the operation of the National Vaccine Injury Compensation Program.

To meet its charge and carry out the reporting requirements under the Act, the TFSCV:
1. reviewed and summarized previously identified safety issues regarding vaccines currently in use
2. reviewed current policies and procedures to ensure the safety of vaccines
3. determined options for improving existing structures to ensure vaccine safety.

As a result, the NIAID Task Force provided the Secretary with a series of recommendations designed to further enhance vaccine safety.

It was not until 1998 that the Task Force's first report was officially published. The document, while not specifically addressing the safety concerns and significant knowledge gaps that had been identified by previous IOM reports, maintains that "new vaccines are extensively studied for safety and are unlikely to proceed through lengthy development steps to licensure if there is evidence of severe adverse reactions."

The report continues by stating that nearly "all childhood vaccines are administered on multiple occasions during the first year of life, a time when rare neurological, immunological, and other disorders may manifest themselves. Vaccination is a nearly universal practice so that controlled evaluations to compare the incidence of such events in vaccinated and unvaccinated children have become increasingly difficult to conduct.

However, lack of definitive case definitions for some of these events, combined with difficulties in controlling for myriad confounding variables, has made these studies virtually impossible to carry out. The cost of such studies has also been considered prohibitive, particularly in the environment of efforts to reduce spiraling health care costs."

Read the full report online

https://books.google.ch/books?id=5PzXTODZ1rMC

The information for Section 2127 is currently found under the NVICP, Mandate for safer childhood vaccines, 42 US Code § 300aa–27.

https://www.law.cornell.edu/uscode/text/42/300aa-27

Box 1.
Section 2127(b) of the Public Health Service Act Created the Task Force on Safer Childhood Vaccines.

Section 2127 of the Act embodies explicit language regarding safety as well as the specific mandate of the Task Force on Safer Childhood Vaccines. It provides in its entirety as follows:

a. General Rule—In the administration of this subtitle and other pertinent laws under the jurisdiction of the Secretary, the Secretary shall:

(1) promote the development of childhood vaccines that result in fewer and less serious adverse reactions than those vaccines on the market on the effective date of this part and promote the refinement of such vaccines; and

(2) make or assure improvements in, and otherwise use the authorities of the Secretary with respect to, the licensing, manufacturing, processing, testing, labeling, warning, use instructions, distribution, storage, administration, field surveillance, adverse reaction reporting, and recall of reactogenic lots or batches of vaccines, and research on vaccines, in order to reduce the risks of adverse reactions to vaccines.

b. Task Force:

(1) The Secretary shall establish a task force on safer childhood vaccines which shall consist of the Director of the National Institutes of Health, the Commissioner of the Food and Drug Administration, and the Director of the Centers for Disease Control.

(2) The Director of the National Institutes of Health shall serve as chairman of the task force.

(3) In consultation with the Advisory Commission on Childhood Vaccines, the task force shall prepare recommendations to the Secretary concerning implementation of the requirements of subsection (a).

c. Report—Within two years after the effective date of this part, and periodically thereafter, the Secretary shall prepare and transmit to the Committee on Energy and Commerce of the House of Representatives and the Committee on Labor and Human Resources of the Senate a report describing the actions taken pursuant to subsection (a) during the preceding two-year period.

Thanks to a 2018-FOIA request by Informed Consent Action Network (ICAN), it was discovered that no HHS Secretary has EVER submitted a formal report to either committee since 1989.
The first and only report submitted by the TFSCV was published in 1998.

"Vaccines undergo rigorous and extensive testing to determine their safety and effectiveness."

—US Food and Drug Administration

"Vaccines are held to the highest standard of safety."

—US Department of Health and Human Services

"Observing vaccinated children for many years to look for long-term health conditions would not be practical and withholding an effective vaccine from children while long-term studies are being done wouldn't be ethical."

—CDC Parent's Guide to Childhood Immunization 2015

Afterword

Before I began investigating vaccination, I believed that vaccines were life-saving, safe, and effective.
My last vaccine was against tetanus, given before traveling to Nepal. I did not ask any questions nor read anything before rolling up my sleeve upon my doctor's recommendation. It hurt like hell and left me with a dead arm all day—which I shrugged off as "normal." Never did my doctor inform me of the risks of injecting both thimerosal and aluminum into my system, nor that these ingredients had never been adequately tested for safety. A year later in 2006, after over thirty years of use, that vaccine, Anatoxal Te® by Berna, was withdrawn from the Swiss market, with all other mercury-containing vaccines still in circulation in Switzerland. The thimerosal-containing vaccines for children were removed in 2000, after 60 years of widespread worldwide use.

I began questioning vaccination in 2015 when my 70-year-old non-smoking father died. Within two months of his flu shot, he was hospitalized (NHS-UK) and diagnosed with stage IV lung cancer. He had no respiratory issues and was treated for brain inflammation with steroids, which I found rather bizarre. By that time I was much more critical of modern medicine and its allopathic reductionist approach to health. His death triggered my deep-dive investigation into, and obsession with, vaccines. Spending days at libraries, months reading and learning, I attended online courses, a week-long vaccinology course at Oxford University, and spoke to medical professionals, scientists, and parents. Over the years, I experienced unpleasant cognitive dissonance before fully digesting the fact that vaccines by definition can never be truly safe.

Critical analysis and discourse in science is necessary, especially in fields that use empiricism, outdated models and prevailing layers of assumptions, and are sensitive to conflicts of interest, ideologies, limited perspective, and willful dismissal of safety signals. Vaccine safety is not scientifically settled; even public heath experts admit behind closed doors to the inadequacy of safety science, yet publicly reject the numerous concerns regarding mid- to long-term health complications. Instead of making vaccines safer, public health authorities push them even harder on the population while maintaining their insistence of vaccines being "safe and effective." To add insult to injury, the WHO, government authorities, and media tech-giant Google censor information to limit the spread of any criticism of vaccination—misleadingly sold as "immunization." Critics are labeled anti-vaxxers, and the solution to the crisis in confidence is censorship and cancel culture.

I take this opportunity to declare that I am pro-health, pro-science, and highly respect the many skilled and compassionate people working in the field of medical science and healthcare.
Life-saving trauma medicine, with its ability to put people back together, alleviate pain, and replace limbs, is incredible. Without emergency surgery, I would no longer be on this planet, so let me reassure the reader that I have no intention of undermining the scientific medical progress made in the field of health. I have much respect and admiration for those working in healthcare and science, who operate with integrity and humility. Thank you to all those souls who have dedicated their lives to be of service to others.

Acknowledgments

I would specifically like to express my sincere gratitude to Lisa Renberg; Dr. Robert W. Malone; Dr. Jill Glasspool-Malone; Tony Lyons; Hector Carosso; and especially to Zoey O'Toole. These folks have been instrumental to getting this publication out—thank you! Much appreciation also goes out to the team working behind the scenes at Children's Health Defense and Skyhorse Publishing.

I extend my gratitude to vaccinologist Dr. Geert Vanden Bossche for his time, feedback, and input; to Swiss doctors Dr. Léna Gorgé Giancola, Dr. Jean-Paul Ecklin, Dr. François Choffat, Dr. Françoise Berthoud, Dr. Nathalie Calame, Dr. Pascal Büchler, and French infectious disease professor Christian Perronne MD, PhD, for the informative discussions, their sharing of firsthand experience, and for being courageously outspoken on vaccine-related matters. A warm thank you also goes out to Christopher Shaw, PhD, and James Lyons-Weiler, PhD, for their valuable review and feedback, and for their ongoing vaccine research.

Many thanks goes to Patrick Jordan for the extensive timeline in his 2015 book *ICD-999: Vaccine Induced Diseases – The Chronic Serum Sickness Postulate* and to the websites of Immunize.org and The College of Physicians of Philadelphia.

Thanks to the Swiss and UK National Archives, the British Library, the Swiss National Library, Alexandra Elbakyan of Sci-Hub, the Medical History Library at the College of Physicians of Philadelphia, the National Museum of American History, the Smithsonian, the Science History Institute, the Museum of Health Care at Kingston, Health Heritage Research Services in Canada, Informed Consent Action Network, the Wayback Machine, Google Books, numerous scholars, researchers, medical professionals, parents, teachers, warrior moms, activists, and scientists, not to mention the friends and acquaintances who have tolerated my vaccine obsession and contributed to many conversations over the years.

Last, but certainly not least, I would like to extend my deepest gratitude to my wonderful mother, Silvia, and stepfather, Serge, for their unwavering support, confidence, encouragement, trust, and love. A massive "merci amour" to my partner, Fabien, for his patience and music that nourishes my soul, and thanks to my dear big brother, Nawaz, for always listening with an open heart and mind.

Thank you for reading!

Check out the website for an extensive bibliography, references, and suggested reading list.
www.TheUltimateVaccineTimeline.com

Index

References and source materials are available on
www.TheUltimateVaccineTimeline.com

About the Author

Shaz Khan is a London-born, Swiss-Indian creative designer, independent researcher, information junkie, and critical thinker, somewhat obsessed with vaccines.

She graduated from Central St. Martins College of Art & Design with a BA (Hons) in Product Design and has complemented her education with certifications and courses in nutrition, marketing and communications, anatomy and physiology, immunobiology, and vaccinology.

Her persistent curiosity along with an inquisitive mind, interest in history, and thirst for knowledge has expanded her perception and understanding of the world, notably vaccination. Following the untimely death of her seventy-year-old father in 2015, she began asking questions and initiated an immersive, deep-dive investigation into the history and development of vaccines from official sources and medical academia.

After years of research and months spent in libraries and national archives, she was disturbed to discover the enormous amount of history regarding vaccine safety and the scope of injuries observed following their administration that was acknowledged by authorities but never disclosed to the public.

Her commitment to the truth, freedom, and the preservation of health, for present and future generations, is the motivation behind this publication.

She self-published a small illustrated book called *The Virus,* available on Amazon in English and French, publishes content on HealthScienceSimplified.com, and spends her free time on the beautiful banks of Lake Geneva, Switzerland.

HealthScienceSimplified
www.HealthScienceSimplified.com

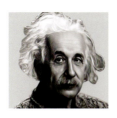

"A foolish faith in authority is the worst enemy of truth."

—Albert Einstein, PhD (1879–1955)

German-born Swiss theoretical physicist, celebrated as one of the world's most influential scientists and awarded the 1921 Nobel Prize in Physics. This quote is taken from a 1901 letter written to his friend, Jost Winteler, found in *The Collected Papers of Albert Einstein: The Early Years, 1879-1902.*

"Science is not really about truth, it's about power. The real aim of science as a project—as an establishment—is not truth, it's power."

—Prof. Yuval Noah Harari, PhD

Israeli historian, philosopher, and author of *Sapiens: A Brief History of Humankind, Homo Deus: A Brief History of Tomorrow, 21 Lessons for the 21st Century,* and *Nexus: A Brief History of Information Networks from the Stone Age to AI.* He is a regular WEF contributor and is quoted as saying the above in a 2015 interview by Intelligence Squared on "The Myths We Need to Survive," filmed at the Royal Geographical Society.